Lecture Notes in Physics

Edited by H. Araki, Kyoto, J. Ehlers, München, K. Hepp, Zürich
R. Kippenhahn, München, H. A. Weidenmüller, Heidelberg,
J. Wess, Karlsruhe and J. Zittartz, Köln

Managing Editor: W. Beiglböck

298

Mladen Georgiev

F' Centers in Alkali Halides

Springer-Verlag
Berlin Heidelberg GmbH

Author

Mladen Georgiev
Institute of Solid State Physics, Bulgarian Academy of Sciences
1784 Sofia, 72 Lenin Blvd., Bulgaria

ISBN 978-3-662-13673-7 ISBN 978-3-540-39076-3 (eBook)
DOI 10.1007/978-3-540-39076-3

Originally published by Springer-Verlag Berlin Heidelberg New York in 1988
Softcover reprint of the hardcover 1st edition 1988

2158/3140-543210

IN MEMORIAM

MY BROTHER

Preface

These Notes are intended to offer a comprehensive discussion of the experimental and theoretical work done so far on the F' centers in alkali halides. They may thereby be expected to fill a gap in the review literature on color centers. Indeed, more than two decades have passed since a survey of a similar type was published by Luty (1961b). Yet, the selected topic is not a subject of merely historical interest, which is perhaps best demonstrated by the number of experimental and theoretical investigations during the period since then that were inspired by some highly controversial features of the F-F' conversion.

Indeed, since the early findings by Pick in 1938 many subsequent works have disclosed the existence of several "unorthodox" and still unexplained properties of what is "otherwise believed to be the most simple color center conversion in the alkali halides". (The author is thankful to Professor J.Z. Damm for suggesting this brief but well-conceived formulation, while using it to justify his own devotion to the problem.)

The arrangement of these Notes can be seen from the Contents. The F' center (also designated F^-) is defined as a two-electron single-anion-vacancy entity that has no other anion vacancies in its immediate neighborhood. Although automatically excluding the M center, this definition provides for the inclusion of F' centers perturbed by more distant anion vacancies, as well as F'_A centers or Ohkura's Z_2 center, which is an F' center perturbed by a nearest-neighbor impurity--cation-vacancy dipole. It may also be seen that considerable space is devoted to both theoretical and experimental contributions. A central position is occupied by the pair interactions between color center species. While undoubtedly reflecting the author's personal preference, this also emphasizes the heuristic aspect of the present F' problem. Theoretical approaches, based on quantum mechanics, to deriving quantities that control or characterize F'-related physics, such as the F-F' conversion efficiency, the F' thermal lifetime, the F' band shape, etc., are also given due space. Most of the presented material is arranged so as to reflect the gradual improvement of the theory. While much of the theoretical work is complementary, no final judgement has been made on the precise nature of the mechanisms involved. In addition to purely F'-related matter, a number of "interdisciplinary items" pertaining to F centers are also touched upon, without which the basic problem could not be set. However, it should be stressed that this is not a textbook on color centers, much better examples of which are listed in appropriate reviews or current literature [see Schulman and Compton (1963), Pick (1965,1972), Fowler (1968), Klick (1972), Farge and Fontana (1974), Stoneham (1975)]. Finally, a recently revived attempt to describe the F-F' conversion in terms of the electron transfer via (bound) polarons is described, albeit incompletely, work on the matter still being in progress. Generally, the present Notes are hopefully expected to stimulate further experimental and theoretical investigations, some options for future research being summarized at the end.

The text is divided into 20 chapters and a number of sections. Mathematical equations are designated by Arabic numerals preceded by the chapter number, while color center reactions are lettered, and numbered analogically. Tables and figures throughout the text are marked with numerals. Two general tables appearing at the end are designated by Roman numerals. A list of references arranged alphabetically can be found at the end.

Acknowledgements

During the course of my own research or while preparing these Notes, I have benefited from collaboration with my co-authors: Professors S.G. Christov and G. Dechev, Drs. A. Gochev, G. Todorov, T. Todorov, and Y. Vassilev, and with M. Baltova, M. Karamikhailova, N. Koralov, A. Kyuldjiev, G. Mladenov, M. Mladenova, K.D. Nierzewski, M. Staikova, and N. Tomova, to all of whom I now extend my thanks. My research was carried out at successively the Central Laboratory of Biophysics (Bulgarian Academy of Sciences), Faculty of Physics (Sofia University), Institute of Solid State Physics (Bulgarian Academy of Sciences), Institute of Low Temperature and Structure Research (Polish Academy of Sciences), and Instituto de Fisica (Universidad Nacional Autonoma de Mexico).

I feel privileged to have met the late Prof. Heinz Pick, pioneer of F'-center research, during the fall of 1976 in West Berlin. Due credit is also given to Professor St. Petroff and the Physics Department at VIMMESS for creating the "color center atmosphere" in this country. Useful talks with and/or material kindly provided by Professors M.A.

Aegerter, A.A. Berezin, L. Bosi, R. Capelletti, J.D. Comins, J.Z. Damm, U.M. Grassano, N. Itoh, G. Jacobs, F. Agullo-Lopez, F. Lüty, M. Manfredi, C. Ruiz-Mejia, G.W.A. Newton, I.A. Parfianovich, St. Petroff, D. Popov, J.L. Alvarez-Rivas, D. Schoemaker, W.A. Sibley, A.M. Stoneham, K. Terzijski, M. Wagner, and many others, are all gratefully acknowledged. D. Popov has also been kind in placing his own F' data and reference material at my disposal. Fritz Lüty's and Zbigniew Damm's scepticism is particularly appreciated for having cooled down my enthusiasm for speculation. Particular thanks are due to Professor S.G. Christov, who introduced me to his reaction-rate method, which later formed the basis of my understanding of radiationless transitions at color centers, for recommending these Notes to Springer. It would not be fair to conclude without also acknowledging the influence of his recent book: S.G. Christov, Collision Theory and Statistical Theory of Chemical Reactions, Lecture Notes in Chemistry # 18, Springer-Verlag, Berlin-Heidelberg-New York (1980). It is hoped that its contribution to our understanding of phonon-coupled solid-state processes will soon be appreciated. Finally, I am grateful to A. Manov for his skilful assistance during the preparation of the manuscript.

Last, but not at all least, I am greatly indebted to both my mother and my family for their patience.

CONTENTS

ABBREVIATIONS

CB - conduction band
cc - cubic centimeter (cm^3)
CC - configurational coordinate
ENDOR - electron nuclear double resonance
ESR - electron spin resonance
IR (ir) - infrared
IV dipole - impurity-vacancy dipole
LAirT - liquid air temperature
LCO_2T - dry-ice sublimation point
LHeT - liquid helium temperature
LMO - local mode oscillator
LNT - liquid nitrogen temperature
LO - longitudinal optic
MCD - magnetic circular dichroism
nir - near infrared
NMR - nuclear magnetic resonance
nn - nearest neighbor
nnn - next nearest neighbor
nonrad - nonradiative
ODESR - optical detection electron spin resonance
PPAE - photoplastic aftereffect
PPE - photoplastic effect
PSTL - photostimulated thermoluminescence
rad - radiative
RES - relaxed excited state
rf - radio frequency
RT - room temperature (300 K)
RTT - radiative tunneling transition
STE - self trapped exciton
STT - spontaneous tunneling transition
TB - thermal bleaching
TBE - temporary bleaching effect
TEE (TSEE) - thermostimulated exoelectron emission
TL - thermoluminescence
TO - transversal optic
TSC - thermally stimulated currents
uv - ultraviolet

2. Historical background

Early experimental work by Gudden and Pohl in 1925 found an increase of the optical absorption spectrum to the red of the F band following the absorption of light by the F centers. This phenomenon was called "Erregung". Subsequently the Erregung was studied by Smakula (1930), who measured the quantum efficiency of the process in X-rayed NaCl and KCl crystals at different temperatures. Subsequent work by the Göttingen School revealed various aspects of the phenomenon. However, it was not until 1938 that the first systematic step towards understanding the physical nature of the Erregung was undertaken by Pick (1938). He assigned the label F' to the new, more loosely bound, absorbing centers giving rise to the longwavelength increase in absorption. Moreover, he considered the optical conversion of F to F' centers to be the simplest case of a photochemical reaction in alkali halides. To prevent possible complications introduced by X-raying, Pick made his measurements on additively colored crystals. Indeed, indications of the occurrence of such complications in X-rayed crystals were found by Smakula, who observed, in addition to the reversible Erregung, a second parallel bleaching process leading to irreversible changes in absorption, the so-called Entfarbung.

Pick made his measurements on samples of KCl, KBr, and NaCl and found that the maximum percentage of F centers that can be converted to F' - 72% (KCl), 75% (KBr), 46% (NaCl) - depends on the compound and is attainable at different temperatures for the various compounds. He also measured the dependence of the number of F centers destroyed on the number of light quanta absorbed. The slope of the curve obtained gives directly the quantum yield of the F-F' process. The experimental data point to a strong temperature dependence of the initial quantum yield $\eta_{F-F'}$. The latter dependence was measured for all three salts, which exhibit the same qualitative behavior: $\eta_{F-F'}$ first increases steeply with temperature but gradually saturates at the highest temperatures. The oscillator strength of the F band had been accurately determined for KCl only, which enabled the absolute determination of $\eta_{F-F'}$ for this salt alone. Under these conditions the maximum value of $\eta_{F-F'}$ at saturation was estimated to be 2. Pick interpreted this to mean that two F centers are destroyed after the absorption process, in which only one electron is ejected by the absorbed photon. These two centers are destroyed after the stable trapping of the photoelectron, resulting in two F' centers with an excess electronic charge. The trapping site was believed to be in the vicinity of an F center. Pick had apparently identified a center with <u>one</u> trapped electron; now the F' center appeared to be something like a diatomic molecule. The first F center disappears after the ejection of the photoelectron, while the second one is destroyed when that electron is stably trapped in the vicinity of another F center, leading to its disappearance, too. Pick's idea of an F center was in line with the Göttingen model: an electron localized on an alkali lattice ion.

Pick also measured the temperature dependence of the maximum number of F centers converted to F': ΔF_{max} . This quantity increases sharply as the temperature is increased in the lower temperature range but slows down gradually to attain a maximum at about -80°C in KCl, after which ΔF_{max} decreases with the temperature. The fall of ΔF_{max} at the higher temperatures is attributed to the decreasing F' thermal lifetime. At such temperatures the F' centers decompose thermally to gradually restore the F centered state of the crystal. The thermal decay of the F' band (the recovery of the F band) was found not to obey an exponential law. For this reason Pick measured the F' halflife (again in terms of the number of F centers revived) and found that it follows the Arrhenius dependence on the absolute temperature for all the three salts. He also pointed out, correctly, that the limited F-F' conversion efficiency $\Delta F_{max}/F_0$ (always below 100 percent) results from the overlap of the F and F' bands. Consequently, the efficiency must depend on the ratio of the F to F' absorption constants, as well as on the quantum yields of the F-F' process and the reverse F'-F process.

The conversion of F' to F was investigated in a second piece of work by Pick (1940). These measurements were delayed for some two years because of the necessity to cool down to low temperatures in order to achieve significant conversion efficiencies. By plotting the number of rebuilt F centers vs. the number of absorbed light quanta in the F' band, he obtained the initial quantum yield $\eta_{F'-F}$ of the conversion. This quantity proved independent of the wavelength within the F' band. It was found to rise steeply as the temperature was lowered, attaining the ultimate value of 2 at the lowest temperatures (KCl). This was to be expected in view of the model proposed earlier. Pick also mentioned for the first time the possibility that F' could, in fact, be a two-electron center.

In discussing the obtained temperature dependences of $\eta_{F-F'}$ and

$\eta_{F'-F}$ Pick argued that an activation energy is needed for the F-F' conversion (either for ejecting the F electron after the absorption of a light quantum, or for creating the F' binding), while the reverse F'-F process occurs with full quantum efficiency throughout the entire temperature range. However, while the end product of the F' to F process is two F centers at low temperature, the result is increasingly two F' centers as the temperature is raised. Although Pick did not elaborate, he forsaw the future interpretation and also suggested that F' is a potential photoconductivity source at the lowest temperatures.

The conventional F-F' conversion model based on Pick's ideas was in excellent agreement with most, if not all, the experimental observations until the early sixties. It is best described in Lüty's paper (1961b). Among other things it predicts an intensity-independent F' density at saturation within the temperature range where the F' center is thermally stable. However, experiments by Costikas and Grossweiner (1962) revealed that the saturated F' density depends de facto on the excitation-light intensity at high intensities. This has been interpreted in terms of an intensity-dependent electron-trapping cross-section σ_F of the F center. More recently, Georgiev and Todorov (1976b) analyzed their data on the F-F' conversion induced by flash-light excitation at room temperature and concluded that not only σ_F, but also the trapping cross-section of the anion vacancy σ_α^* may be intensity-dependent. Alternatively, the possibility has been discussed that the deexcitation rate of the excited F electron may vary with the intensity at high excitation levels. These conclusions strongly suggest that our conventional standpoint on the conversion mechanism should be revised.

At sufficiently high temperatures the F' center decomposes thermally, its extra electron being ejected into the conduction band. The F' lifetime decreases steadily as the temperature is raised. In a couple of experiments $\tau_{F'}$ was found to be both concentration (Schmid and Wolf (1962)) and intensity (Goldberger and Owens (1971)) dependent. It has been suggested that the observed behavior may result from the corresponding dependence of σ_F (Georgiev and Todorov (1976b)). Other experiments seem to point to the existence of several decay channels for the F' center, depending on the temperature and, perhaps, on the purity of the sample (Cordovilla and Alvarez-Rivas (1974)).

Another series of experiments revealed that trapping of conduction electrons by F centers is not the only way to produce F' centers following the absorption of light quanta in the F band (Chiarotti and Grassano (1966b), Ruedin, Schnegg, Jaccard and Aegerter (1972,1973)). It has been verified that excited F electrons can tunnel, even at very low temperatures, to neighboring F centers to form short-lived F' centers, which decompose by way of another tunneling transition to the ground state of the initial center. At F center densities as high as 10^{17} cm^{-3}, and with a random F center distribution, the percentage of tunnel-produced F' centers is rather low, but as the concentration is increased these processes become more important. In particular, they are responsible for the observed concentration quenching of both the luminescence and the photoionization of the F center. The effectiveness of these tunneling phenomena is also enhanced at moderate F center densities by F band bleaching at room temperature, which presumably leads to the formation of clusters of high local F-center density (Delbecq (1963)).

The above-mentioned deviations from the conventional model give rise to the important question as to the occurrence of short-range interactions between neighboring species which can influence both the transition and trapping probabilities of the F center electron.

3. The F' band

Throughout a large part of this paper we shall be dealing with additively colored alkali halide crystals that have initially been in the 'freshly quenched state'. It is achieved by rapidly cooling the samples down to the desired temperature following thermal anneal at about 400°C. Crystals in the freshly-quenched state are believed to contain atomically-dispersed F centers almost exclusively. On illuminating such a crystal with light that falls spectrally within the F band range at a temperature above LNT, a new band labelled F' appears at the expense of the F band. The F' band is very wide (halfwidth about 2 eV), the larger portion of it usually extending on the long wavelength side of the F band (Fig.1). A notable exception seems to be NaI (and possibly NaBr too), where the F' band lies chiefly on the high-energy side of the F band (Baldacchini, Pan and Lüty (1981)). As the temperature gets higher, the F' becomes increasingly unstable thermally. In many alkali halides, F' occurs just as a transient band at RT. F' is also optically unstable: light in the F' band range bleaches that band at arbitrarily low temperatures, the F band recovering in the process. The F' band shape is believed to be temperature-insensitive. Although this conclusion is mainly based on low-temperature measurements (within the F' thermal-stability range), it has also been confirmed at RT in transient-absorption experiments (Todorov, Koralov and Georgiev (1975)).

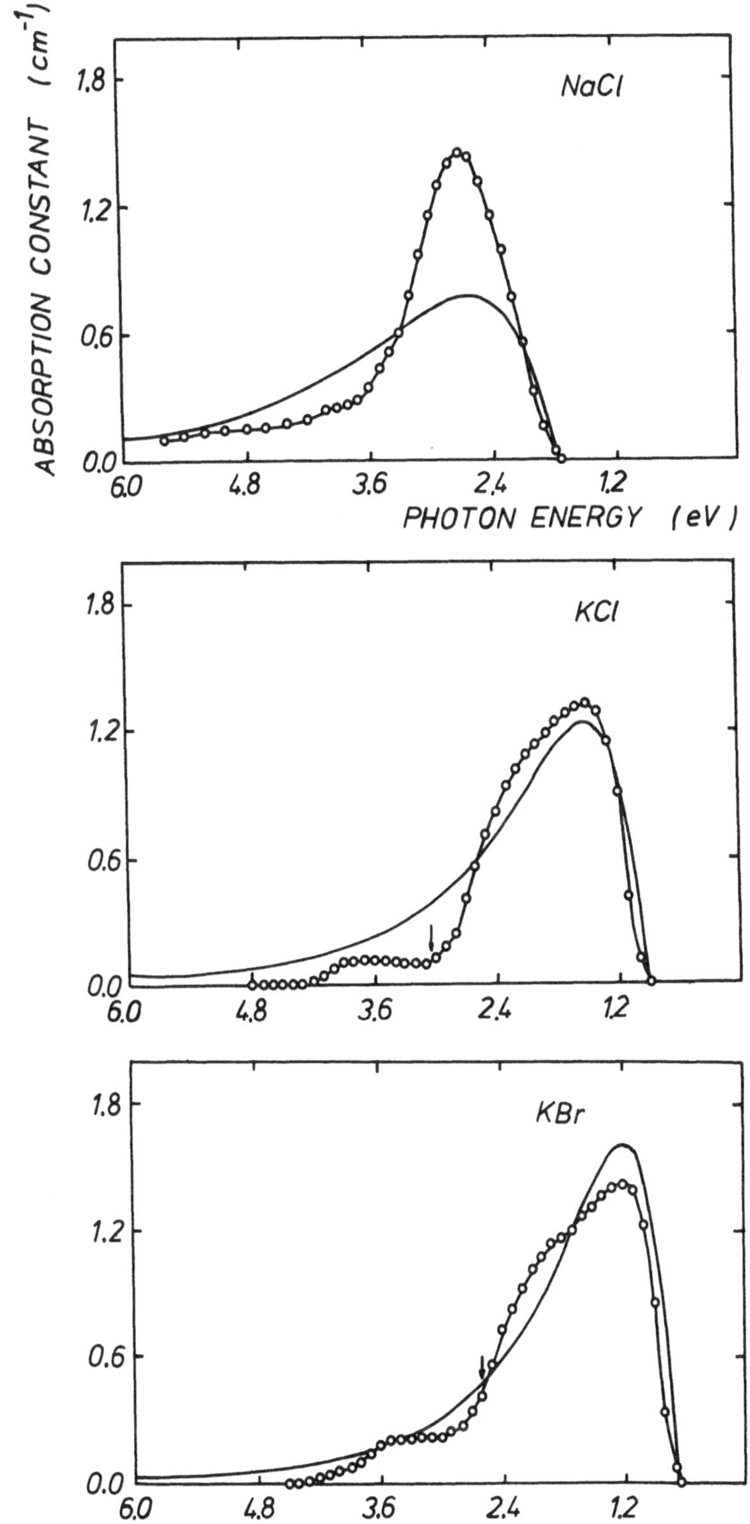
ABSORPTION CONSTANT (cm^{-1})
NaCl
1.8
1.2
0.6
0.0
6.0
4.8
3.6
2.4
1.2
PHOTON ENERGY (eV)
KCl
KBr

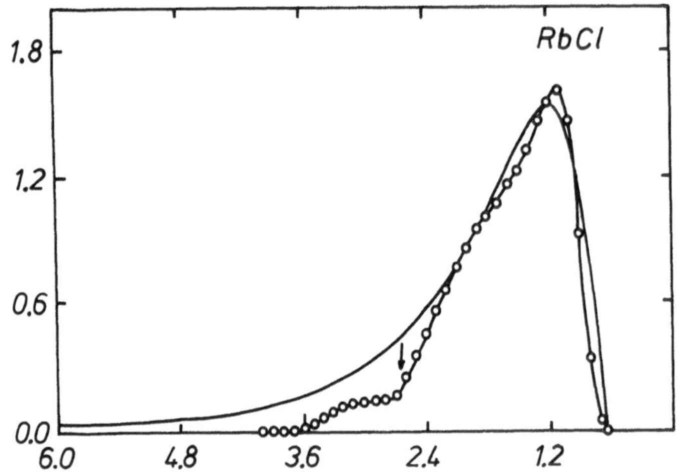
RbCl
1.8
1.2
0.6
0.0
6.0
4.8
3.6
2.4
1.2

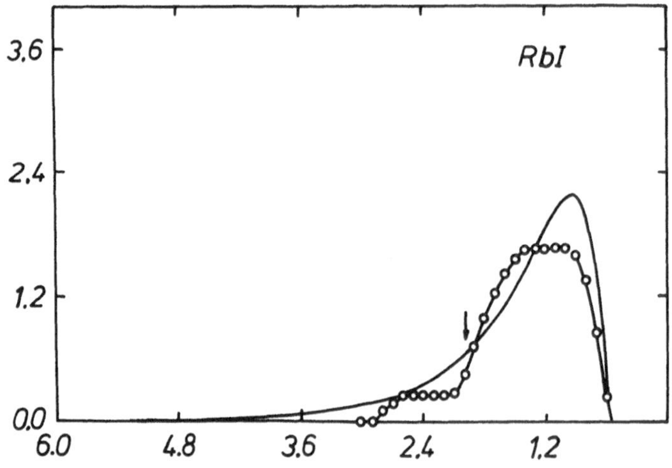
RbI
3.6
2.4
1.2
0.0
6.0
4.8
3.6
2.4
1.2

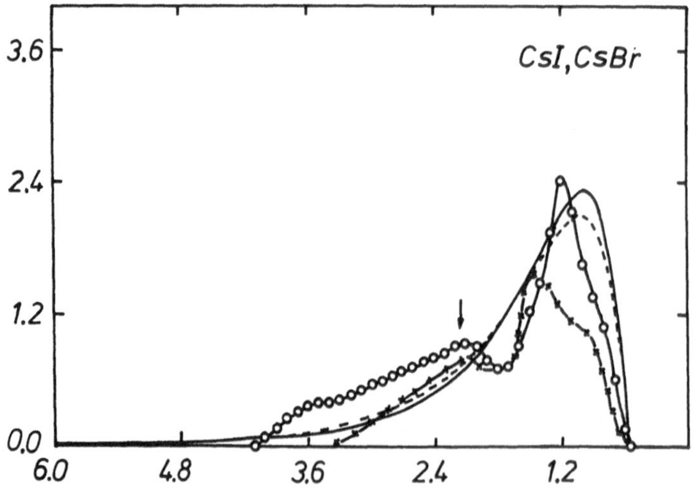
CsI,CsBr
3.6
2.4
1.2
0.0
6.0
4.8
3.6
2.4
1.2

Figure 1: The F' bands in several alkali halides. The circles are experimental points (o–o), while the solid curves (—) are calculated theoretically using a semicontinuum approach. The arrows mark the presumed energies beyond which the F' photoionization may follow reaction (5.a) rather than the "classical ionization path (4.h)" at lower energies. (After Lynch and Robinson (1968)).

The temperature insensitivity and the optical bleachability of the F' band at low temperatures, as well as its large halfwidth, suggest strongly that the band corresponds to optical transitions from a bound state to the conduction band. Accordingly, the F' band shape reflects in some way or the other the energy distribution of the density of states in the conduction band.

Again, NaI and NaBr constitute two notable exceptions from this general rule. The F' band in these materials is somewhat narrower and more bell-shaped, and its halfwidth is strongly temperature-dependent. It has been suggested that the F' band in the above crystals may arise from optical transitions to a bound excited F' state which is autoionized during the relaxation of the lattice (Baldacchini, Pan and Lüty (1981)).

Data on the optical threshold energies, peak positions and halfwidths of the F' band in various alkali halides are presented in Table I.

4. F-F' conversion via the conduction band

The nature of the F' center, the absorbing species responsible for the F' band, as well as the general features of the F-F' conversion have mainly been understood on the basis of the following experimental observations:

(i) The photoconductivity in the F band is inversely proportional to the F center concentration at low excitation levels in the temperature range where the F' band is thermally stable (Glaser).

(ii) The F' center is apparently photoionized on illumination in the F' band at low temperature. Prolonged F' band illumination leads to the complete recovery of the original F band at such a temperature (Pick (1940)).

(iii) The ultimate quantum yields of the F-F' and F'-F conversions, that is, the ultimate initial changes of the F center concentration per one absorbed quantum in the F and F' bands, respectively, are both equal to 2. This implies that two F centers are ultimately destroyed per each absorbed photon in the production of F' centers from F centers, and that two F centers per quantum are ultimately recreated in the reverse optical reaction (Pick (1940)).

Inasmuch as photoelectrons from F centers are mainly trapped by non-ionized F centers (i), combined the above evidence comes to show that the F' center is the result of adding an extra electron to an F center: two electrons eventually trapped in the same anion vacancy.

Based on the available experimental evidence, a mathematical description of the conversion kinetics, as well as of the related physical quantities, can be made by first considering the elemental reactions (see Lüty (1961b)), the rate of each individual reaction being given on its right-hand side:

$h\nu_F + F \rightarrow F^*$	$q_F F$	(4.a)
$F^* \rightarrow F$	F^*/τ_r	(4.b)
$F^* \rightarrow \alpha + e$	F^*/τ_i	(4.c)
$e + \alpha \rightarrow F^*$	$\gamma_\alpha^* n \alpha$	(4.d)
$e + \alpha \rightarrow F$	$\gamma_\alpha^0 n \alpha$	(4.e)
$e + F \rightarrow F'$	$\gamma_F n F$	(4.f)
$F' \rightarrow F + e$	$F'/\tau_{F'}$	(4.g)
$h\nu_{F'} + F' \rightarrow F + e$	$q_{F'} F'$	(4.h)

Here F, F^*, and α stand for the concentration of F centers in ground, relaxed excited, and ionized state, respectively; F' is the F' center density, while n is the concentration of conduction band electrons e. $h\nu_F$ and $h\nu_{F'}$ denote light quanta in the F and F' bands, respectively. Should one deal with photoexcitation in the F band alone (F-F' conversion), $h\nu_{F'}$ has to be replaced by $h\nu_F$, while in considering the F'-F conversion one must set $h\nu_F = 0$ and cancel reaction (4.a). q_F and $q_{F'}$ stand for the photon absorption rates per center for F and F', respectively. Both q's are proportional to the excitation light intensity I, $q_{F,F'} = c_{F,F'} I$, assumed uniform throughout the illuminated part of the sample, the c's being the corresponding photon absorption cross-sections. τ_r^{-1} and τ_i^{-1} are the rates per center for F^* deexcitation and thermal ionization, respectively, $\tau_{F'}^{-1}$ is the thermal ionization rate of an F' center. The γ's are the electron trapping coefficients, reactions (4.d) and (4.e) accounting for excited state pre-trapping and direct ground state trapping in the field of the anion vacancy, respectively.

Combining the rates of the above reactions, one obtains the following rate equations for F^* and n:

$$\dot{F}^* = q_F F - F^*/\tau_F + \gamma_\alpha^* n\alpha \qquad (4.1)$$

$$\dot{n} = F^*/\tau_i - (\gamma_\alpha^* + \gamma_\alpha^o) n\alpha - \gamma_F nF + \beta_{F'} F' \qquad (4.2)$$

where

$$\tau_F^{-1} = \tau_r^{-1} + \tau_i^{-1} \quad . \qquad (4.3)$$

τ_F is the total F^* lifetime, while

$$\beta_{F'} = q_{F'} + \tau_{F'}^{-1} \quad . \qquad (4.4)$$

$\beta_{F'}^{-1}$ is the total F' lifetime against both optical and thermal destruction. Under the conditions of a conventional F' experiment, F^* is kept in photoequilibrium with all related species, giving $\dot{F}^* = 0$. This yields

$$F^* = (q_F F + \gamma_\alpha^* n\alpha)\tau_F \qquad (4.5)$$

Inserting into (4.2) one obtains

$$\dot{n} = \eta_i q_F F - \gamma_\alpha n\alpha - \gamma_F nF + \beta_{F'} F' \qquad (4.6)$$

where the following new important quantities have been introduced:

$$\eta_i = \tau_F/\tau_i \quad , \qquad (4.7)$$

the F ionization quantum yield, and

$$\gamma_\alpha = \gamma_\alpha^o + (\tau_F/\tau_r)\gamma_\alpha^* = \gamma_\alpha^o + (1 - \eta_i)\gamma_\alpha^* \quad , \qquad (4.8)$$

the electron-trapping coefficient of an anion vacancy. The latter is composed of a direct term γ_α^o plus a via-the-excited-state term, given by the product of γ_α^* times the probability τ_r^{-1}/τ_F^{-1} for deexcitation of the F^* electron.

The complete set of rate equations for the species under consideration can be complemented by two conservation conditions which occur after integration and are not independent of the equations:

$$\alpha = n + F' \quad , \qquad (4.9)$$

the charge-balance condition, and

$$F + F^* + F' + \alpha = F_0 \quad , \qquad (4.10)$$

the vacancy-conservation condition. Here F_0 is the F center density in the freshly-quenched state of the sample. Under the usual conditions the free-electron density is kept low compared to the color-center concentrations leading to $F' = \alpha$. It is also implied that there are no anion vacancies in any significant amount other than the ones originating from the photoionization of F centers.

In addition, we shall also assume photoequilibrium for the conduction electrons when dealing with the conversion kinetics. Setting $\dot{n} = 0$ in

(4.6) we obtain the photoequilibrium electron density

$$n = (\eta_i q_F F + \beta_{F'} F')/(\gamma_\alpha \alpha + \gamma_F F) \qquad (4.11)$$

This is a quantity proportional to the photocurrent. Now the rate equations for F and F' are

$$\begin{aligned} \dot{F} &= -\eta_i q_F F + \beta_{F'} F' + (\gamma_\alpha \alpha - \gamma_F F)n \\ &= 2(-\eta_i q_F \gamma_F F^2 + \gamma_\alpha \beta_{F'} \alpha F')/(\gamma_\alpha \alpha + \gamma_F F) \end{aligned} \qquad (4.12)$$

and

$$\begin{aligned} \dot{F}' &= \gamma_F n F - \beta_{F'} F' \\ &= (\eta_i q_F \gamma_F F^2 - \gamma_\alpha \beta_{F'} \alpha F')/(\gamma_\alpha \alpha + \gamma_F F) \end{aligned} \qquad (4.13)$$

4.1. F' thermal-stability range

Within the temperature range where the F' center is stable thermally $\beta_{F'} = q_{F'}$. From eq.(4.12) one obtains the initial quantum yield of the F-F' conversion, neglecting the F' containing terms at $t = 0$:

$$\eta_{F-F'} = (-\dot{F}/q_F F)_{t=0} = 2\eta_i \qquad (4.14)$$

Similarly, we obtain the initial quantum yield of the reverse F'-F conversion by setting $q_F = 0$ in eq.(4.12), giving

$$\eta_{F'-F} = (\dot{F}/q_{F'} F')_{t=0} = 2\gamma_\alpha \alpha/(\gamma_\alpha \alpha + \gamma_F F) \qquad (4.15)$$

It is therefore equal to twice the probability that a conduction electron is trapped by an anion vacancy in ground state. Experimentally, the F-F' conversion preceding the $\eta_{F'-F}$ measurements at various temperatures is kept constant, e.g. $F/F' = 0.5$.

Pick investigated the change in the F center concentration as a function of the number of absorbed quanta at several conversion temperatures for the F-F' (Pick (1938)) and the F'-F (Pick (1940)) conversion. In either case the kinetic curves tended to gradually saturate after rising initially steeply, apparently as a result of the increasing role of the reverse conversion. This general behavior is in qualitative agreement with eq's (4.12) and (4.13), as solved under the appropriate initial conditions. The initial slopes in both cases, that is, the quantum yields $\eta_{F-F'}$ and $\eta_{F'-F}$, respectively, are plotted in Fig.2 vs. the corresponding conversion temperatures. According to eq.(4.14) the temperature dependence of $\eta_{F-F'}$ reflects the temperature variation of η_i attaining the ultimate value of 2 when η_i attains 1. From equation (4.15), the temperature variation of the quantum efficiency

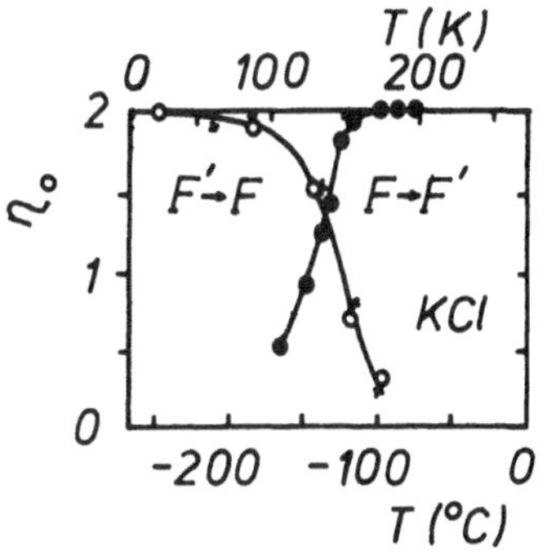

Figure 2: Temperature dependences of the quantum yields of the F-F' and F'-F photoconversions in KCl. (After Pick (1940)).

$\eta_{F'-F}$ arises from γ_α/γ_F, provided F/F' has been kept constant. Physically this means that while the F-F' conversion efficiency increases as a result of the increasing F center photoionization probability, the F'-F yield drops due to the decreasing probability for electron-vacancy recombination as the temperature is raised. $\eta_{F'-F}$ attaining the ultimate value of 2 at the lowest temperatures would imply that γ_α/γ_F has exceeded largely the pre-selected value of F/F'.

The ultimate F-F' conversion at a given temperature can be obtained from eq.(4.13) setting $\dot{F}' = 0$ and using $F' = \alpha$, $F^* \ll F$:

$$F'_{max} = F_0 \,/\, (2 + (\gamma_\alpha \, \beta_{F'}/\gamma_F \eta_i \, q_F)^{\frac{1}{2}}) \qquad (4.16)$$

F'_{max} is thus predicted to be strongly temperature-dependent. At the lowest temperatures only a negligible fraction of the original F centers can be converted into F', while as the temperature is raised F'_{max} increases steeply. The temperature dependence arises from

$$(\gamma_\alpha/\gamma_F)/\eta_i = (\gamma_\alpha^0/\gamma_F)/\eta_i + (\gamma_\alpha^*/\gamma_F)(\tau_i/\tau_r) \qquad (4.17)$$

for $\beta_{F'} = q_{F'}$. This quantity decreases as the temperature is raised within the F' thermal-stability range. Fig.3 shows the experimentally observed temperature variation of F'_{max}. The fall at the highest temperatures arises from the thermal instability of the F' center ($\beta_{F'} = \tau_{F'}^{-1}$)(Pick (1938)).

A photocurrent is initiated on optical excitation in either the F or the F' band, prior to or after the F-F' conversion, respectively. The photoexcitation is usually chosen sufficiently low to prevent the accumulation of a space charge (Pohl (1937)). The steady-state photocurrent in both cases can be obtained from eq.(4.11). For historic reasons we prefer to derive an expression for the photoconductive yield ζ, the

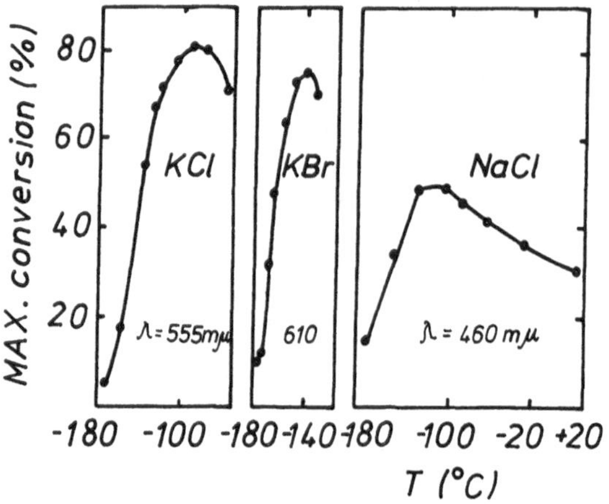

Figure 3: Temperature dependences of the maximum number of F centers bleached in the F-F' conversion in several alkali halides (percent data). The wavelengths of the bleaching light are also indicated. (After Pick (1938)).

photoconductivity per absorbed quantum. Accordingly, we obtain

$$\zeta_F = e\mu_e n/q_F F = e\mu_e \eta_i \tau_e = \tfrac{1}{2} e\mu_e \tau_e \eta_{F-F'} \qquad (4.18)$$

for the F-photocurrent, and

$$\zeta_{F'} = e\mu_e n/q_{F'}F' = e\mu_e \tau_e/(\gamma_\alpha \alpha/\gamma_F F + 1)$$
$$= e\mu_e \tau_e(1 - \tfrac{1}{2}\eta_{F'-F}) \qquad (4.19)$$

for the F'-photocurrent. Here

$$\tau_e = (\gamma_F F)^{-1} \qquad (4.20)$$

is the electron-trapping time by F centers, while μ_e is the electron mobility. The temperature dependences of ζ_F and $\zeta_{F'}$ obtained experimentally by Glaser in 1937 and Domanic (1943), respectively, are shown in Fig.4. According to eq's (4.18) and (4.19) they arise for different reasons: As the temperature is raised, ζ_F increases due to the increase of η_i, while the rise of $\zeta_{F'}$ is due to the parallel drop of $\eta_{F'-F}$. Both yields tend to level out at the highest temperatures attaining the ultimate value of $e\mu_e\tau_e$, e-times the 'displacement distance' per unit field.

Equations (4.14), (4.15), (4.18), and (4.19) show just how measurements of $\eta_{F-F'}$, $\eta_{F'-F}$, ζ_F, and $\zeta_{F'}$, can be used to calculate η_i, γ_α/γ_F, and τ_e. In particular, ζ_F-based calculations at known μ_e have yielded $\gamma_F = 7.5 \times 10^{-8}\ cm^3 s^{-1}$ at 200 K for KCl

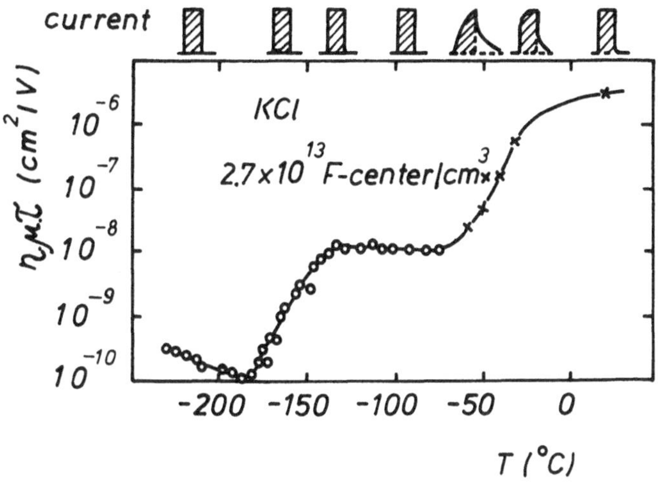

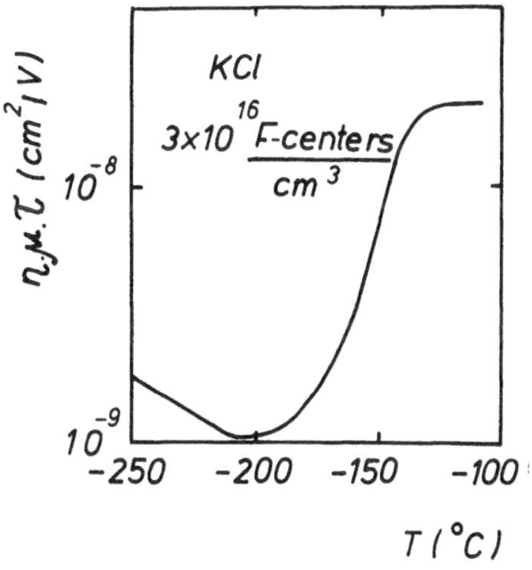

<u>Figure 4:</u> Temperature dependences of the F- (top) and F'- (bottom) photocurrents. The experimental data are by Glaser (1937) and Domanic (1943), respectively. After Lüty (1961b).

(Markham (1966)). This quantity is not expected to be strongly temperature-dependent, as we shall see it shortly. Inasmuch as $\gamma = \sigma u$, where σ is the trapping cross-section and u is the electron thermal velocity, taking $u = 0.95 \times 10^7$ cm/s one obtains $\sigma_F = 7.5 \times 10^{-15}$ cm^2, which is the order of magnitude of the geometrical cross-section of the F center. By similar arguments the value of $\gamma_\alpha^* / \gamma_F$ has been estimated at about 20 at 80 K (Markham (1966)). This gives $\gamma_\alpha^* = 1.5 \times 10^{-6}$ cm^3s^{-1} and $\sigma_\alpha^* = 1.5 \times 10^{-13}$ cm^2. As the temperature is raised, the pre-trapping term decreases due to the increasing F^* thermal-ionization probability, in agreement with eq.(4.8). An early interpretation of Pick's quantum yield vs. temperature dependences is due to Markham (1952).

A continuum theory of diffusion-controlled electron trapping which derives expressions for the γ's has been developed by Pekar (1951). The general form of the trapping coefficient is

$$\gamma = 4\pi D_e \Big/ \int_0^{1/r_0} \exp(eV/k_BT)\, d\frac{1}{r} \tag{4.21}$$

where $V = V(r)$ is the potential of the force exerted on the electron at distance r from the trapping center, $D_e = (k_BT/e)\mu_e$ is the electron diffusion coefficient, and r_0 is the radius of the bound state. Assuming a semicontinuum model for the field of the anion vacancy (Fowler (1968)) we get

$$V(r) = -e/\varepsilon r \qquad \text{for } r > r_0 \tag{4.22}$$

and set $e^2/\varepsilon r_0 k_BT > 1$ to obtain

$$\gamma_\alpha^* = 4\pi e \mu_e/\varepsilon \tag{4.23}$$

for the excited-state pre-trapping coefficient. Equation (4.23) has apparently been derived much earlier (see Crandall (1965)). It is therefore well-familiar. To estimate γ_α^o we set

$$V(r) = V_0 \quad (r_0 < r < R), \quad = -e/\varepsilon r \quad (R < r < \infty) \tag{4.24}$$

which gives

$$\gamma_\alpha^o = \gamma_\alpha^* / (1 + (e^2/\varepsilon k_BT)(r_0^{-1} - R^{-1})\exp(eV_0/k_BT)) , \tag{4.25}$$

that is, $\gamma_\alpha^o \lesssim \gamma_\alpha^*$ for $R \gtrsim r_0$ and $-eV_0/k_BT > 1$. R is the radius of the vacancy. The field in the case of trapping by F centers is assumed to be of the monopole-induced dipole type:

$$V(r) = -\tfrac{1}{2}(\alpha_F e/\varepsilon^2 r^4) \tag{4.26}$$

where α_F is the F center polarizability. Inserting into (4.21) we obtain for $\alpha_F e^2/2\varepsilon r_0^4 > k_BT$,

$$\begin{aligned} \gamma_F &= 4.4\pi\mu_e(k_BT)^{3/4}(\alpha_F/2e^2\varepsilon^2)^{1/4} \\ &= 1.1\,(k_BT)^{3/4}(\alpha_F\varepsilon^2/2e^6)^{1/4}\gamma_\alpha^* \end{aligned} \tag{4.27}$$

All the three trapping coefficients considered depend on the temperature by way of the electron mobility mainly.

Equation (4.27) works surprisingly well at low temperatures: As a matter of fact, setting $\alpha_F = 2 \times 10^{-23}\ cm^3$, $\varepsilon = 4.84$ we obtain $\gamma_F/\gamma_\alpha^* = 0.014$ at 80 K for KCl which is not far from an experimental value of 0.016 (Costicas and Grossweiner (1962)). On the other hand,

Crandall (1965) has measured the F' photocurrent for a direct experimental determination of the electron trapping coefficients. This allows an experimental check-up of eq's (4.23) and (4.27). He plots

$$(e\mu_e / \zeta_{F'})/\alpha = \gamma_\alpha + \gamma_F(F/\alpha) \qquad (4.28)$$

vs. F/α at given temperatures in KBr to obtain straight lines of slope γ_F and intercept γ_α. Between 40 to 80 K the intercept is found to follow eq.(4.23) rather well, while above 80 K there is an increasing F^* thermal ionization and the more general equation (4.8) turns on. However, there are considerable deviations from eq.(4.23) below 40 K. It seems likely that the conditions under which eq.(4.21) has been derived may be violated below that temperature. As a matter of fact, the general expression for the electron trapping coefficient, as derived by Pekar's theory is

$$\gamma = \gamma_b \exp(-eV(r_o)/k_BT)/(1 + (\gamma_b/4\pi D_e)\exp(-eV(r_o)/k_BT) \times \int_o^{1/r_o} \exp(eV(r)/k_BT)d\tfrac{1}{r}) \qquad (4.29)$$

Here γ_b is a rate constant for the transition of the electron into a bound state. Equation (4.21) obtains when that probability is high and the electron migration to the trapping center is rate-determining. As the temperature is lowered, the electron mobility increases and the situation may reverse: now, the transition into a bound state may become rate-determining; consequently,

$$\gamma = \gamma_b \exp(-eV(r_o)/k_BT) \qquad (4.30)$$

Crandall's data are shown in Fig.5. Best agreement with eq.(4.8) is obtained for

$$\eta_i = (1 + 3.45 \times 10^{-8} \exp(0.13 \text{ eV}/k_BT))^{-1} \quad \text{(KBr)}$$

In calculating the left-hand side of (4.28) the electron-mobility data by Ahrenkiel and Brown (1964) have been used, while the α-center concentration has been determined by measuring the F center density and assuming eq's (4.9) and (4.10), F^* and n neglected. A similar temperature plot has later been reproduced by Hoffmann (1973) for KCl using data by Hunger, again manifesting the applicability of eq's (4.8) and (4.23) between 50 to 160 K. The slope of eq.(4.28) in Crandall's experiment has also been found to be temperature-dependent. These data are represented in Fig.6. However, the agreement between theory and experiment now seems to be less satisfactory. In no case has there been any experimental indication of a sizeable γ_α^o term.

Prior to Crandall's measurements a comprehensive study of the photoconductive and interconversion properties of F and F' centers in KCl

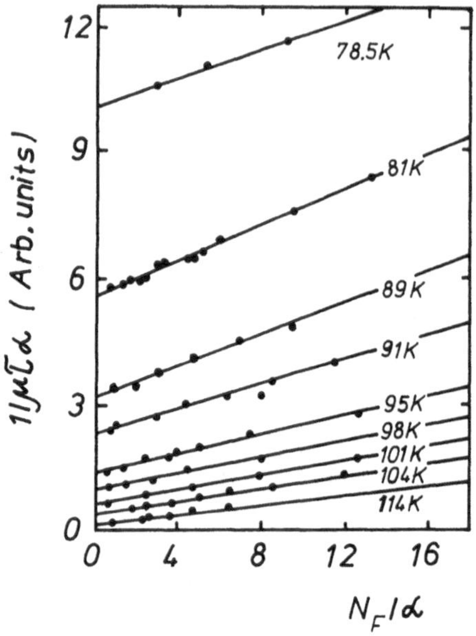

Figure 5: Experimental data from KBr showing the validity of eq.(4.28) at not too low temperatures or high α center densities. The slope of a straight line determines the electron trapping coefficient γ_F of the F center, while the intercept with the ordinate yields the electron trapping coefficient γ_α of the anion vacancy. (After Crandall (1965)).

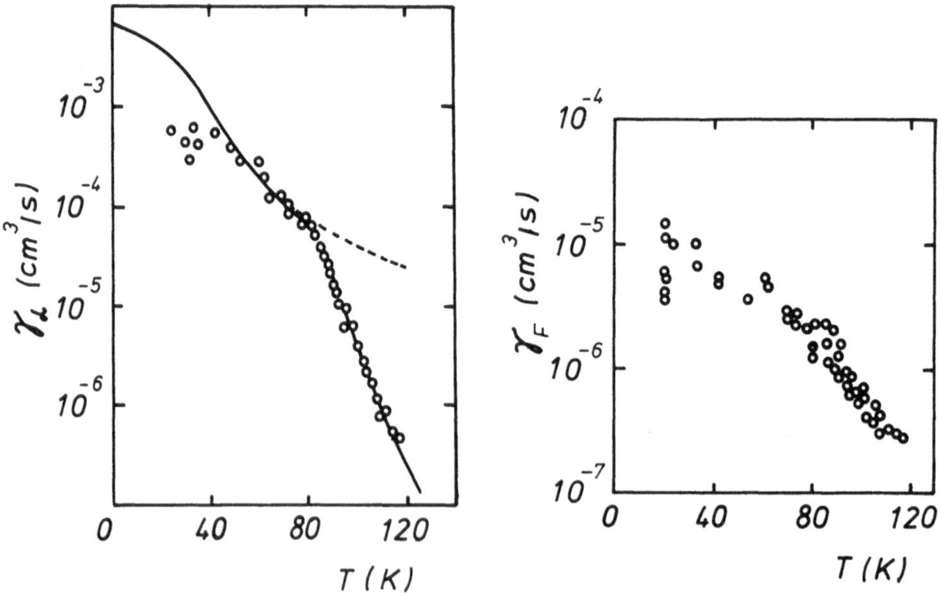

Figure 6: Temperature dependences of γ_α (left) and γ_F (right) in

KBr measured as explained in Fig.5. The solid line in the drawing on the left follows eq.(4.8) with $\gamma_\alpha^o = 0$ and γ_α^* given by eq.(4.23)(extrapolated by a dashed line).(After Crandall (1965)).

has been made by Fedders, Hunger, and Lüty (1961). Among the quantities measured are: $\eta_{F-F'}$, $\eta_{F'-F}$, and $\zeta_{F'}$ between 90 to 180 K. The first of these has been used to find the temperature dependence of the F ionization quantum yield η_i (see eq.(4.14)), while experimental data on the third have been exploited to calculate γ_α/γ_F from the intercept-to-slope ratio according to eq.(4.28). Apparently the lack of any reliable mobility data has prevented separating slope from intercept, as done in Crandall's case. The ionization yield data have been found to fit well

$$\eta_i = (1 + 2.3 \times 10^{-7} \exp(0.16\ \mathrm{eV}/k_BT))^{-1} \quad \text{(KCl)},$$

while eq.(4.8) has proved applicable with $\gamma_\alpha^o/\gamma_F = 0$ and $\gamma_\alpha^*/\gamma_F = 6$. In turn, using the latter data and eq.(4.15) has led to a good agreement with the direct $\eta_{F'-F}$ measurements. Although this implies an apparent triumph for the conventional F-F' phenomenology, it is not at all clear whether a constant γ_α^*/γ_F has resulted from any insufficient experimental accuracy or from any inadequacy of Pekar's theory. The system studied has been fairly dilute ($F_0 = 10^{16}$ cm^{-3}) to avoid interference by color center pair interactions.

The ionization quantum efficiency η_i has also been derived from experimental measurements of the F^* lifetime in KCl, KBr, KI, and NaCl (Swank and Brown (1963)). Assuming $\tau_r = \tau_R = 0.577 \times 10^{-6}$ s , the radiative lifetime, one obtains for KCl below 150 K

$$\eta_i = (1 + 4.3 \times 10^{-7} \exp(0.15\ \mathrm{eV}/k_BT))^{-1} \quad \text{(KCl)} .$$

More recently Popov and Terzijski (1973) have investigated the F' formation kinetics in KCl below 150 K. The shape of the experimental curves has been found to follow closely the predictions of the conventional F-F'conversion model. In particular, this conclusion also applies to the dependence of the growth curves on the excitation-light intensity, investigated at 130 K. In a subsequent work (Terzijski and Popov (1974)) the steady state F' density (4.16) is shown to be independent of the excitation-light intensity I below 150 K. The temperature dependence of the relative F-ionization quantum yield has followed closely Fedders et al.'s data (Popov (1979)).

An obscurity occurs regarding the predictions of eq.(4.16) when high excitation-light intensities are used. A study by Costicas and Grossweiner (1962) at 80 K in KCl reveals that the photoequilibrium F' density

F'_{max} depends as log(I) rather than being independent of I , as required by eq.(4.16) for $\beta_{F'} = q_{F'}$. It is suggested that the inconsistency may arise from an intensiry dependence of the trapping cross-section σ_F at high I , σ_F increasing with I because of the increasing probability that the F center potential field V(r) is overlapped by the field of an α center. Later, an intensity dependence of F'_{max} has also been reported by Tubbs and Wright (1971) for KCl at 135 K.

A series of highly informative experiments have dealt with the field-assisted ionization of the excited F center. The field effect can be observed in alkali halides because of the relatively low F^* thermal-ionization energy, the order of 0.1 eV. Field-assisted ionization can materialize in two ways: at higher temperatures a lowering of the thermal-ionization barrier by the applied field (Schottky emission) would mainly occur, while at lower temperatures a narrowing of the barrier would predominate leading to an increased ionization rate via tunneling. The applied field can therefore be expected to enhance the ionization quantum yield η_i and to lower the electron-trapping coefficient γ_α at a given temperature. In agreement, field-assisted experiments have indeed shown $\eta_{F-F'}$ to increase (Lüty (1958; 1961b)) and $\eta_{F'-F}$ to decrease with the field strength (Lüty, 1961b). For a discussion of the field-effect theory see Fowler (1968) and literature cited therein.

An electric-field modulation of η_i has been performed by Reguzzoni and Samoggia (1968). An a.c. electric field of frequency ν is applied to an additively colored KCl crystal at LNT which is simultaneously illuminated with white light to excite the F center. The field modulates the F^* ionization probability η_i . The combined photoelectric effect initiates presumably the following reaction:

$$F + F + h\nu \rightarrow F^* + F \rightarrow \alpha + e + F \rightarrow \alpha + F' \rightarrow F + F \ .$$

Inasmuch as the F' center is thermally stable at LNT, the last step should occur in a way other than the thermal ionization of the undisturbed F' center. As a result of the above photoelectric cycle the concentrations of F and F' vary periodically with time around their average values and the monitoring light through the sample is consequently amplitude-modulated at 2ν . This is a technique of great sensitivity since a coherent amplification technique is used. The resulting spectrum of $\Delta I / I$ contained an F band (negative) and the F' band (positive), F_0 being about 10^{16} F/cc. The F' peak was at about 9000 Å. ΔI was found to drop as ν increased, while the shape of the spectrum remained unchanged. A time constant the order of 10^{-3} s has proved characteristic of the particular F-F' process. A linear dependence of the signal on the field strength has been established for fields up to 50 kV/cm.

Later Nosenzo, Reguzzoni, and Sammogia (1970) extended these measurements to additively colored (5×10^{16} F/cc) KCl and KBr at LNT. Transmission changes $\Delta I / I$ of less than 0.001 % have easily been detected by integrating the output of a lock-in amplifier. The spectral distribution of $\Delta I / I$ for a freshly-quenched sample has again shown both the F and F' bands. The F' band has also occurred in the presence of the F aggregate bands, formed by RT F band bleaching.

The F-F' conversion in KCl, KBr, and KI has been investigated under illumination in the higher-energy bands of the F center as well (Lüty (1960)). The L_3 band, most prominent of the L's, has varied along with the F band during the F-F' conversion. In turn, the L-F' and K-F' conversions have been found to bring about a reduction of the F band. At -180 ^oC and $F_0 = 5.3 \times 10^{16}$ F/cc the corresponding quantum yields increased in the order: $\eta_{F-F'}$, $\eta_{K-F'}$, $\eta_{L-F'}$. In a subsequent paper Hunger and Lüty (1965) have determined the absolute quantum yield $\eta_{L-F'}$ to demonstrate that the L bands are in fact transitions of the F center. Investigating the F band bleaching under L_2 light reveals a constant initial rate of F center destruction independent of the temperature between -193 ^oC to -120 ^oC. Consequently L_2 is a temperature-independent source of photoelectrons captured by F to form F'. $\eta_{L-F'}$ has been determined in two ways by comparing with the well-known values of $\eta_{F-F'}$:

(i) A direct comparison with the F band reduction under F and L_2 light was made and the number of photons absorbed in both cases obtained from the ratio of F to L_2 absorptions through calibrating the relative light intensity used. The result was $\eta_{L-F'}/\eta_{F-F'} = 1.15 \pm 0.23$.

(ii) In another approach the F' photocurrent was utilized. If N_F and N_L are the numbers of photons absorbed in F and L, and Δi_F and Δi_L are the corresponding increases of the photocurrent, $\eta_{L-F'}/\eta_{F-F'} = \Delta i_L N_F / \Delta i_F N_L = 1.25 \pm 0.13$. At the temperature of measurement (150 K), $\eta_{F-F'} = 1.75$ giving $\eta_{L-F'} = 2.0$ or 2.2, respectively, by the two methods.

Two peculiar examples of unusual F' centers are apparently provided by NaBr and NaI (Baldacchini and Lüty (1977), Pan and Lüty (1977), Baldaccini, Pan, and Lüty (1981)). These crystals have been colored with great difficulties. On irradiation with light in the F band, a F' band develops at the expense of F which peaks on its high-energy side, at odds with their disposition in most alkali halides.(Fig.7). In addition, the F' halfwidth in both materials is temperature-dependent through

$$H(T)^2 = H(0)^2 \coth(h\nu/2k_BT)$$

with $\nu = 152$ cm^{-1} (NaI) and 133 cm^{-1} (NaBr). The optical F to F' con-

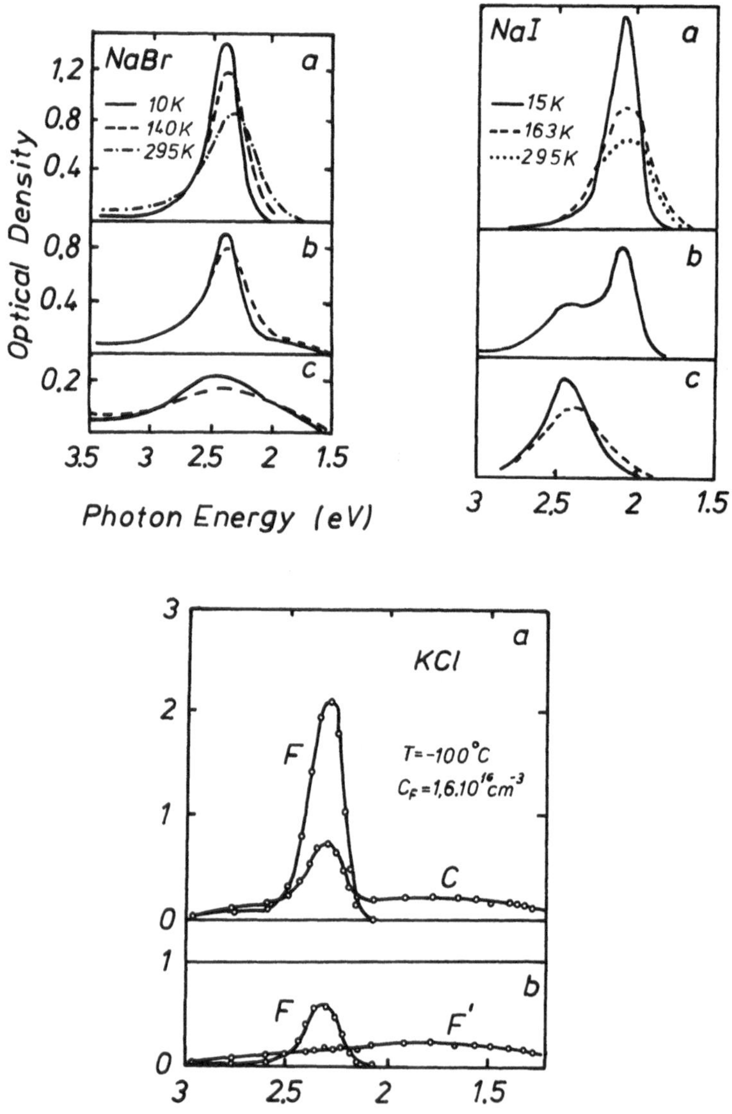

Figure 7: The F and F' bands in NaBr and NaI: (a) pure F bands; (b) absorption following an F-F'conversion; (c) pure F' bands. The situation in KCl is also shown for comparison, after Pick (1940). In contrast to KCl, the F' bands in NaBr and NaI peak on the high-energy side of the F band. (After Baldacchini, Pan, and Lüty (1981)).

version can be completely inverted by irradiation into the F' band. The temperature dependence of $\eta_{F-F'}$ in a dilute F center system exhibits the usual behavior attaining an ultimate value of 2, as verified for NaI assuming the same oscillator strength for the F band as in KBr.

An Arrhenius plot of $\eta_i^{-1} - 1$ where $\eta_i = 0.5\,\eta_{F-F'}$ yields

$$\eta_i = (1 + 1.8 \times 10^{-2} \exp(0.030\ \mathrm{eV}/k_BT))^{-1} \quad \text{(NaI)}$$

$$\eta_i = (1 + 5.56 \times 10^{-4} \exp(0.061\ \mathrm{eV}/k_BT))^{-1} \quad \text{(NaBr)}$$

At the same time, $\eta_{F'-F}$ has been found to be temperature-independent (virtually 2) for NaI between 10 to 120 K. This implies that the recapture of the F' photoelectron by an ionized F center may effectively occur into the ground state directly leading to a high F'/F ratio (cf. eq.(4.15)). On the other hand, the symmetric and rather narrow bandshape suggests transition into a bound singlet F' state which may be autoionized during lattice relaxation, to account for the absolute $\eta_{F'-F}$ value measured. Studies of Raman scattering have also been made (Pan and Lüty (1977)) using a two-beam technique: one strong pump-beam initiates and shifts the F-F' photoequilibrium, while another weaker probe-beam measures the Raman effect at a different laser frequency. By this method, the Raman responses of F and F' could be separated. Both Raman spectra have been found to mainly consist of a sharp A_{1g} line (at 113 cm^{-1} for F and at 121 cm^{-1} for F'). The F' line has exhibited a larger thermal broadening.

Baldacchini, Gallerano, Grassano, and Lüty (1981) have carried out interesting magneto-optical experiments aimed at verifying the assignment of the F' center absorption in NaI and NaBr. They have first studied the presumed F-F' conversion in a high magnetic field where it might be inhibited because of the spin polarization of the F center electron. Confirming these expectations, the F-F' quantum yield was found to drop to one half without getting completely quenched as the magnetic field increased up to 80 kGauss. The partial quenching was attributed to the loss of spin memory during an optical pumping cycle, the spin-mixing parameter being estimated at $\epsilon = 0.26$ (NaI), 0.13 (NaBr). To check this hypothesis the authors measured ϵ independently using the magnetic circular dichroism (MCD) of the F band. The MCD signal is proportional to the ground-state spin polarization of the F center. The latter can be changed through optical pumping in one of the dichroic peaks, while monitoring in the other. (Spectrally, the MCD signal composes of two peaks of opposite polarity centered symmetrically on both sides of the F band peak.) The rise-time of the MCD signal due to the pumping light is inversely proportional to ϵ, while the decay-time (with the pumping light switched-off) is proportional to the spin-lattice relaxation time t_1. The following values were obtained: ϵ = 0.12 (NaI), 0.03 (NaBr), in partial disagreement with the ones obtained from the F-F' yield. It appears that more spin memory is lost in the optical cycle leading to F' center formation which suggests a second mechanism of F' production, probably via an excited triplet state. Indeed, a new metastable and field-dependent absorption was found to occur during the F-F' conversion which could be due to transitions from the F' triplet to the conduction band.

More recently Takiyama, Fujita, Nojima, and Nishi (1983) seem to have also observed an "odd F' band" in LiCl lying on the short wavelength side of the F band. The samples have been colored by ^{60}Co (3×10^8 R) gamma rays at LNT, and the optical absorption spectrum measured at 80 K from 2.8 to 8.25 eV. The F band occurred at 3.27 eV. On excitation with F light at 80 K, a new band appeared at 3.65 eV which was ascribed to some type of F' center, based on the reversibility of the presumed F-F' conversion and the measured F-F' quantum efficiency of 2 (to be reported in detail elsewhere). Another band at 7.6 eV is attributed to the β center, while the alpha band is found to be at 7.45 eV. It is to be noted that LiCl falls well within the group of materials expected to exhibit "odd F' bands" based on Luty's criterion $r_+ / r_- < 0.5$, the cation-to-anion-radii ratio (Baldacchini, Gallerano, Grassano, and Lüty (1981)).

The electron Hall-mobility in alkali halides containing F, F', and α centers has been investigated down to LHeT by Ahrenkiel and Brown (1964). For a KBr sample in which the mobility at 7.7 K was 11000 cm^2/Vs before the F-F' conversion, a 40 % conversion reduced the mobility to 1800 cm^2/Vs. The mobility was also found to gradually increase to its original high value upon bleaching in the F' band at 7.7 K. This implies that there are no other traps to compete effectively with α and F' even when the latter are reduced to low concentration by bleaching. Consequently, the presence of F' and α, strong scatterers due to the Coulombic charge, leads to a marked decrease of the electron mobility. An attempt was made to describe the dependence of μ_{Hall} on the α concentration in terms of the Conwell-Weisskopf formula, giving a fair agreement. The following empirical relationships have been obtained:

$$\mu_{Hall} = A\, F^{-0.41}$$

$$\mu_{Hall} = B\, \alpha^{-0.52}$$

Some highly informative investigations have been made of the effect of a magnetic field on the photoconductivity of F centered alkali halides. By studying the longitudinal magnetoconductivity, evidence has been obtained for the dependence of the electronic transport at low temperature on the orientation of the conduction-electron spins (Hodby, Borders, and Brown (1970)). The experimental results have been understood in terms of the interaction of spin-polarized conduction electrons with spin-polarized F centers. It is concluded that the transport of photoelectrons at LHeT in KCl, KBr, and KI containing 10^{16} to 10^{17} F/cc is likely to occur by way of a diffusive motion and that a spin-dependent scattering or trapping time may be defined. These studies have later been extended by Jenkin, Stacey, Crowder, and Hodby (1978). In a zero magnetic field the carrier lifetime before final trapping approached the scattering time at 4 K when the F center density exceeded 10^{17} cm^{-3}. Now the trapping process is important in reducing the carrier mobility. In a high magnetic field the ratio of the cross-section for trapping by F centers in a triplet state to the singlet cross-section was found to be less than 5 % for KBr, KI, RbCl, and RbI, and about 30 % for KCl. The reason for the unusually high KCl value is unknown, since it is generally believed that the F' center does not have a bound triplet state. Such an orbitally antisymmetrical state would be very diffuse and with little binding to the central positive charge, since the singlet itself is bound by only half an electron volt. Photoexcitation from F' centers in a magnetic field enables one to compare between the transport of electrons, initially polarized (those originating from F centers), and non-polarized (coming out from F' centers). At high magnetic fields, F' conduction electrons attain an average polarizability of about 20 % as compared with over 90 % for F electrons. Evidence for a weak spin exchange in scattering has also been obtained.

A study of the low-temperature photoconductivity of KCl between 10 to 90 K by Ursula Haupt (1963) has apparently been aimed at checking the hypothesis whether the residual photocurrent in the F or F' band is related to colloids, as suggested frequently (Seitz (1946), (1954), see Fig.4). Among other things, the temperature dependence of the Schubweg $\zeta_{F'}/e$ was found to be weak with a sudden drop at the lowest temperatures. The Schubweg in a zone-refined crystal was always a bit higher than in an air-grown crystal. Electric glow curves were obtained yielding three main peaks. The highest-temperature glow peak at 198 K is ascribed to F'. It is worth noting that the F' peak has appeared in a colloid-containing crystal as well, the peak being much higher in an air-grown sample than in a zone-refined sample. Although the low-temperature photoconductivity in colloid-containing crystals has been about an order of magnitude higher than in F centered crystals, no definite conclusions are drawn on the nature of the residual photocurrent.

In an attempt to produce F' centers, additively colored CsF samples were bleached optically in the K band at 140 K (Cases, Alcala, and Orera (1982)). This resulted in a broad band (followed between 750 to 1600 nm) that peaked at about 925 nm at 140 K which also exhibited a shoulder at about 1050 nm. The band has been assigned to F' based on the overall information collected: The "F' band growth curve" was found strongly dependent on the temperature of bleaching. Measurements at 75 K and at 140 K at the same F_0 showed that both $(dF'/dt)_{t=o}$ and F'_{max} were the larger the higher the temperature. Only about 25 % of the F centers were destroyed at saturation at 140 K. Bleaching with "F' light" led to the decrease of F' and the increase of F. The F' bleaching rate was higher at 75 K than it was at 140 K. The F' were found to become unstable thermally at higher temperatures.

4.2. Range of F' thermal instability

Quite a few experiments have been made in the temperature range where the F' center is thermally unstable. Now the reverse F'-F conversion occurs spontaneously in the dark in a crystal pre-irradiated with F light to produce F' centers. The F' band thermal decay has been found not to follow the first-order reaction type. Nevertheless, Pick (1938) measured the temperature dependence of the F' halflife in KCl, KBr, and NaCl; activation energies of the order of half an electron volt can be estimated from the Arrhenius plots.

The deviation of the F' band thermal decay from the exponential type is physically clear. Indeed, having been ejected to the conduction band, the F' extra electron has a chance of not only recombining with an ionized F center but also of getting trapped by an F center to restore an F' center. Consequently, the first-order reaction rate $F'/\tau_{F'}$ has to be multiplied by the probability that the ejected F' electron will be captured by an empty anion vacancy, to obtain

$$\dot{F}' = -(F'/\tau_{F'}) \frac{\gamma_\alpha \alpha}{\gamma_\alpha \alpha + \gamma_F F} \tag{4.31}$$

which also follows directly, as it should, from eq.(4.13) at $q_F = 0$ and $\beta_{F'} = \tau_{F'}^{-1}$. Now using (4.9) and (4.10) (F^* and n neglected) we obtain

$$\dot{F}' = -(F'/\tau_{F'}) \frac{F'}{(\gamma_F/\gamma_\alpha)F_0 - (2(\gamma_F/\gamma_\alpha) - 1)F'} \tag{4.32}$$

whose solution for $F' = F'_0$ at $t = 0$ is

$$(\gamma_F/\gamma_\alpha)F_0(F'(t)^{-1} - F_0'^{-1}) + (2(\gamma_F/\gamma_\alpha) - 1)\ln(F'(t)/F'_0)$$

$$= g(t) \tag{4.33}$$

with

$$g(t) = \int_0^t dt/\tau_{F'} \quad (= t/\tau_{F'}, \text{ for } \tau_{F'} = \text{const}) \qquad (4.34)$$

In principle, fitting eq.(4.33) to an experimental F' decay curve can yield the values of γ_F/γ_α and $\tau_{F'}$.

The F' decay kinetics has been investigated experimentally by a number of authors. Ikezawa, Hirai, and Ueta (1962) have found a bimolecular decay type (second-order reaction) for the F' band in KCl between -85 and -36 ^{o}C which is predicted by eq.(4.33) at $\gamma_F/\gamma_\alpha = 0.5$ and constant $\tau_{F'}$. The bimolecular decay constant $A = 1/(\gamma_F/\gamma_\alpha)F_0\tau_{F'}$ has been found inversely proportional to the initial F center concentration, in agreement with the F-F' conversion model. From the temperature dependence of A, following fairly the Arrhenius type, an activation energy of 0.50 eV has been determined and interpreted accordingly as the thermal ionization energy of the F' center. These experiments have also revealed the shape of the F' band, transient in the above range of temperatures, in agreement with the low-temperature data.

In an investigation carried out by monitoring the variations of both the optical absorption and the ESR signal due to F centers in KCl during optical bleaching in the F band at -90 ^{o}C, Schmid and Wolf (1962) have obtained a close agreement with eq.(4.33) assuming a time-dependent $\tau_{F'}$ of the form

$$\tau_{F'} = \tau_0 + (\tau_\infty - \tau_0)(1 - \exp(-bt)) \qquad (4.35)$$

where b is a constant (nearly independent of F_0), τ_0 and τ_∞ are are the F' thermal lifetimes at the beginning and at the late stage of of the decay, respectively. While τ_∞ has not been found concentration-dependent, τ_0 has, dropping steeply as F_0 and F'_0 increased. To extract the lifetime, a constant value of $\gamma_F/\gamma_\alpha = 11$ was assumed, as extrapolated from Luty's data (1961b). The observed concentration dependences, as well as the time-dependent lifetime, suggest that pair interactions may have been involved. Schmid & Wolf's experiment has later been extended by Goldberger and Owens (1971) between -100 and -58 ^{o}C with the aim of obtaining the temperature dependence of τ_∞. This was done by assuming $\gamma_F/\gamma_\alpha = 1/6(1 - \eta_i)$ and extrapolating the data by Fedders, Hunger, and Lüty (1961) to the above temperature range. An experimental dependence of the form

$$\tau_{F'} = 7.1 \times 10^{-13} \exp(0.53 \text{ eV}/k_BT) \quad , \text{ s}$$

was obtained for the late-stage lifetime. The pre-exponential factor, however, has not been determined very convincingly. It is nearly two orders of magnitude lower than the one by Ikezawa et al.: 6×10^{-11} s.

The implications of assuming a time-dependent $\tau_{F'}$ are that the F' thermal lifetime is in fact dependent on the F' concentration, that is, on the density of the ionized species, which exert, through their local electric fields, a stimulating effect on the thermal escape of the F' electron. If so, τ_{∞} should be expected to be the thermal lifetime of an isolated F' center, when the density of the ionized species has dropped down to a low level and the mean distance between F' and α centers has increased beyond some critical interaction distance. Although the idea itself is plausible, care should be taken in appraising the significance of these conclusions, insofar as they have been drawn by assuming values for γ_F/γ_α based on extrapolation of low-temperature data. In this respect the direct evaluation of γ_F/γ_α and $\tau_{F'}$ by fitting eq.(4.33) to an experimental decay curve may be more reliable. Such a procedure has first been used by Kitada, Kakui, and Tomura (1968) for the F' center in KBr between -140 and -20 ^{o}C assuming a time-independent $\tau_{F'}$ (Fig.8).

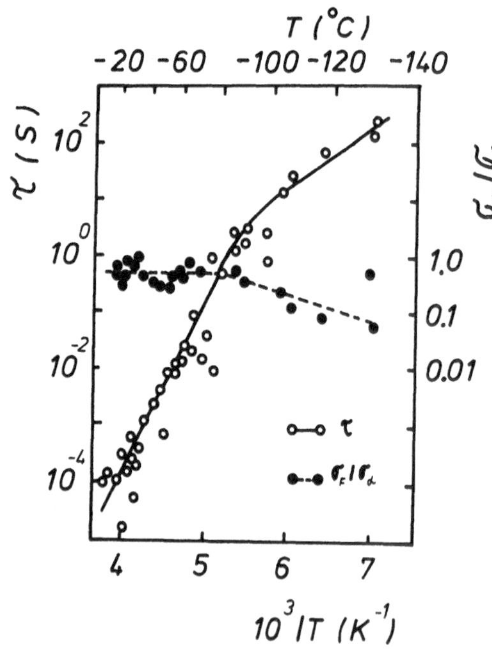

Figure 8: Temperature dependences of σ_F/σ_α and $\tau_{F'}$ in KBr obtained by fitting eq.(4.33) to experimental F' decay curves. (After Kitada, Kakui, and Tomura (1968)).

An intense flash-light F band excitation has been used. Under these conditions the obtained values of $\sigma_F/\sigma_\alpha = \gamma_F/\gamma_\alpha$ are generally lower by an order of magnitude than the ones estimated from Crandall's data, the two temperature ranges matching each other in the vicinity of 140 K. It is argued that the observed discrepancy may result from the preferred drift rather than random motion of the conduction electron to the anion vacancy at high excitation levels. This would lead to an increased σ_α in Kitada's experiments.

The direct evaluation of γ_F/γ_α and $\tau_{F'}$ has also been the method employed in a series of works by Georgiev and Todorov (1976). Light

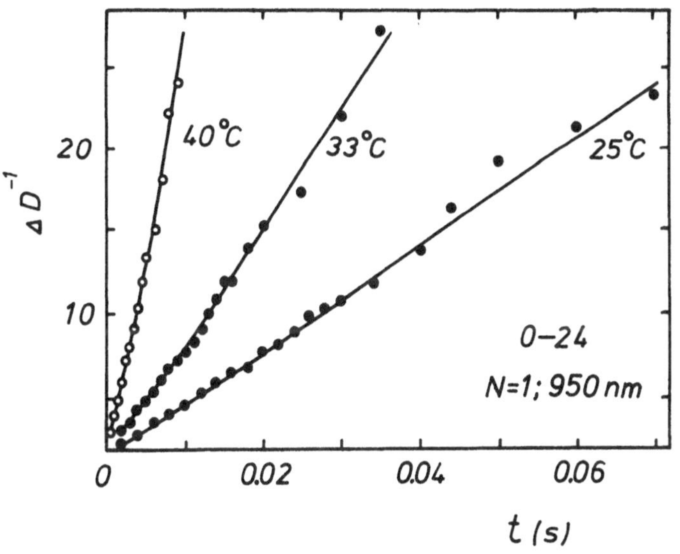

Figure 9: F' band decay curves in KCl near room temperature. The solid lines are best fits of eq.(4.33) to the experimental points marked by circles. A constant F' thermal lifetime has been assumed. (After Todorov, Tomova, and Georgiev (1974)).

flashes of considerable intensity have been used to produce F' centers in additively colored KCl crystals between 290-350 K. A good agreement has been found between eq.(4.33) at time-independent $\tau_{F'}$ and the experimental decay curves within the entire temperature range, as shown in Fig.9 (Todorov, Dechev, Tomova, and Georgiev (1973), Todorov, Tomova, and Georgiev (1974)). From such fits the temperature dependences of γ_F/γ_α and $\tau_{F'}$ have been deduced. The obtained F' thermal ionization energy 0.42 ± 0.21 eV agreed with data by other authors. However, due to uncontrollable flash-light intensity variations, the dispersion was too large to permit any reliable determination of the frequency factors. On the other hand, γ_F/γ_α was also found to decrease as the temperature was raised and some 0.44±0.27 eV were extracted from the Arrhenius plot. Undoubtedly, the latter temperature dependence arises from the time-constants entering into eq.(4.8) for γ_α.

It has reasonably been assumed that in the temperature range under study $\tau_F = \tau_i$. Inasmuch as the thermal ionization energy of the F^* center in KCl is 0.15 eV (Fowler (1968)), it follows that some 0.59 eV have to be ascribed to τ_r. This is about the height of the potential barrier for non-radiative cross-over transitions of the F^* electron to the ground state, as estimated from optical absorption and emission data (Russell and Klick (1956)). It follows that thermal cross-over de-excitation becomes important enhancing the trapping coefficient of the anion vacancy in the vicinity of room temperature. Again, there has been no indication of the existence of a direct ground-state trapping of the conduction electron.

In parallel measurements, the flash-produced yield F'_m of F' centers has been followed as a function of temperature, as well as of the

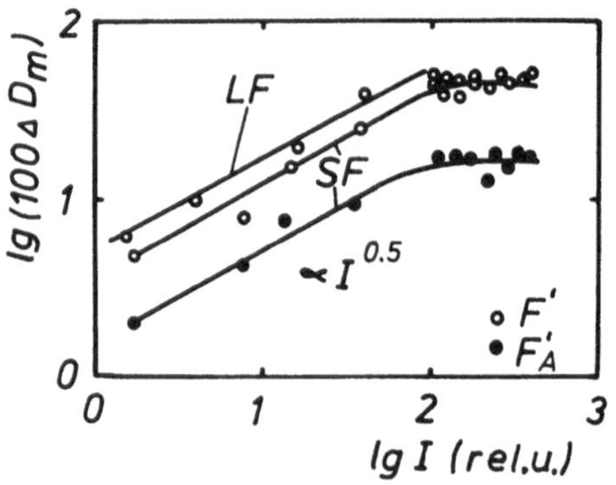

Figure 10: F' yields vs. excitation-light intensity in KCl at RT. Two types of light flashes have been used to excite the F centers labelled SF and LF, respectively. Hollow circles refer to F', filled ones to F'_A(Na). (After Georgiev and Todorov (1976b)).

excitation-light intensity I at RT (Todorov, Dechev, and Georgiev (1972), Todorov, Baltova, and Georgiev (1975), Georgiev and Todorov (1976b)). Two sorts of light flashes, labelled SF and LF, of duration t_{SF} = 40 μs and t_{LF} = 1 ms, respectively, have been used. In both cases F'_m was found to decrease Arrhenius-like as the temperature was raised. However, the obtained thermal slopes were in a 2:1 ratio: 0.41

eV (SF) and 0.21 eV (LF). On the other hand, both yields varied as $I^{\frac{1}{2}}$ before saturating with the excitation-light intensity I (Fig.10). Assuming that while the SF yield corresponds to the initial stage of F' formation (described by eq.(4.13) in which all F' and α containing terms are neglected)

$$F'_{mSF} = \eta_i \, q_F \, F_0 \, t_{SF} \quad , \qquad (4.36)$$

the LF-yield corresponds to a F-F' photoequilibrium, given by eq.(4.16) at $\beta_{F'} = q_{F'}$ (a high excitation level),

$$F'_{mLF} = F_0(\eta_i(\gamma_F/\gamma_\alpha)(q_F/q_{F'}))^{\frac{1}{2}} \quad . \qquad (4.37)$$

Now the obtained temperature- and intensity- dependences can be explained, provided

$$\eta_i \propto I^{-\frac{1}{2}} \exp(0.41 \text{ eV}/k_BT) \; , \qquad \gamma_F/\gamma_\alpha \propto I^{3/2} \; .$$

In the light of the data from the decay kinetics, the latter assumption can only be understood if during an intense flash-excitation $\gamma_\alpha = \gamma_\alpha^*$ ($\eta_i \ll 1$), while $\gamma_\alpha < \gamma_\alpha^*$ ($\eta_i \sim 1$) after the system has relaxed somewhat during the F' dark decay. We shall later speculate on the reasons why the F^* ionization probability may drop when the crystal is highly excited. More recently the flash-light data have generally been confirmed in an investigation using intense photoexcitation pulses by a mode-locked Nd-glass laser (Vassilev, Georgiev, Todorov, and Todorov (1978)).

At very high excitation levels, such as those attained in flash-light or pulsed-laser experiments, reactions (4.a) to (4.h) of the conventional F-F' conversion model may be complemented by

$$e + F^* \longrightarrow F' \qquad \gamma_F^* \, n \, F^* \qquad (4.i)$$

which may be significant at high F^* densities during the excitation. Now eq.(4.5) for F^* turns into

$$F^* = \tau_F(q_F \, F + \gamma_\alpha^* \, n\alpha)/(1 + \gamma_F^* \tau_F \, n) \qquad (4.38)$$

and consequently

$$\dot{n} = \hat{\eta}_i \, q_F \, F + \beta_{F'} F' - \hat{\gamma}_F \, n \, F - \hat{\gamma}_\alpha n\alpha \qquad (4.39)$$

where

$$\begin{aligned} \hat{\eta}_i &= \eta_i/(1 + \gamma_F^* \tau_F \, n) \\ \hat{\gamma}_F &= \gamma_F + \gamma_F^* \tau_F \, q_F/(1 + \gamma_F^* \tau_F \, n) \\ \hat{\gamma}_\alpha &= \gamma_\alpha^o + \gamma_\alpha^* (1 - \hat{\eta}_i) + \tilde{\gamma}_\alpha \\ \tilde{\gamma}_\alpha &= \gamma_\alpha^* \gamma_F^* \tau_F \, n/(1 + \gamma_F^* \tau_F \, n) \end{aligned} \qquad (4.40)$$

Accordingly

$$\dot{F}' = \hat{\gamma}_F n F - \beta_{F'} F' + \tilde{\gamma}_\alpha n\alpha \quad (4.41)$$

$$\dot{\alpha} = \dot{F}' + \dot{n} = \hat{\eta}_i q_F F - (\hat{\gamma}_\alpha - \tilde{\gamma}_\alpha) n\alpha$$

At photoequilibrium we have from (4.41)

$$\hat{\eta}_i q_F F (\hat{\gamma}_F F + \tilde{\gamma}_\alpha \alpha) = (\hat{\gamma}_\alpha - \tilde{\gamma}_\alpha) \beta_{F'} F'\alpha \quad (4.42)$$

It can be assumed that $\gamma_F^* \tau_F n \ll 1$. Now, eq.(4.42) converts into

$$\eta_i q_F \hat{\gamma}_F F^2 = \gamma_\alpha \beta_{F'} F'\alpha \quad (4.43)$$

leading to eq.(4.16), provided $F^* \ll F'$, with γ_F replaced by

$$\hat{\gamma}_F = \gamma_F(1 + (\gamma_F^*/\gamma_F) q_F \tau_F) \quad (4.44)$$

The above equations demonstrate clearly that the inclusion of new color center reactions effective at high excitation levels can lead to intensity-dependent values of the apparent electron-trapping coefficients.

The conclusion that thermally-activated non-radiative de-excitation becomes important in KCl near RT (Georgiev and Todorov (1975)) has a major significance for our understanding of the electron transition and trapping probabilities at the F center. It is now clear why greater care should be taken in extrapolating low-temperature data to the F' thermal-instability range. As a matter of fact, the inclusion of the cross-over transitions will gradually lead to a continuous drop of η_i as the temperature is raised. Consequently γ_α will increase tending to attain the ultimate value of γ_α^* at sufficiently high temperature. The presumed temperature variation of η_i and γ_F/γ_α based on all available data, even though incomplete, is shown in Fig.11. The net effect of an intense light excitation is displacing both temperature curves to the left: the crystal seemingly resists to the external force which tends to completely ionize the F center system (Georgiev and Todorov (1976b)).

Gudat, Scott, and Wagner (1974) have estimated the F ionization quantum yield η_i to be between 0.2 to 0.5 in KCl at RT. They have done it from measurements of the saturation behavior, as the electric field strength is increased, of a charge transferred across the crystal following the shining of a F-light pulse. Apparently, such smaller-than-unity η_i values within the thermal instability range of F' in KCl would lend support to the predictions in Fig.11.

In a comprehensive study of the photocurrent kinetics in KCl and KBr within the F' thermal-instability range, Hoffmann (1973) has obtained important results. The basic equations used for interpreting his data have been derived in an earlier paper by Hoffmann, Stöckmann, and Tödheide-Haupt (1973), based on the conventional F-F' phenomenology. Approximate solutions of the conversion equations have been derived:

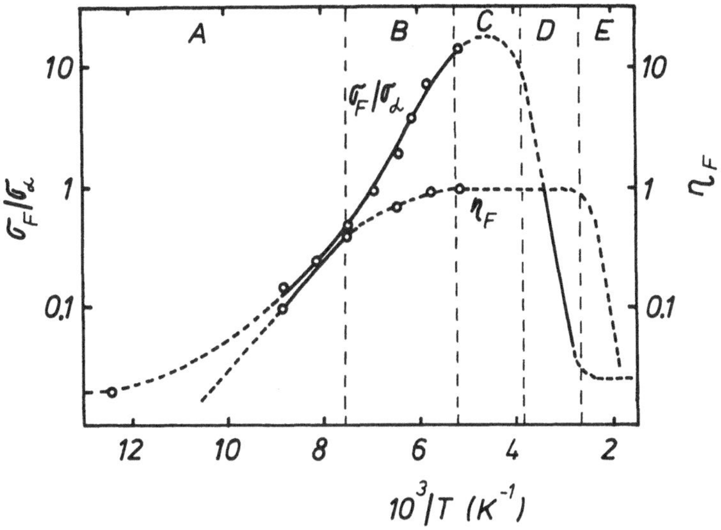

Figure 11: Temperature variations of γ_F/γ_α and the F photoionization quantum yield η_i in KCl. The solid curves and the circles are experimental, while the dashed ones are speculative. Various temperature ranges labelled A through E are separated by vertical dashes. (After Georgiev and Todorov (1976b)).

$$n(t) = n_{st} \tanh(t/\theta) \tag{4.45}$$

for the growth of the free-electron density under 'light-on', and

$$n(t) = n_{st} / (1 + t/\theta) \tag{4.46}$$

for the free-electron density under 'light-off' conditions. Here n_{st} is the photoequilibrium electron density given by

$$n_{st} = (\eta_i\, q_F\, \beta_{F'}/\gamma_\alpha \gamma_F)^{\frac{1}{2}} \tag{4.47}$$

which can be obtained from (4.11) and (4.13) at $\dot{F}' = 0$ and $F = F_0$. θ is a time-constant:

$$\theta = (\gamma_F/\eta_i\, \gamma_\alpha\, q_F\, \beta_{F'})^{\frac{1}{2}} \tag{4.48}$$

It follows that

$$\gamma_\alpha = (n_{st}\,\theta)^{-1} \tag{4.49}$$

The quantities n_{st} and θ can be easily measured from the photocurrent kinetics. Both n_{st} and $1/\theta$ have varied as $I^{\frac{1}{2}}$ which obtains from eqs.(4.47) and (4.48) for $\beta_{F'} = \tau_{F'}^{-1}$. Now, using (4.49) one can evaluate γ_α, which has been found to follow eq.(4.23) between 220 and 340 K (KCl) and 140 to 260 K (KBr). Hoffmann argues that the values measured are those of γ_α^o , expected to exhibit nearly the same temperature dependence (see eq.(4.25)), because $\eta_i = 1$ in the temperature range studied and the second term in eq.(4.8) is practically vanishing. However, he has overlooked the cross-over transitions. In our opinion, the agreement between Hoffmann's γ_α data and eq.(4.23) may rather reflect the fact that the thermally-activated deexcitation leading to $\gamma_\alpha = \gamma_\alpha^*$ has been predominating in his crystals. Further, from the temperature dependences of n_{st} and θ^{-1} Hoffmann has evaluated the F' thermal-ionization energy. Both sets of data on $E_{F'}$ agreed well with each other indicating that not only γ_α but also η_i must be nearly temperature-independent. However, the latter statement does not cast any doubt on the cross-over option in the light of the Fig.11 data.

In a subsequent work, Dokter and Hoffmann (1976) have investigated the reversible bleaching kinetics of the F band in several additively colored alkali halides in the temperature range where F' is thermally unstable. The growth and decay of the F' band at low excitation levels are interpreted by means of equations derived earlier (Hoffmann, Stöckmann, and Tödheide-Haupt (1973)) for $t \gg \tau_e$, $F' \ll F_0$:

$$F'(t) = F'_{st} \tanh(t/\theta) \qquad \text{(growth)} \tag{4.50}$$

$$F'(t) = F'_{st} \,/\, (1 + t/\theta) \qquad \text{(decay)} \tag{4.51}$$

where θ is given by eq.(4.48), while

$$F'_{st} = (\eta_i \gamma_F \tau_{F'} q_F / \gamma_\alpha)^{\frac{1}{2}} F_0 \ , \tag{4.52}$$

which follows directly from eq.(4.16) for $F \sim F_0$ and $\beta_{F'} = \tau_{F'}^{-1}$. The intensity and temperature dependence of F'_{st} and θ have been measured experimentally. The expected square-root dependence on I was confirmed nearly perfectly. Assuming η_i and γ_α to be temperature-independent, the Arrhenius plots, almost perfectly straight lines, have presumably yielded $E_{F'}$, the F' thermal ionization energy. γ_α has also been calculated from the obtained data.

A more recent paper by Hoffmann (1981) again deals with the F-F' conversion kinetics within the F' thermal-instability range. In contrast to his earlier papers, he now considers the case when electron trapping via the excited state dominates over the direct ground-state trapping in electron-anion vacancy recombination. Hoffmann argues that the opposite case $\gamma_\alpha^o \gg \gamma_\alpha^*(1 - \eta_i)$ suits better the data on NaCl, KBr, RbI and perhaps KCl, while the $\gamma_\alpha^o = 0$ case is better adaptable to RbCl and especially to RbBr.

Experimental data on the F' thermal ionization energies $E_{F'}$ are presented in Table II. The following Table 1 lists available data on the electron trapping coefficients within the F' thermal-instability range. Thermal-stability range data can be found in Luty's paper (1961b).

Table 1

Electron-trapping coefficients in the F' thermal-instability range (cm^3/s)
u (cm/s) is the electron thermal velocity

Crystal	γ_α	γ_F
LiF		$1.7 \times 10^{-16}u$ [a]
NaF		$1.5 \times 10^{-16}u$ [a]
NaCl	2×10^{-4} [b]	$2.1 \times 10^{-14}u$ [a]
KCl	10^{-9} [b]	
KBr	10^{-7} [b]	
RbCl	7×10^{-10} [b]	
RbBr	10^{-10} [b]	
RbI	10^{-8} [b]	

[a] From Levin, Berggren, and Honnold (1967)
[b] From Dokter and Hoffmann (1976)

Goode and Simpson (1977) have studied both the fraction of F centers that can be converted to F' by steady F band light and the F' decay in the dark, all between 77 and 230 K in additively colored KCl. The decay curves have been analyzed by means of eq.(4.33) put in the form

$$L(t) = 2\ln(F'(t)/F'(0)) + F_0(F'(t)^{-1} - F'(0)^{-1})$$
$$= (\gamma_\alpha/\gamma_F)(t/\tau_{F'} + \ln(F'(t)/F'(0))) \qquad (4.53)$$

Plots of $L(t)$ vs. t have yielded straight lines between 180-210 K with slopes found to be Arrhenius functions of the temperature with an activation energy of 0.50 ± 0.02 eV. The latter is interpreted as the difference between $E_{F'}$, the F' thermal ionization energy, and E_i , the F^* thermal ionization energy, assuming eq.(4.8) with $\gamma_\alpha^o = 0$, $\tau_F = \tau_i$, $\tau_r = \tau_R$. Below 180 K the decay curves proved to be nonlinear. Above 210 K the results have not been reproducible: repeated F' formation + decay cycles have caused the F' conversion efficiency to progressively fall down and the F' decay to get steeper. The occurrence of an additional F' + α tunneling channel, enhanced by gradual clustering during the recycling is suggested. The F center concentration in these experiments has been less than 7×10^{16} F/cm^3.

Several authors have reported that the F' decay kinetics is a single exponential or a superposition of several first-order components, described by exponentials with different lifetimes. For example, Hoffmann (1973) has observed a first-order F' decay at low F center densities (below 10^{16} F/cc) and very low excitation levels. He attributed this to the contribution of a residual concentration α_0 of anion vacancies, frozen-in during quenching, which are otherwise overwhelmed by the large amount of vacancies originating from the photoexcitation. The first-order decay type follows directly from eq.(4.31) for $\gamma_\alpha \alpha \sim \gamma_\alpha \alpha_0 >> \gamma_F F$. α_0 was estimated at 10^{11} to 10^{12} cm^{-3}. In an earlier investigation, Tomura, Murase, Takebayashi, and Kitada (1964) have used light flashes to produce F' centers in KCl at RT. The subsequent F' decay in the dark has been resolved by means of three exponentials. Parallel measurements of the photocurrent under the same conditions have revealed a close relationship to the F' decay. The three different exponentials have been attributed to decay-times in spatially inhomogenious regions of the crystal. However, it is not clear whether any eq. (4.33)-based analysis has been attempted in addition to the multiexponential fit. A more recent work by Cordovilla and Alvarez-Rivas (1974) has disclosed a series of at least six F' thermal annealing processes. This conclusion has been drawn from a parallel investigation of the decay of the F' band and of the luminescence induced as an ionized F' electron recombine with an anion vacancy via the excited state. Both additively-colored and RT gamma-rayed KCl samples have been investigated from 80 to 300 K. The F center density ranged from 10^{15} to 10^{17} F/cc. Only two of the annealing steps have been of the second-order reaction type, while the remaining ones have been first-order. While sum-of-several-exponentials conclusions must always be drawn cautiously, the authors's results may be regarded as an evidence for a strong F'-defect interaction in their crystals leading to correlated first-order decay processes.

Further experimental verification of the predictions of the conventional F-F' phenomenology has come from an investigation by Nierzewski, Todorov, and Georgiev (1978). Both the F' yield and the decay kinetics induced by flash-light excitation of a pure KCl crystal at RT have been studied as a function of the initial F center density F_0 . The crystal has been colored by RT gamma raying; subsequent thermal annealing at about 150 ^{o}C has led to stabilization of the color and in many respects the samples behaved like additively colored ones. The important point in using a gamma-rayed rather than an additively colored crystal was the possibility of controlling the initial F center density F_0 by varying the annealing conditions. Taking into account the inhomogeneous excitation of the sample resulting from light attenuation, the F' yield has been found to follow closely the dependence on the initial F band optical density D_{F_0} based on an $F'_m \propto F_0$ assumption; this has virtually confirmed the validity of eq.(4.36). The proportionality between the bimolecular F' decay parameter and D_{F_0} has also been established in agreement with the first term on the left-hand side of eq.(4.33). From some decay kinetics $\gamma_F/\gamma_\alpha = 2$ and $\tau_{F'} = 0.5$ ms have been obtained, eventually the RT values characteristic of a pure KCl crystal.

Alvarez-Rivas (1970) has investigated the temporary bleaching induced by a spark-light excitation of a gamma-irradiated NaCl crystal at RT. He measured the optical densities in the F and F' bands (465 nm and 525 nm, respectively), as well as the photocurrent while the spark-light was on, at various initial F_0. This enabled him to determine F, F', and n in photoequilibrium at given F_0, the aim being to estimate the capture cross-section $\sigma_\alpha = \gamma_\alpha / u$ of an anion vacancy according to

$$\gamma^*_\alpha = \eta_i (F^*/\tau_F)/nF' \quad , \qquad (4.54)$$

which follows from eqs.(4.5), (4.11), and (4.13) with $\dot{F}' = 0$, as well as from eq.(4.8) for $\gamma^o_\alpha = 0$. F^* was determined by eq.(4.10) assuming $F' = \alpha$. It was also assumed that $\tau_F = \tau_i$, leading to $\eta_i = 1$, and $\tau_r = \tau_R$. The following numerical data have been used: $\tau_R = 10^{-6}$ s and $\tau_i = 1.5 \times 10^{-10}$ s, based on an extrapolation of the low-temperature data by Swank and Brown (1963). For F_0 between 2×10^{15} and 3×10^{16} F/cc, γ^*_α was between 0.2 to 0.8 cm^3/s. It gives $\gamma_\alpha = 6 \times 10^{-5}$ cm^3/s for $\gamma^*_\alpha = 0.4$. At the same time, γ^*_α has been found to increase for $F_0 > 3 \times 10^{16}$ F/cc. In this latter F_0-range, F' decreased on increasing F_0, as did the free-electron concentration n. It is concluded that another process, possibly electron tunneling from F' to α, may occur at high F_0. One should note the enormously high γ^*_α values at the excitation levels employed! This implies that some of the assumptions made may not have been justified. The free electron density n was about 2.5×10^{10} cm^{-3} at low F_0.

The F-F' conversion in additively colored KF has been investigated by Collins and Schneider (1976). A broad F' absorption band was obtained extending to about 1000 nm with a plateau peak at 700 nm. The photoequilibrium F' density increased with the temperature above 100 K to attain a maximum $\Delta F/F_0$ ratio of about 40 % at about 160 K. This ratio decreased above 200 K because of the F' thermal instability. Pulsed annealing experiments were made to examine the F' thermal decay, found to occur at an exponential rate.

Onuki and Ohkura (1960) have measured the pulsed photoconductivity of additively colored KBr crystals. The space charge effects were eliminated by applying an a.c. electric field and a pulsed illumination synchronized with the plus and minus peaks of the field. The time decay of the current after switching off the illumination has been studied and analyzed to conclude that three kinds of electron traps are present. The longest of the three time-constants has been attributed to an F'-electron release. From its temperature dependence between 180-300 K the $\tau_{F'}$ vs. temperature curve was deduced. The trapping cross-section σ_F was estimated to be between 10^{-16} and 3×10^{-16} cm^2. These estimates, however, have to be regarded as being very rough due to the approximations made in the analysis of the data. Additional data on KCl are reported in a subsequent paper (Onuki and Ohkura (1961).

F center bleaching data on KBr and KI obtained by means of a Q-switched ruby laser pulse of 20 ns duration at RT and LNT are reviewed by Walker and May (1974). Most of the experimental information has been collected in an earlier work by the authors (May and Walker (1973)). The crystals have been additively colored or gamma rayed to below 6×10^{16} F/cc. At full intensity, the RT laser excitation at 694 nm induced an almost complete transparency, while the original absorbance recovered within microseconds thereafter. The intensity dependence of the fractional bleaching $(D_0 - D)/D_0 = (F_0 - F)/F_0$ during the pulse increased steeply above 10^{23} photons/cm^2s showing saturation trends above 10^{26} photons/cm^2s. However, this quantity saturated earlier, at about 4×10^{24} photons/cm^2s, when measured 10 ns after the pulse. F' absorption has presumably been created which was monitored at 1152 nm during the excitation pulse. It decayed over microseconds concurrent with the recovery of the F band. By estimating the rate of repopulation of the F ground state at low bleaching levels, the mean lifetime for deexcitation is found to be 1×10^{-9} s at RT. Another important conclusion is that the F band is homogeneously broadened, since it has bleached equally when monitored at different wavelengths within the band. The F^* ionization rate has been estimated at about 10^8 s^{-1} at 295 K. Data are also mentioned of the complete bleaching of the F' absorption at 633 nm by a ruby laser pulse at 183 K in KCl and KBr. Comments are made regarding the broadness and the red shift of F' relative to F: in solid media the e-e repulsion is apparently not compensated by any enhanced polarization energy due to the double electronic charge. To the contrary, in liquids the solvent can be further polarized and the absorption band of solvated dielectrons is blue-shifted. These suggestions can be found useful in trying to understand the "odd" disposition of F and F' in NaBr and NaI, as mentioned in Section 4.1. The reported data also suggest that the F' centers form in about 1 ns following the excitation at RT. It is also remarkable that the F^* deexcitation rate has been estimated to exceed the ionization rate at RT: this lends support to the assumption by Georgiev and Todorov (1976b) regarding the high excitation levels near RT.

Werner, Stradowski, and Sugier (1980) have investigated the RT kinetics of flash-stimulated changes of the F center absorption in gamma-irradiated KCl crystals. The authors followed both the F' decay kinetics at 650 nm and the kinetics of F band recovery at 560 nm, after excitation by short (15-20 μs) flashes of white light from a Xe lamp. The F' kinetics was found to be only slightly faster than the F band recovery indicating that only a small number of F' centers decay to form species other than F, presumably F-aggregates. Apparently, the overwhelming number of F' centers, formed as primary products of the F band excitation at RT, decay to recover the original F centers. Within the millisecond delay range (0.25-3.5 ms), the F' decay kinetics was found to follow eq.(4.33), in agreement with earlier data (Nierzewski, Todorov, and Georgiev (1978)). However, the F' decay type within the microsecond delay range (2.0-3.5 μs) was logarithmic, indicative of a tunneling process. This process is interpreted as $F' + \alpha \rightarrow 2F$ step occurring between distributed $F'-\alpha$ pairs. We shall discuss this point in greater detail later.

A synchronous detection technique has recently been used by Hoebeeck, Provoost, and Jacobs (1983) to trace out the F' band (highly transient) during a reversible F-F' optical conversion in RbBr. The following method was employed: First, the intensity I_0 of the probe beam was measured at different wavelengths without simultaneously exciting the crystal with F band light. Following this I was scanned over the same wavelength range with the F light on (under saturation conditions), and the absorption change calculated as $\log(I_0/I)$. The decay of the F' band at given wavelengths was also followed after switching off the F light. The crystals used were gamma-colored (^{60}Co) to 4×10^{15} F/cc, and samples were cleaved to $3 \times 15 \times 13$ mm^3. Appropriate photodetectors were used to cover a wide spectral range from 0.37 to 1.6 eV. The optimum conversion temperature was found to lie between 110-120 K, while the F' thermal decay was measured at 80, 95, and 110 K to the effect of establishing its bimolecular (second-order kinetics) character. A decay time τ was determined from $OD(t) = OD_{max}/(1 + t/\tau)$, and the three temperature measurements were found to fall off rather well along the τ vs. $1/T$ line reported previously by Dokter and Hoffmann (1976).

A reversible optical F-F' conversion has been observed in KI crystals at LNT, either colored additively or x-rayed at RT (Delbecq, Pringsheim, and Yuster (1951)). The alpha band has been closely watched and found to correlate with the F' band changes. An assignment of the former is made for the first time as a fundamental band perturbed by empty anion vacancies. The F' and α bands have later been used as effective optical probes to separately follow the F' and α concentrations in a conversion experiment.

For example, Onaka and Fujita (1962) have investigated the isothermal decay of the α-band in NaCl, KCl, and KBr, and found it to be of the bimolecular form $\alpha(t) = \alpha_0 /(1 + t/\tau)$. The halflife τ has proved to be Arrhenius temperature-dependent in crystals colored additively and freshly quenched. Setting τ = 1000 s, the following temperatures are obtained: +22 °C (NaCl), -71 °C (KCl), -138 °C (KBr), similar to the data reported by Pick (1938) on the F' decay: + 20 °C (NaCl), -70 °C (KCl), -140 °C (KBr). This suggests that α-centers decay by recombining with thermally released F' electrons which is also confirmed by a linear relationship observed between the induced absorption changes in the α- or F'- bands and the F band.

4.3. Band theory approach to the thermal ionization rates

Based on band theory, a simple expression can be derived for the thermal-ionization rate of traps which may be useful in explaining some but not all the features of the temperature dependences observed. Suppose there are N_t traps per unit volume, n_t of which are filled, and let n be the free electron concentration in the conduction band (CB). The thermal ionization rate will be defined by the number of untrapping events per unit time per filled trap. Since in equilibrium the overall trapping and untrapping rates must be equal to each other, we obtain

$$\tau_{ion}^{-1} = \gamma\, n\, (N_t - n_t) / n_t \tag{4.55}$$

We further assume Fermi and Boltzmann-tail statistics for the discrete trap level E_t , and the CB bottoming at E_c to calculate n_t and n

$$n_t = N_t / (1 + \exp(E_t - E_f)/k_BT) \tag{4.56}$$

$$n = 2(2\pi m^* k_BT / h^2)^{3/2} \exp(-(E_c - E_f)/k_BT) \tag{4.57}$$

(Stasiw (1959)) where E_f is the Fermi energy, m^* is the electron effective mass. The pre-exponential factor in (4.57) is called an 'effective density of states in the CB" and labelled N_c . Inserting into

(4.55) we get

$$\tau_{ion}^{-1} = 2\gamma(2\pi m^* k_B T/h^2)^{3/2} \exp(-(E_c - E_t)/k_B T) \qquad (4.58)$$

$$= \gamma N_c \exp(-(E_c - E_t)/k_B T)$$

Equation (4.58) applies equally well to F' and F^* centers, giving

$$\tau_{F'}^{-1} = \gamma_F N_c \exp(-E_{F'}/k_B T) \qquad (4.59)$$

$$\tau_i^{-1} = \gamma_\alpha^* N_c \exp(-E_i/k_B T) \qquad (4.60)$$

However, its use in color-center kinetic models is only justified when the electron-phonon coupling can be ignored.

5. F' luminescence?

When a crystal containing F, F', and α centers is illuminated with F' light at low temperature, it emits infrared light. At least to a first approximation the emission spectrum coincides with the one characteristic of the intrinsic F center luminescence (Lüty (1961b)). Moreover, measurements of the radiative lifetime have also indicated that the same center is responsible for emitting both the F and the F' luminescence (Benci and Manfredi (1973)), the same lifetime being experimentally obtained whether the crystal is irradiated with F or with pure F' light. Therefore, it is concluded that the F' luminescence is in fact a phosphorescent process in which electrons released optically from traps (F' centers) recombine radiatively with ionized activator centers (α centers). However, there seem to be two remarkable exceptions from the above rule, long believed to have a general validity: NaBr and NaI again (Baldacchini, Pan, and Lüty (1981)). Both crystals exhibit an infrared emission of double spectral structure. While the excitation spectrum of the longer-wavelength band resembles the F absorption band, the higher-energy component seems to be excited by illumination in the F' band. It has been concluded that the former results from radiative F^* deexcitation (as in the "usual" F center systems), while the latter is due to a radiative recapture process from a higher state, possibly in the conduction band, to the F ground state.

The phosphorescent quantum yield $\eta_{F'}$ has been measured in detail by Lüty and co-workers (Lüty (1961b)), as well as by Miehlich (1963). Analytic expressions for the luminescent yields, the number of quanta emitted in a given process per one quantum absorbed, can be obtained by means of the basic phenomenological equations of the F-F' conversion. We define the following quantities:

(i) F fluorescent yield

$$\eta_F = \eta_R = (F^*/\tau_R)/q_F F = \tau_F/\tau_R \qquad (5.1)$$

under F light illumination of a crystal containing F centers only.

(ii) Equilibrium luminescent yield under F light excitation at F-F' photoequilibrium

$$\eta_{F+F',e} = (F^*/\tau_R)/(q_F F + q_{F'} F')$$
$$= \eta_R \, (q_F F + \gamma_\alpha^* n\alpha)/(q_F F + q_{F'} F')$$

Now, since in photoequilibrium

$$\dot{\alpha} = F^*/\tau_i - (\gamma_\alpha^* + \gamma_\alpha^o) n\alpha = 0$$

it follows

$$\gamma_\alpha^* n\alpha = \frac{\eta_i}{1 - \eta_i + (\gamma_\alpha^o/\gamma_\alpha^*)} q_F F$$

and, consequently,

$$\eta_{F+F',e} = \eta_R \frac{1 + (\gamma_\alpha^o/\gamma_\alpha^*)}{1 - \eta_i + (\gamma_\alpha^o/\gamma_\alpha^*)} (1 + (q_{F'}/q_F)(F'/F))^{-1} \qquad (5.2)$$

(iii) Phosphorescent yield under F' light

$$\eta_{F'} = (F^*/\tau_R)/q_{F'} F' = \eta_R \, (\gamma_\alpha^* n/q_{F'})$$

Inserting n from eq.(4.11) at $q_F = 0$ and using eq.(4.8) one obtains for $F' = \alpha$:

$$\eta_{F'} = \eta_R \frac{1}{1 - \eta_i + (\gamma_\alpha^o/\gamma_\alpha^*)} (\beta_{F'}/q_{F'}) \, \tfrac{1}{2} \eta_{F'-F} \qquad (5.3)$$

In photoequilibrium, the luminescence under F light illumination is always a mixture of F' phosphorescence and F fluorescence. Inasmuch as

$$F' = F \, (\eta_i \gamma_F \, q_F/\gamma_\alpha \beta_{F'})^{\frac{1}{2}} \quad , \qquad (5.4)$$

the mixture is strongly temperature-dependent. Nevertheless, the same emission band is obtained in most alkali halides. This is the sound basis for the statement that both processes are related to the same deexcitation transition: the F^* radiative deexcitation (Lüty (1961b)). The observation that the same emission lifetime is obtained in both photoequilibrium luminescence and pure F' phosphorescence seems to rule out the existence of any F' bound excited state active in emission (Benci and Manfredi (1973)), unless that emission exhibits the same spectral and lifetime characteristics as the F^*-F radiative process. However, the emission lifetime τ_F obtained under fluorescent conditions (purely F centered system) is considerably smaller (Benci and Manfredi (1973)). The temperature dependences of η_F, $\eta_{F+F',e}$, $\eta_{F'}$ in KCl are shown in Fig.12 after Fedders, Hunger, and Lüty (1961) and Lüty (1961b). The decrease of η_F above -180 °C is due to F^* thermal ionization which becomes increasingly important as the temperature is rai-

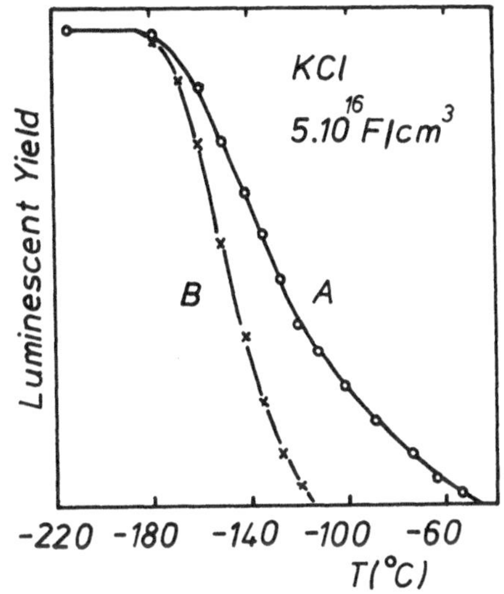

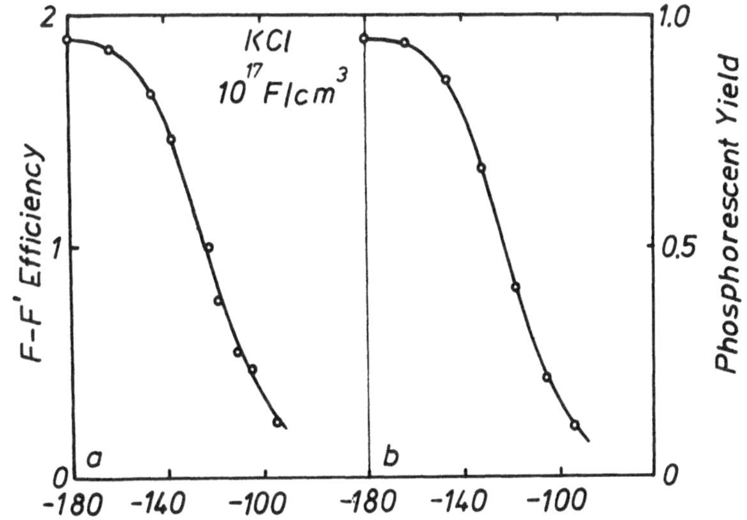

Figure 12: Experimental temperature dependences of: the F-fluorescent yield η_F (B) and the equilibrium luminescent yield $\eta_{F+F',e}$ (A) (top), and the F'-phosphorescent yield $\eta_{F'}$

(bottom right), as compared with the quantum efficiency $\eta_{F'-F}$ of the F'-F conversion (bottom left). The latter comparison reveals that both the nonradiative deexcitation of F^* and the direct ground-state electron capture by the anion vacancy are insignificant in KCl (see eq.(5.3)). (After Fedders, Hunger, and Lüty (1961)).

sed, in accordance with eqs.(4.3) and (5.1). It has also been found that for KCl below -90 °C $\eta_R = 1 - \eta_i$, which implies that non-radiative F^* deexcitation is negligible in this material at lower temperatures. Under these conditions eq.(5.3) predicts $\eta_{F'} = \frac{1}{2}\eta_{F'-F}$ within the F' thermal stability range, provided $\gamma_\alpha^0/\gamma_\alpha^*$ is also negligible. This is exactly the situation depicted in Fig.12. On the other hand, $\eta_{F+F',e}$ also decreases, more slowly than η_F though, as the temperature is raised. This follows directly from eq.(5.2) under the above conditions: apparently, F'/F increases more slowly with the temperature than η_R drops.

(iv) Equilibrium luminescent yield under F light excitation when only the F band absorption is accounted for: -

$$\eta_{F,e} = \eta_{F+F',e}\frac{q_F F + q_{F'} F'}{q_F F} = \eta_R \frac{1 + (\gamma_\alpha^0/\gamma_\alpha^*)}{1 - \eta_i + (\gamma_\alpha^0/\gamma_\alpha^*)} \qquad (5.5)$$

Under the same assumptions as above, $\eta_{F,e} = 1$ independent of the temperature. Experiments by Miehlich (1963) have indicated this to be

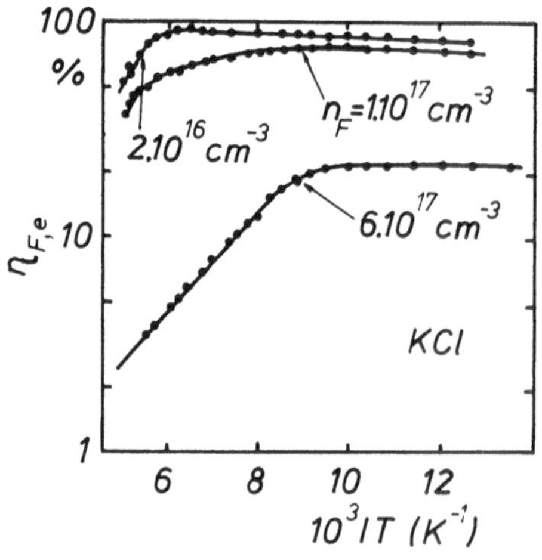

<u>Figure 13:</u> Temperature dependence of the corrected F equilibrium luminescent yield $\eta_{F,e}$ in KCl. (After Miehlich (1963)).

very nearly the case below 170 K in KCl at low F center density, as shown in Fig.13. The observed drop above that temperature has led Hoffmann (1973), (1981) to the conclusion that γ_α^o may play an important role at higher temperatures. Alternatively, it is even more likely that the thermally-activated deexcitation rate (Georgiev and Todorov (1976b)) is involved. If so,

$$\tau_r^{-1} = \tau_R^{-1} + \tau_{nR}^{-1} \tag{5.6}$$

and, consequently, $1 - \eta_i = \eta_R + \eta_{nR}$, where

$$\eta_{nR} = \tau_F / \tau_{nR} \tag{5.7}$$

is the nonradiative deexcitation efficiency. Now, from (5.5) we have

$$\eta_{F,e} = \eta_R / (\eta_R + \eta_{nR}) = (1 + \tau_R/\tau_{nR})^{-1} \tag{5.8}$$

so that the drop of $\eta_{F,e}$ may reflect the increase of the nonradiative rate τ_{nR}^{-1} as the temperature is raised. Recently the thermally-activated nonradiative deexcitation has been shown to effectively "switch on" in KCl above 200 K (Popov (1979)). Consequently, an excited-state pre-trapping may in fact be the prerequisite for electron - anion vacancy recombination in KCl. Moreover, the direct trapping of a conduction electron by the vacancy in ground state through a nonradiative process seems highly unlikely due to the large number of phonons involved. An example of a radiative transition of that kind may have been found recently by Baldacchini, Pan, and Lüty (1981) for NaBr and NaI.

An implicit assumption made in suggesting reaction (4.h) of the F-F' conversion model is that following the photoionization of an F' electron the remaining electron goes non-radiatively to the ground state of the F center formed. Alternatively, one can assume a radiative transition through the intermediate step

$$h\nu_{F'} + F' \longrightarrow F^* + e \qquad q_{F'}^* F' \tag{5.a}$$

This could lead to phosphorescent yields $\eta_{F'}$ in excess of 100 %. Such yields have been measured experimentally by Miehlich (1963) and explained accordingly. Later reaction (5.a) has also been proposed by Lynch and Robinson (1968) to account for a shoulder on the high-energy tail of the F' band. Reaction (5.a) may be enhanced by the presence of a high concentration of ionized species following a strong F-F' conversion.

The inclusion of (5.a) in the thermal-stability range leads to

$$\begin{aligned} n &= (q_{F'} + q_{F'}^*)F'/(\gamma_\alpha \alpha + \gamma_F F) \\ F^* &= \tau_F (\gamma_\alpha^* n\alpha + q_{F'}^* F') \end{aligned} \tag{5.9}$$

at $q_F = 0$. Now, redefining the phosphorescent yield one obtains

$$\eta^*_{F'} = (F^*/\tau_R)/(q_{F'} + q^*_{F'})F'$$

$$= \eta_{F'} + \eta_R(1 + q_{F'}/q^*_{F'})^{-1} \quad (5.10)$$

setting $\gamma^o_\alpha = 0$. At the same time, $\eta_{F'-F}$ remains unchanged since the inclusion of (5.a) merely substitutes $q_{F'} + q^*_{F'}$ for $q_{F'}$ in both the numerator and the denominator of eq.(4.15). The spectral distribution of the phosphorescent yield at 90 K following a strong F-F' conversion at 180 K is shown in Fig.14 at three different F center densities in KCl. From these data one obtains $c^*_{F'} = c_{F'}$ at 700 nm assuming

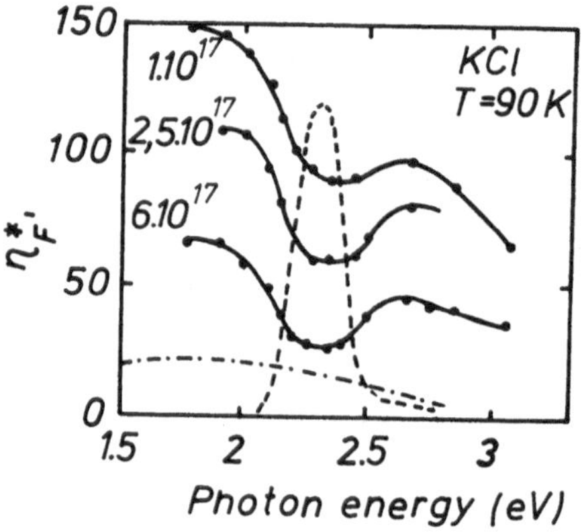

Figure 14: Excitation spectrum of the F' phosphorescence in KCl. The high quantum yield values in excess of 100 % indicate that reaction (5.a) could play a significant role at the lower photon energies (cf. eq.(5.10)), in contrast to the conclusion drawn by Lynch and Robinson (1968) from the higher-energy part of the optical absorption spectrum (see Fig.1). (After Miehlich (1963)).

$\eta_R = 1 - \eta_i = 1$ at the lowest F center density. It is remarkable that the high yields obtain in the low-energy range, contrary to Lynch and Robinson's suggestion. This gives room to speculations that reaction (5.a) may be assisted through some form of F'-α interaction at high α densities which makes it more favorable at low photon energies. Accordingly, the F' band shape may be expected to be somewhat different at low and high α . Unfortunately, no deliberate measurements seem to have been made so far to clarify this point.

Measurements by Becker and Pick (1956) and by Miehlich (1963) have disclosed a rather unusual temperature behavior of the quantum yield

$\eta_{F\,F',e}$ at low F center densities: As the temperature is increased above 70 K, the yield first rises to pass through a maximum before falling down with the temperature, in what appears to be a qualitative agreement with Lüty's data (Lüty (1961b)). The rise before the maximum at 100 K does not follow from eq. (5.2). From the descending portion, represented as $\eta^{-1} - 1$ in an Arrhenius temperature plot, an activation energy of 0.115 eV has been determined which decreased down to 0.059 eV as the F center density increased from 2.5×10^{16} to 6×10^{17} F/cc. At the same time, the temperature slope anticipated on the basis of eq.(5.4) is 0.08 eV, half the F^* thermal ionization energy. The reason for this poor agreement between theory and experiment is not known, all the more that eq.(5.2) is expected to hold better at lower F center densities.

5.1. Intracenter quenching

We will now touch briefly some quenching mechanisms of the F center luminescence which do not involve ionization to either polaron or CB states. Consider a weak coupling situation, as in Fig.15 (left): Upon

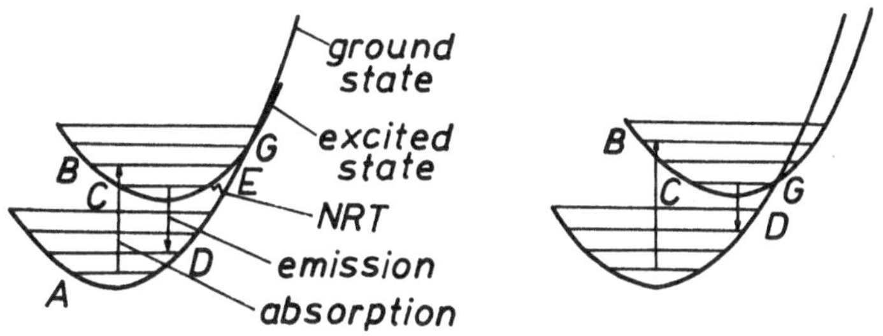

Figure 15: Configurational-coordinate diagrams showing various processes at the F center following the absorption of a light quantum. (After Georgiev (1985a)).

absorption of a photon of appropriate energy the system is lifted from A to B. Suppose now that the crossover barrier at G is above B on the energy scale. The system will then relax to the point at C dissipating the excess energy $Sh\nu$ (ν - the phonon frequency coupled to F^*, S -

the Huang-Rhys factor) to the lattice. Once at C, there are two distinct paths for the system to return back to the ground electronic state:

(i) radiative CD (rate τ_R^{-1}) followed by nonradiative relaxation from D to A,

(ii) nonradiative CE through quantum-mechanical tunneling through the crossover barrier (rate τ_{nR}^{-1}), followed by nonradiative vibronic relaxation to A. In fact, τ_{nR}^{-1} is the total rate involving energy-conserving transitions between various vibronic levels in the respective wells. These start from the lowest level at C, a thermal equilibrium being established as the vibronic system has relaxed from B to C.

Consider next a situation in which the crossover energy at G is below point B reached in absorption, as shown in Fig.15 (right). There now is another conceivable effective path for the vibronic system to deexcite to the ground electronic state (Dexter, Klick, and Russell (1955), Bartram and Stoneham (1975)): Immediately following the optical absorption the system performs violent vibrations at energies near B to reach the vicinity of G before relaxing down to C. There now is a finite probability that the system will undergo an overbarrier transition to the ground electronic state. Eventually, this will be a process competitive to the nonradiative deexcitation of rate τ_{nR}^{-1}, which starts from the relaxed excited state at C once thermal equilibrium has been established during the vibrational relaxation from B. Apparently, transitions of the nonequilibrium type (Dexter-Klick-Russell (DKR) transitions) are conceivable at any given vibronic level, either above or below the crossover barrier, as the system relaxes towards C, leading to the virtual nonattainability of the relaxed excited state. The physical condition for a DKR process to materialize is that it should be faster than the vibronic relaxation to C. The overall DKR deexcitation rate τ_{DKR}^{-1} will mainly be determined by the overbarrier transitions, since as the system steps down the staircase towards C the DKR underbarrier transitions, occuring through tunneling, will grow less probable and, therefore, less competitive to the vibrational relaxation, which tends to bring the system down to the level at C. From the condition that B lies higher than G, the following inequality derives from the configurational parabolae in Fig.15 (right) to represent the criterion for the occurrence of an F luminescence (inefficiency of the DKR deexcitation):

$$\Lambda = S\, h\nu \,/\, E_{abs} \leqslant \tfrac{1}{4}$$

(Bartram and Stoneham (1975)). E_{abs} is the optical absorption energy.

Clearly, both τ_{nR}^{-1} and τ_{DKR}^{-1}, if sufficiently effective, may bring about an intracenter quenching of the F luminescence. An experimental evidence for the former seems to have been provided by Podini (1966). He investigated the temperature dependences of the F ionization and emission efficiencies in NaF. The samples were x-rayed at LNT and then annealed at RT. While the relative emission yield was measured in the usual way (the emission band peaking at 785 nm), the ionization efficiency was estimated from the intensity of a thermoluminescent glow peak at 291 K, following a photostimulation in the F band (peaked at 338 nm at LNT) at various temperatures. The TL peak intensity was assumed proportional to the number of photoelectrons ejected from F centers. Two thermal activation energies have been found from the Arrhenius slopes: 0.06 eV from the photoluminescence and 0.12 eV from the photoionization. The former is assumed to give the height of the cross-over barrier for radiationless deexcitation of the F^* center responsible for the observed drop with the temperature (thermal quenching) of the fluorescence. The latter is the F* thermal ionization energy. However, no detectable F' band has been observed to occur in NaF colored by x-rays on F band irradiation at any temperature.

For an experimental check up of the DKR hypothesis see Hirai (1983).

5.2. F' bound excited singlet?

Crandall (1965) has made an interesting suggestion with regard to his experimental data in Fig.6: He proposed that the temperature dependence of γ_F might in fact follow the type predicted by eq.(4.8) resulting from pre-trapping by some F' bound excited state. The alternative is Pekar's equation (4.27). As a matter of fact, this equation does predict an increase of the electron trapping coefficient as the temperature is lowered because the electron mobility within the phonon-scattering range

$$\mu_e = 15(\exp(240/T) - 1) \quad , \mathrm{cm}^2/\mathrm{Vs} \qquad (\mathrm{KBr})$$

(Markham (1966)) increases more steeply than $T^{-0.75}$. Nevertheless, the dependence calculated from eq.(4.27) is somewhat less inclined than the one shaped by the experimental points in Fig.6. However, the scatter is too large to allow drawing any definite quantitative conclusions. In the light of this uncertainty Crandall's suggestion for the existence of an F' excited state as deep as 0.06 eV below the conduction band does not seem to be conclusive.

In another experimental investigation, Harris, Haven, and Richards (1968) have compared lifetimes of the F' center obtained under steady illumination with relaxation times from modulated absorption measurements. The former were found by following the F' absorption at 650 nm (after F band irradiation) as a function of time, while in the latter the modulated part of the 650 nm absorption was measured using a lock-in amplifier and illuminating with chopped light in the F band. This investigation has been carried out at 80 K and 203 K in KCl containing from 2 to 20 x 10^{16} F/cc. Two relaxation times were found: a long one of the order of 10 s, and a short one of the order of 10 ms. The former is believed to be related to the F' ground state, while the latter is attributed to a higher excited F' state (^{3}F'?). From an analysis of the F' decay curve $\gamma_\alpha / \gamma_F = 6$ has been obtained, however, it is not reported whether this value has been dependent on the temperature. Instead, it is stated that electron trapping at both α and F occurs via an excited state. The quantum yield $\eta_{F-F'}$ was also measured, additively colored and x-rayed crystals giving agreeable results, both by the steady- and modulated- absorption techniques. However, the values obtained at 200 K are somewhat too low (0.07 on the average).

Gorbenko (1964), (1966) also claims to have obtained some evidence confirming the presumed existence of an F' bound excited state. He reports an F' peak at 740 nm for KCl at -180 ^{o}C. Illumination of an F' containing crystal with F' light gives rise to an emission band at 1040 nm which is shifted to the longer wavelengths relative to the F emission band (1000 nm). Illuminating such a crystal with F light brings about a widening of the F emission band and shifts it to the longer wavelengths. It is concluded that energy transfer from F* to F' takes place which excites the F' emission (1040 nm) in addition to the direct photoexcitation of the F emission (1000 nm).

However, Gorbenko's results have been questioned by Benci, Capelletti, Fermi, and Manfredi (1975). They studied very carefully the emission spectrum of a KCl crystal on photoexcitation in either the F band (pure F centered system) or in the F' band (after a strong F-F' conversion). Only one emission band has been obtained, the emission spectra being identical to within 4 % indicating that the initial state for the radiative transition has been the same in both cases (the F* state). This makes the existence of an excited F' state highly unlikely: F' apparently ionizes rather than deexciting radiatively. The possibility that an F' bound excited state active in emission forms in KCl is also ruled out by experimental results of Benci and Manfredi (1971) on the luminescent lifetime, found to be practically the same whether excited by F or L_1 light from 20 to 140 K. The temperature dependence of the relative L_1 emission yield coincided with the one under F band excitation.

These results suggest that the fraction of L_1 states which yield luminescence relax directly to the relaxed excited state F*.

5.3. Bound polaron?

Park and Faust (1966) have reported the occurrence of an induced absorption band (peaked at 0.27 eV, halfwidth 0.16 eV) in KI at 77 K apparently accompanying the F' band produced by a ruby laser pulse. This former band was transient in about 60 ms after the excitation. It had given rise to many speculations on whether it could not be ascribed to the optical absorption of an excited F' state. The unidentified band is also similar to the one predicted by Cheban (1963) (peaked at 0.48 eV, halfwidth 0.18 eV) to result from the photodissociation of an F' center into an F center and a free polaron in NaCl.

The mystery surrounding Park & Faust's infrared band has been unraveled to a certain extent in an experimental study by Borms and Jacobs (1971). They have observed the occurrence on photoexcitation of new temporary IR bands in several alkali halides. In NaCl a short lived center absorbing at 4 μm was found to form each time an electron-excess center was photoexcited at low temperature. This band peaked at 0.31 eV and had a halfwidth of 0.07 eV. On photoexcitation in e.g. the M band at 81 K, both the 4 μm band and the M^+ band formed which then both decayed following the bimolecular reaction type $Y = A/(1 + t/\tau)$. Both decays, as well as the M band recovery were found bimolecular with 1.7 s at 81 K. The temperature dependence of τ was also measured: After running essentially constant (about 2 s) from 30 to 90 K, the time-constant decreased following $\exp(E/k_BT)$ above the latter temperature, yielding $E = 0.2$ eV . Simultaneous absorption measurements were made in the F, M, and the IR band after a ruby laser excitation in the F band, and in the F, F', and the IR band following laser excitation in the F' band. It was concluded that the shallow trap, which gives rise to the IR band when filled, is the F center itself. A strong support for this identification is the observation that the IR band forms each time photoelectrons appear in the CB independent of sample purity and F aggregation. The authors have suggested a metastable state of the F' center (possibly $^3F'$) to form when CB electrons are pre-trapped by F centers. Similar IR bands have been observed in: KCl (6 μm), NaF (2.4 μm), and in KBr and KI (both above 4 μm). The activation energy E is interpreted as the thermal depth of the shallow level below the CB.

Brandt (1972) has attributed the mysterious IR band to the absorption of a Frohlich polaron bound to an F center. A further experimental insight into the puzzle has been made by Carlier and Jacobs (1978a), (1978b). The transient IR band has been excited in F centered NaCl and KBr samples, respectively, by ruby laser pulses (see Fig.16). The following peak energies and band halfwidths are now measured (all in eV): 0.29 and 0.16 (NaCl), 0.34 and 0.13 (KBr), at low temperature. What is remarkable is that both IR bands have exhibited a vibronic structure with an average separation between the lines of 0.89 $h\nu_{LO}$ (NaCl) and

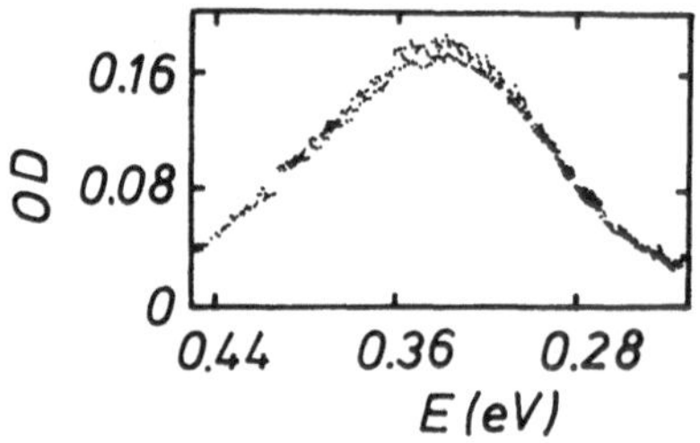

Figure 16: Bound polaron absorption in KBr at 90 K. (After Carlier and Jacobs (1978b)).

0.96 $h\nu_{LO}$ (KBr). This time the decay of the induced temporary band has been tentatively described as logarithmic: $OD(t) = C(1 - \ln(t/\tau))$, apparently to better comply with Brandt's polaron concept. The relaxation time τ is reported to be temperature-independent up to about 50 K followed by a steep Arrhenius-like branch as the temperature is raised. The authors report that the low-temperature portion is strongly influenced by optical bleaching, since in that range τ depends on the intensity of the probe beam. Using the measured band halfwidths and phonon energies, Huang-Rhys factors of 5.2 (NaCl) and 5.7 (KBr) are determined. The phonon energy being close to $h\nu_{LO}$, lends strong support to the bound-polaron interpretation. However, a very recent study by Jacobs (1985) revives the bimolecular decay type for the IR band in NaCl, KCl, RbCl, and KBr. The band has not appeared in RbBr. The following thermal ionization energies are derived from the Arrhenius portions of the temperature dependences of the bimolecular time-constant: 0.22 (NaCl), 0.33 (KCl), 0.28 (RbCl), 0.17 (KBr), all in eV.

It appears that revealing the exact decay type would require a better accuracy of the transient absorption measurements which might not be easily attainable. The problem is thus still open to speculations.

One such attempt made quite recently takes the bimolecular decay type for granted and explains the time-constant versus temperature dependence, as reported by Borms and Jacobs (1971), in terms of a simple quasi-chemical reaction (Georgiev and Mladenov (1986)). The latter involves an endothermic electron transfer from a bound polaron to an excited F center state combined with an exothermic charge charge-transfer from that polaron to the ground-state F center. The exothermic step is virtually temperature-independent and gives rise to the low-temperature time-constant, while the endothermic path depends on the temperature strongly leading to the Arrhenius-like higher-temperature portion. Insofar as the residual low-temperature time-constant is also dependent on the probe-beam intensity, the exact magnitude of the exothermic-path rate remains unknown. One way or the other, given credit to the polaron concept, it would be very interesting to learn more on the nature of the optical transitions at the bound polaron through a more detailed study of data already available (Jacobs (1985)).

6. Pair interactions between color center species

6.1. F*-α interactions

Crandall's photoconductivity experiments in KBr (Crandall (1965)) have brought about a substantial support to the conventional F-F' phenomenology by yielding nearly perfect straight lines at low F center densities ($F_0 = 1.02 \times 10^{16}$ cm^{-3}), in agreement with eq.(4.28). However, the apparent accord between theory and experiment breaks down at higher concentrations: The result for $F_0 = 1.13 \times 10^{17}$ cm^{-3} is reproduced in Fig.17. It should be noted that since $F/\alpha = F_0/\alpha - 2$, any given numerical abscissa is obtained for proportionally larger α when F_0 is increased. Therefore, the reason for the occurrence of the curved portions in Fig.17 can well be the higher α center densities rather than the higher F_0 by itself. This is supported by the plots in Fig.5 and Fig.6 where some of the points have been obtained by extrapolating the linear portions beyond the non-linear range: Thus values of γ_α and γ_F at high F_0 obtained by extrapolation fall in well along the curves shaped by the low F_0 data.

Accordingly, there seems to be a critical vacancy concentration α_c beyond which the experimental points in Fig.17 curve down beneath the expected theoretical line. This critical concentration is markedly temperature-dependent, increasing as the temperature is raised. Two possib-

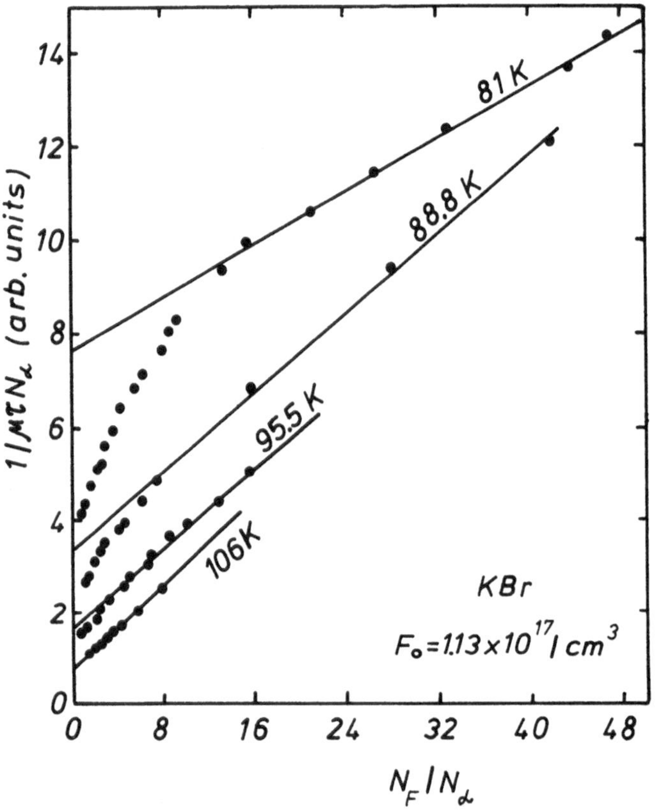

<u>Figure 17:</u> Non-linear deviations from the predictions of eq.(4.28) which have been observed to occur in KBr at lower temperatures and higher anion vacancy concentrations. (After Crandall (1965)).

le non-linear effects can be considered: either the increase of γ_F or the decrease of γ_α as α is increased beyond α_c. Crandall has chosen the latter arguing that it is the cross-sections' overlap which gives rise to the observed non-linearity. It is implied that eq.(4.23) for γ_α^*, derived for an isolated Coulomb center, is no more valid for $\alpha \geqslant \alpha_c$ when the average center separation becomes too small. Assuming that the overlap begins at α_c, the critical concentration can be estimated from the trapping volume of a vacancy using $\frac{4}{3}\pi r_t^3 \alpha_c = 1$. To estimate r_t, we note (Rose (1963)) that the probability of finding an electron at a distance r from a Coulomb center, $p(r) \propto 4\pi r^2 \exp(e^2/\varepsilon r k_B T)$, is at minimum for

$$r_t = e^2/2\varepsilon k_B T \qquad (6.1)$$

This gives

$$\alpha_c = \frac{6}{\pi} (\varepsilon k_B T/e^2)^3 \qquad (6.2)$$

Nevertheless, eq.(6.2) even though correctly predicting the general trend, is not expected to give any quantitative agreement with the experimental data due to the approximations involved.

A good fit to Crandall's data at 81 K is obtained by the following empirical relation

$$\gamma_\alpha = \gamma_\alpha^* (\bar{r}(\alpha)/\bar{r}(\alpha_c)) \qquad \text{for} \quad \alpha \geqslant \alpha_c \qquad (6.3)$$

where $\bar{r}(\alpha) = (6/\pi\alpha)^{1/3}$ is the average distance between vacancies of concentration α, while α_c is taken from the beginning of the non-linear portion, which gives $F_0/\alpha_c = 16$. Physically, eq.(6.3) implies that the electron-trapping coefficient of the anion vacancy decreases proportionally when the average vacancy half-separation falls below the average trapping radius. Similar corrections to the trapping cross-section of a Coulomb center have been advised by Rose (1963) when the mean free path of the electron exceeds the trapping radius of the center. Unfortunately, no consistent mathematical theory of the electron-trapping coefficients at high trapping-center densities seems to have been developed so far.

Ideologically related to Crandall's experiment is an investigation by Bosi and Nimis (1979a) in which the F emission lifetime τ_F in KBr at about 81 K has been followed as a function of F/α . The result is shown in Fig.18 along with Crandall's photoelectric data. The lifetime, nearly constant at low vacancy concentrations, is seen to rise abruptly as α exceeds some critical value at $F_0/\alpha_c = 10$. This again points to the existence of some strong but short-ranged F^*-α interaction which becomes effective when the pair separation drops below some critical value. It has been suggested that the presence of a nearby vacancy may reduce the overlap between the F relaxed-excited state and ground-state wave functions with a resulting drop in radiative deexcitation rate. If so, this does not seem to be simply related to the field produced by the vacancy, since earlier measurements have revealed a decreasing lifetime in an applied dc field (Stiles, Fontana, and Fitchen (1970)). Based on the observed F^*-α interaction effect, Bosi and Nimis (1979b) have even questioned the intrinsic nature of the emission lifetime in alkali halides on grounds that it may not be characteristic of an isolated center due to the presence of a residual vacancy concentration in a quenched crystal (Hoffmann (1973)). It is implied that the critical vacancy concentration α_c should decrease as the temperature is lowered, as in Crandall's experiment, and may approach the residual

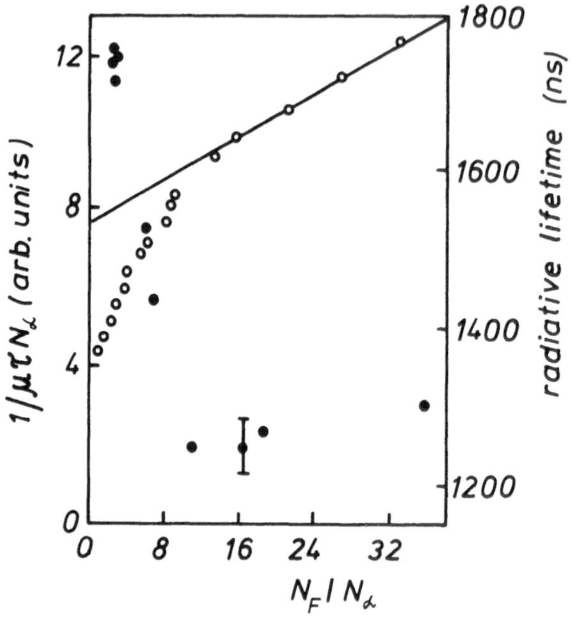

Figure 18: The comparison with a similar effect on the F* lifetime. (After Bosi and Nimis (1979a)).

concentration at low temperatures.

In a related investigation of KCl, Benci and Manfredi (1973) have also studied the effect of the F-F' conversion on the emission lifetime. The luminescent lifetime excited by F light was found to rise proportionally to the α center density for a crystal containing initially 5×10^{16} F/cc. In a subsequent experiment (Benci, Fermi, and Manfredi (1976)) the slope of the straight line obtained was virtually independent of the temperature, being the same at 30 K and at 80 K, for $F_0 = 2 \times 10^{17}$ cm^{-3} . These data are reproduced in Fig.19. The dependence of the lifetime on F_0 at LNT was also studied after 70 % F-F' conversion for each F_0 (Fig.20). Apparently, there are saturation trends at F_0 in excess of 10^{17} cm^{-3}. On the other hand, the temperature dependence from 80 to 160 K of the emission lifetime in a crystal after a given F-F' conversion percentage at given F_0 (that is, at a fixed α) exhibits a good agreement between data obtained by F and F' light excitations (Fig.21)(Benci and Manfredi (1973)). This again rules out the existence of an F' bound excited state. In addition to being longer than the corresponding lifetimes under F band light in a pure

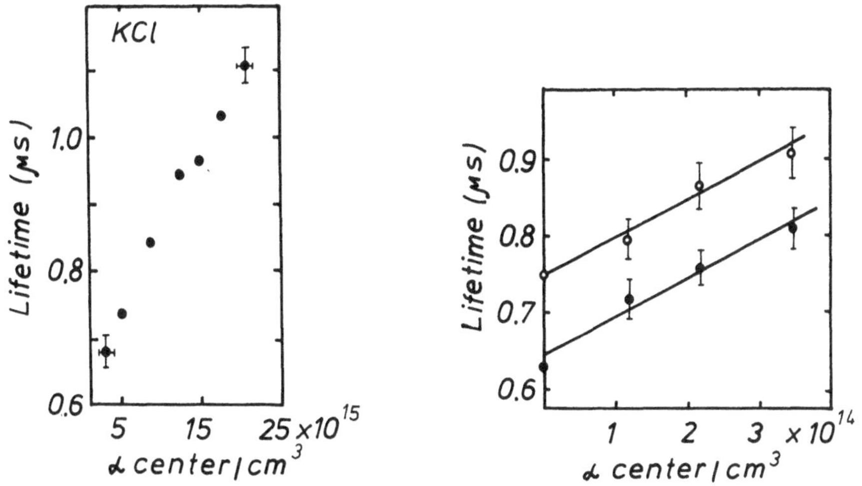

Figure 19: F* lifetime in KCl vs. the α center concentration following a preliminary F-F' conversion. The filled circles are obtained at 80 K, while the open ones are at 30 K. (After Benci and Manfredi (1973)(left) and Benci, Fermi, and Manfredi (1976)(right)).

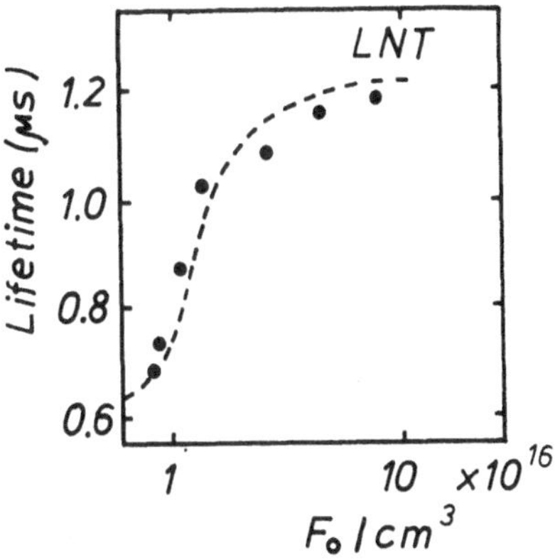

Figure 20: F* lifetime vs. the initial F center density F_0 at LNT (KCl). For each sample about 70 % of the F centers have been converted into F' and α centers prior to the lifetime measurement. (After Benci, Fermi, and Manfredi (1976)).

F centered system, the lifetimes in the presence of F' and α centers decreased more steeply as the temperature was raised, the difference

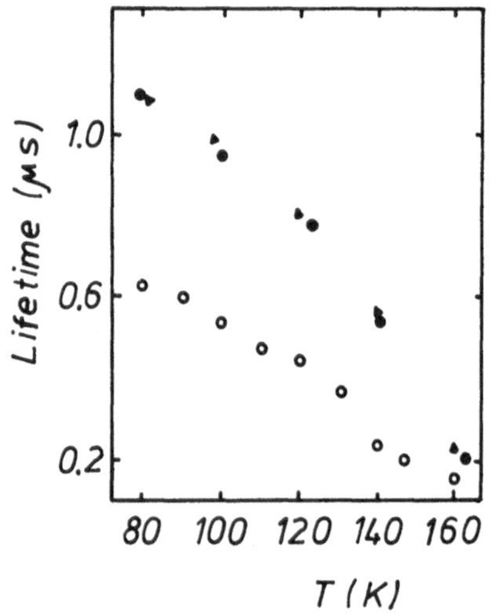

Figure 21: Temperature dependence of the luminescent lifetime in KCl. The F* lifetime has been measured following excitation with F light without (o)(lower set) and after (●)(upper set) a preliminary F-F' conversion (80 %), or with F' light (▲) after 80 % of the original F centers have been converted into F' centers (upper set). (After Benci and Manfredi (1973)).

between the two sets of data tending to vanish above 160 K. Inasmuch as the temperature dependence of the emission lifetime τ_F in KCl above 100 K is attributed to thermal ionization of the relaxed excited state (Swank and Brown (1963)), it would follow that F* ionization becomes effective at lower temperatures when F' and α are present. Consequently, the presumed F*-α interaction brings about either a lengthening of the radiative lifetime τ_R or a decrease of the F* ionization lifetime τ_i. If Crandall's non-linear effect does occur in KCl as well, the latter possibility can be discarded, since from the detailed balance τ_i is inversely proportional to γ^*_α, which decreases as α is increased. Benci et al. attribute the increase in radiative lifetime to the F* electron wave function becoming more diffuse in the presence of a nearby α center which would bring about a reduction of the overlap between excited state-ground state wave functions. However, it is stressed that the effect is not simply dependent on the F*-α separation since it increases as α^1 rather than as $\alpha^{1/3}$. Because of the constant F-F' conversion for the data in Fig.20, the α center concentration therein varies in proportion to F_0. Apparently, saturation occurs when the "interaction spheres" of neighboring centers begin to overlap at high α , that is, when the average F^*-α separation becomes comparable to the critical "interaction distance". The luminescent quantum yield was also measured under various conditions attaining unity below 100 K which indicates once again that it is the radiative lifetime that undergoes a strong F^*-α interaction effect (Benci and Manfredi (1973)). However, no quantitative explanation of the observed dependences has been proposed.

A possible clue may be provided by some earlier observations by Miehlich (1963). He has investigated the luminescent yield on excitation by F band light at various F' densities produced by bleaching the F' band following a strong F-F' conversion. In an η vs. $F/F_0 = 1 - 2\ /F_0$ plot, the yield, apparently independent of the abscissa within the intermediate F/F_0 range, dropped down almost abruptly as F/F_0 approached unity (Fig.22). Under the experimental conditions employed the lumi-

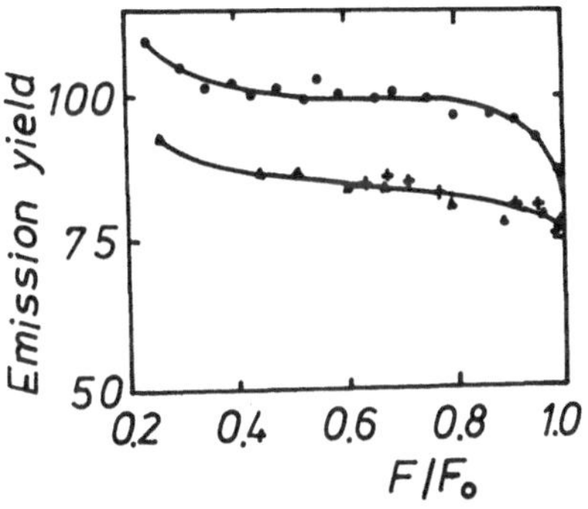

Figure 22: Luminescent yield on excitation in the F band vs. F/F_0 (KCl). The measurements have been made at 90 K following an F-F' conversion at 180 or 130 K. The abscissa has been varied through subsequent bleaching of the F' band. (After Miehlich (1963)).

nescent yield is formally given by eq.(5.2) with $F'/F = \frac{1}{2}(F_0/F - 1)$ not obeying eq.(5.4). However, eq.(5.2) predicts a smoothly increasing yield as F/F_0 is increased, if $\eta_R + \eta_i = 1$. Consequently, to explain Miehlich's data one can introduce some radiationless deexcitation efficiency η_{nR} by means of eq.(5.7) which leads to an yield factor of the form (5.8) before the F'/F containing term in eq.(5.2). Such a step is supported by the lower-than-unity yield values obtained at $F/F_0 = 1$. Now, Miehlich's result will be explained assuming that the luminescence-quenching mechanism is progressively suppressed as the α

center concentration is increased. This would also lead to an increased emission lifetime τ_F due to the decreased deexcitation rate (5.6). Miehlich's data have been taken for F_0 equal to or in excess of 10^{17} cm^{-3}. However, measurements of the relative fluorescent yield in KCl (normalized to unity at 0 K) are claimed to represent the genuine η_R values for $F_0 \sim 10^{16}$ cm^{-3} (Stiles, Fontana, and Fitchen (1969),(1970). In any event, high (virtually 1) relative yield values have been found independent of the temperature up to about 100 K. Now, if the normalization procedure is indeed meaningful physically, this would lend support to Benci and Manfredi's assumption that it is the radiative deexcitation rate τ_R^{-1} which is reduced as a result of the F*-α interaction (Manfredi and Benci). Nevertheless, the above normalization may prove incorrect if there is some radiationless F* decay mode at low temperature.

As F/F_0 decreased beyond the intermediate range (e.g. below 0.5 at $F_0 = 10^{17}$ cm^{-3}), Miehlich's luminescent yield increased above the 100 % level suggesting stimulation of reaction (5.a) at high α center densities, an example of a strong F'-α interaction. Both α-stimulated effects, at high and low F/F_0, respectively, are apparently dependent on F_0, the initial F center density.

6.2. F*-F interactions

Most of the experimental evidence described so far has been collected for dilute F center systems. Some notable exceptions have been mentioned in the preceding Section. In general, the situation changes both quantitatively and qualitatively as the F center density is increased. Before all, the density effects are particularly strong on the F' production rate, the F-F' photoequilibrium, and the luminescent yield of the F center.

There have been some early indications that the F center fluorescence is gradually quenched as the F center density is increased (Becker and Pick (1956), Van Doorn (1958), Pick (1958), Lüty (1961b)). It has been suggested that the concentration quenching may be due to interactions between F* and neighboring F centers (Lüty (1961b)). More systematic investigations have been carried out later by Delbecq (1963) and particularly by Miehlich (1963). Fig.23 presents Miehlich's data at 90 K. The emission yield has been measured under a F-F' photoequilibrium but the contribution of the phosphorescence has been negligible due to the low F' density. Now, since $\eta_i \sim 0$ at 90 K, the luminescent yield

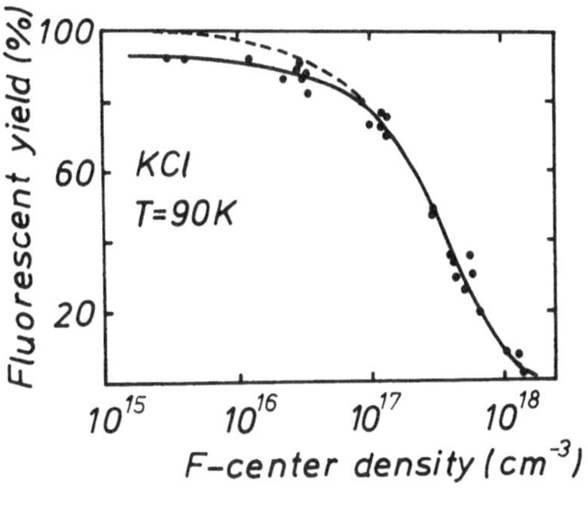

Figure 23: Dependence of the F fluorescent quantum yield in KCl at 90 K on the initial F center density. The dashed line represents eq.(6.7). (After Miehlich (1963)).

(5.2) coincides practically with (5.1). The Figure shows that while at low F center densities the yield is nearly 100 %, it drops down quickly as F_0 is increased, to attain values as low as 1 % for $F_0 \sim 10^{18}$ cm^{-3}. At the same time, the emission spectrum does not change within experimental errors. It can be concluded that concentration quenching is effected through the opening of a new non-radiative deexcitation channel for the F* electron at high F densities.

The initial quantum yield of the F-F' conversion $\eta_{F-F'}$ has also been found to undergo a marked concentration effect (Lüty (1961b)): it drops too, as F_0 is increased, at not too low temperatures, in a way quite similar to the quenching of the luminescence. However, $\eta_{F-F'}$ apparently has a residual non-vanishing low-temperature value (below -180 °C in KCl) which seems to undergo the opposite density effect. According to eq.(4.14) the variations of $0.5\ \eta_{F-F'}$ are those of the F ionization quantum yield η_i .

The effect of the F center density on the F-F' photoequilibrium is shown next in Fig.24 (Lüty (1961b)). While the concentration effect has the character of a suppression above about -150 °C, it plays a stimulating role at lower temperatures. Again, a residual non-vanishing low-temperature value is observed which increases as F_0 is increased. It should be recalled that according to eq.(5.4) the percent conversion is proportional to $\eta_i^{1/2}$; consequently, the drop of η_i can well be the common cause for the observed density effect on the F-F' conversion quantities.

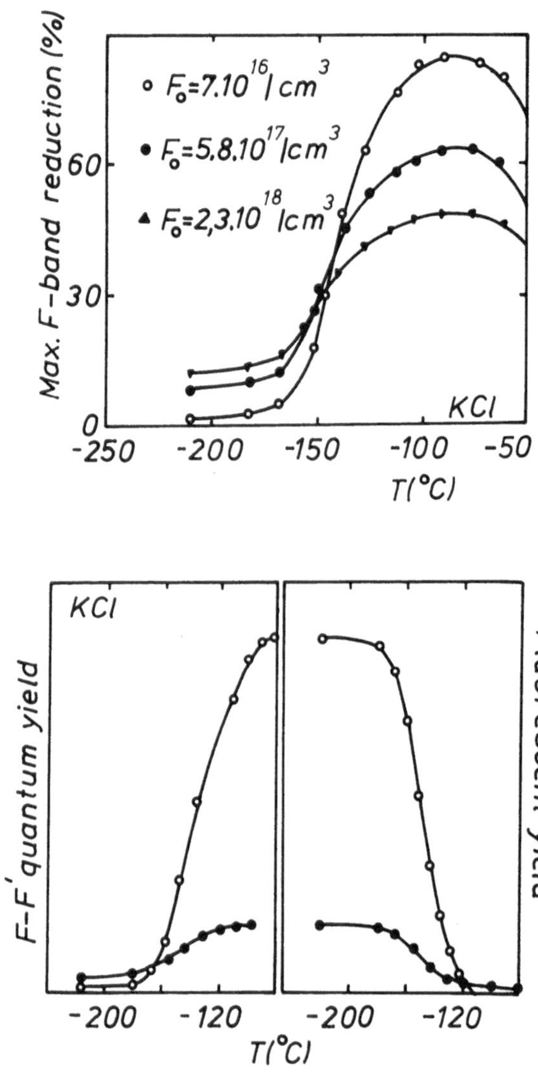

Figure 24: Temperature dependence of the maximum percent F-F'conversion in KCl at three different F center densities (top), and of the F-F' and F fluorescent quantum efficiencies (bottom). (After Lüty (1961b)).

Various quenching mechanisms have been considered in the attempt to disclose the nature of the interaction between F* and F or possibly other defects of the lattice (Lüty (1961b)). Of these, a reaction proposed some time before by Markham, Platt, and Mador (1953) and then revived by Lüty (1961b) has proved to be the genuine one. According to these suggestions the F* electron can tunnel to a nearby F center forming an F' center:

$$F^* + F \longrightarrow F' + \alpha \qquad F^*/\tau_{tf} \qquad (6.a)$$

wherefrom the excess electron can tunnel back to the original center in ground state:

$$F' + \alpha \longrightarrow F + F \qquad F'/\tau_{tb} \qquad (6.b)$$

It is clear that reactions (6.a) and (6.b) would not only provide a radiationless F* deexcitation channel to quench the F luminescence but would also suppress the F photoionization capability.

To derive an expression for the radiative quantum yield, we imagine the F* center to be at the origin of a frame at $r = 0$. Assuming a spherical symmetry, let $n(r)dr$ be the number of F centers pairing with F* at separation between r and $r + dr$, and $w(r)$ be the quenching probability at r . The quantum yield will be given by

$$\eta_R = 1 - \int_0^\infty w(r)n(r)dr \qquad (6.4)$$

For a random F center distribution

$$n(r) = 4\pi r^2 F \exp(-\tfrac{4}{3}\pi r^3 F) \qquad (6.5)$$

(Williams (1968)). Eq.(6.4) thus makes it possible to calculate η_R , provided $w(r)$ is known. Miehlich assumed in effect

$$w(r) = \begin{cases} 1 , & r \leqslant r_c \\ 0 , & r > r_c \end{cases} \qquad (6.6)$$

where r_c is an effective "interaction distance". Inserting into (6.4) one obtains

$$\eta_R = \exp(-\tfrac{4}{3}\pi r_c^3 F) \qquad (6.7)$$

The dashed line in Fig.23 presents eq.(6.7) for $r_c = 13.1$ lattice spacings (KCl). Another example is reported by Compton (1967) yielding $r_c = 12.2$ lattice spacings for KCl at 77 K.

Equation (6.7) works surprisingly well in spite of the very crude approximation (6.6). Another approach to η_R can be based on calculating the total tunneling rate through integration over all pair separations

$$\tau_t^{-1} = \int_0^\infty p(r)n(r)dr \qquad (6.8)$$

where $p(r)$ is the tunneling frequency at r . Now using the quasiclassical WKB expression

$$p(r) = p_o \exp(-r/r_o) \qquad (6.9)$$

and integrating by parts we obtain

$$\tau_t^{-1} = p_o(1 - \frac{1}{r_o} \int_0^\infty \exp(-\frac{4}{3}\pi r^3 F) \exp(-r/r_o) dr) \qquad (6.10)$$

Now the quantum yield can be computed from

$$\eta_R = \tau_F/\tau_R = (1 + \tau_R/\tau_t)^{-1} \qquad (6.11)$$

The average tunneling rate (6.10) has been calculated numerically by Mezger and Jaccard (1981) giving

$$\tau_t^{-1} = 5.7p_o(\frac{4}{3}\pi r_o^3 F) \qquad (6.12)$$

However, inserting (6.12) into (6.11) would not lead to Miehlich's formula for the quantum yield, except at low F , such that $F \ll \frac{3}{4}/\pi r_c^3$; in this case $r_o = r_c(5.7p_o\tau_R)^{-1/3}$.

Bosi, Bussolati, and Spinolo (1970) have checked the effect of the F center concentration F_0 on the emission lifetime τ_F of the KCl F center at 80 K. The luminescent decay was well exponential over 4 decades at three different F_0 : 6.6×10^{16} , 2×10^{17} , and 7.5×10^{17} cm^{-3} , the decay constant being the same, 628 ± 6 ns , in all the three cases. This implies that the F* centers which emit light have been far apart from other F centers and have therefore had a small chance to deexcite non-radiatively through tunneling. Although the reported lifetime is nearly half the value measured by Benci, Fermi, and Manfredi (1976) at LNT within the same F_0 range, it should be recalled that Benci's data have been taken following a strong F-F' conversion.

6.2.1. Tunnel-produced F' centers

The first experimental verification of reactions (6.a) and (6.b) has been sought and found by Chiarotti and Grassano (1966a), (1966b) by means of modulation spectroscopy. The experiment was made at 77 K which is too low a temperature for any sizeable F* ionization in KCl. Nevertheless, a distinct F' band was found to occur (Fig.25) at low chopping rates in crystals containing 1.4×10^{17} F/cc. While the F' signal was apparently independent of the chopping rate at low rates, it dropped down steadily as the rate was increased above 100 Hz. It was concluded that at least one of the above reactions was governed by a time-constant of the order of 10^{-2} s. The modulation index vs. chopping rate curve was analyzed by means of the first-order rate equations corresponding to reactions (6.a) and (6.b):

$$\dot{F}^* = q_F F - F^*/\tau_F \qquad (6.13)$$

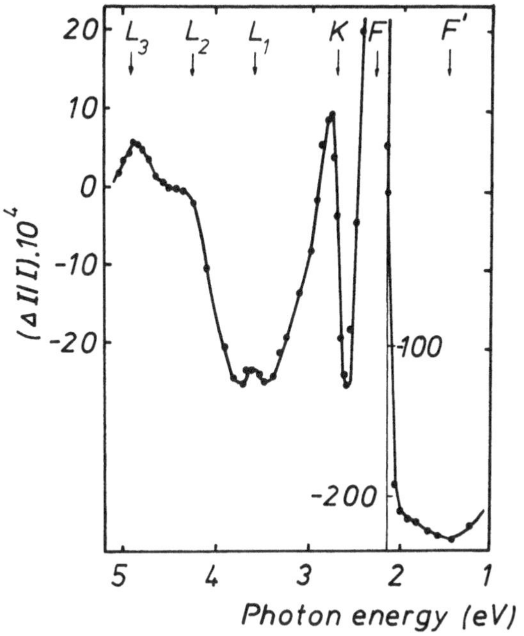

Figure 25: Spectral distribution of the modulation index in KCl showing F', F, K, and the L bands at 77 K. The chopping frequency is 75 Hz. (After Chiarotti and Grassano (1966a)).

$$\dot{F}' = F^*/\tau_{tf} - F'/\tau_{tb} \tag{6.14}$$

where

$$\tau_F^{-1} = \tau_R^{-1} + \tau_{tf}^{-1} \tag{6.15}$$

and $q_F \propto (1 - \cos(\omega t))$. A fair agreement between theory and experiment has been obtained for $\tau_{tf} = 3 \times 10^{-5}$ s , $\tau_{tb} = 4 \times 10^{-3}$ s , taking $\tau_R = 5.7 \times 10^{-7}$ s (Swank and Brown (1963)) which was verified by measuring the K band modulation index in the samples investigated. Increasing F_0 led to an apparent increase of the F' signal, in line with the general expectations for the tunneling process. However, inserting the above values of τ_{tf} and τ_R into eq.(6.11) one obtains $\eta_R = 98$ % which is higher than the luminescent yield measured by Miehlich (1963): about 70 % at the same F_0 .

Another ingenious insight into the nature of the F' formation process is due to Porret and Lüty (1971). Their experiments have been made at very low temperature (1.7 K). Inasmuch as no phonon-assisted F* ionization is possible at this temperature, the authors have effected field-assisted F* ionization to produce F' centers via the conduction band until a photoequilibrium has been established. In equilibrium, according to eq.(5.4),

$$(F'(E)/F(E))^2 = (\gamma_F/\gamma_\alpha)(q_F/q_{F'})\,\eta_i(E) \qquad (6.16)$$

where E is the electric field strength. This produces $\Delta F/F$ of up to 33 % at the highest E . If now a magnetic field is applied to the crystal, spin polarization of the F center electrons will occur. This polarization is known to be almost completely preserved during the optical excitation of the F center. If it is also preserved during the subsequent field-assisted ionization process, we can expect spin-aligned conduction electrons. The polarization is given by

$$P = (n_+ - n_-)/(n_+ + n_-) = \tanh(g\beta H/2k_BT) \qquad (6.17)$$

where

$$n = n_o \exp(\pm g\beta H/2k_BT) \qquad (6.18)$$

are the densities of conduction electrons aligned parallel or antiparallel to the magnetic field H , respectively. Introducing the electron trapping coefficients by F centers in the singlet (γ_F^s) and the triplet (γ_F^t) state, that is, when the conduction- and F center- electrons are spin antiparallel and parallel, respectively,

$$\gamma_F = (n_+\gamma_F^t + n_-\gamma_F^s)/(n_+ + n_-)$$

Now, since

$$n_+/(n_+ + n_-) = \tfrac{1}{2}(1+ P)$$
$$n_-/(n_+ + n_-) = \tfrac{1}{2}(1 - P)$$

we obtain

$$\gamma_F = \tfrac{1}{2}(\gamma_F^t(1+ P) + \gamma_F^s(1 - P)) \qquad (6.19)$$

Consequently, we get inserting (6.19) into (5.4)

$$\frac{F'}{F}(E,H)^2 = \frac{F'}{F}(E,0)^2\,\frac{\gamma_F^t(1 + P) + \gamma_F^s(1 - P)}{\gamma_F^t + \gamma_F^s} \qquad (6.20)$$

Fig.26 shows the magnetic-field dependence of $\Delta F(E,H)/\Delta F(E,0)$, which reveals a decreasing ability to produce F' centers as H (or P) is increased. This suggests a spin-dependent electron trapping by F centers, the declining F-F' conversion resulting from $\gamma_F^t \ll \gamma_F^s$ which leads to a drop in γ_F as P is increased. The solid line in Fig.26 has been obtained from eq.(6.20) at $\gamma_F^t \leq 0.01\gamma_F^s$. It has also been shown that the loss of spin memory during the field-ionization process is negligible. Therefore, triplet trapping of conduction electrons by F centers, if any, is to be neglected, as found later by Jenkin, Stacey,

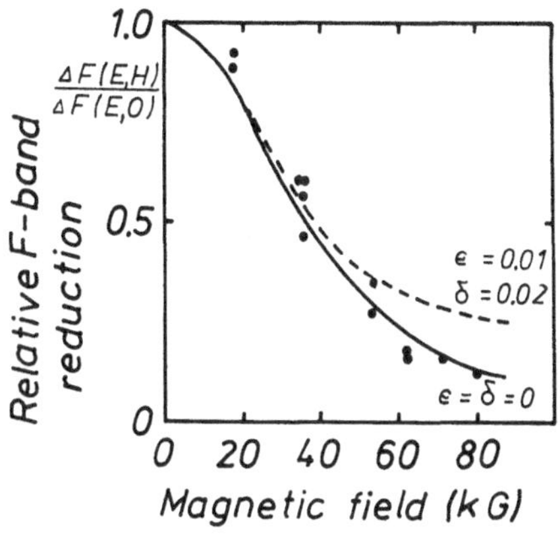

Figure 26: Relative F band bleaching through electric-field assisted photoionization at 1.7 K as a function of the applied magnetic field. The solid line is the expected behavior for full spin memory and zero triplet trapping. (After Porret and Lüty (1971)).

Crowder, and Hodby (1978). Since no spin memory seems to be lost in a relaxation process, the observed lack of triplet trapping is due to the absence of a triplet F' state, as required by the Pauli exclusion principle.

The behavior of the F luminescent yield in a magnetic field has also been studied at 1.7 K (Porret and Lüty (1971)). No electric field has been applied so that only radiative and radiationless F* deexcitation processes have been expected to occur. The result is shown in Fig.27 for two F center densities. The quantum yield η_R is seen to increase steadily with the magnetic field H tending to the ultimate value of 1 as H approaches some 80 kG. The observed behavior presumably results from a spin-dependent tunneling in the non-radiative deexcitation channel (reaction (6.a)), due apparently to the Pauli principle again. In other words, the F*-F electron tunneling, which leads to concentration quenching of the F luminescence in denser F center systems, is in its turn quenched by a strong magnetic field, tunneling between spin-aligned F*-F pairs being prohibited by the exclusion principle. To derive an expression for the quantum yield in a magnetic field, the authors consider a cluster of n F centers within an "interaction sphere" of radius r_c . The quantum yield is given by the probability that all F centers within the cluster are aligned (along H or $-H$) which is

$$\eta_R(n) = 2^{-n}((1+P)^n + (1-P)^n) \qquad (6.21)$$

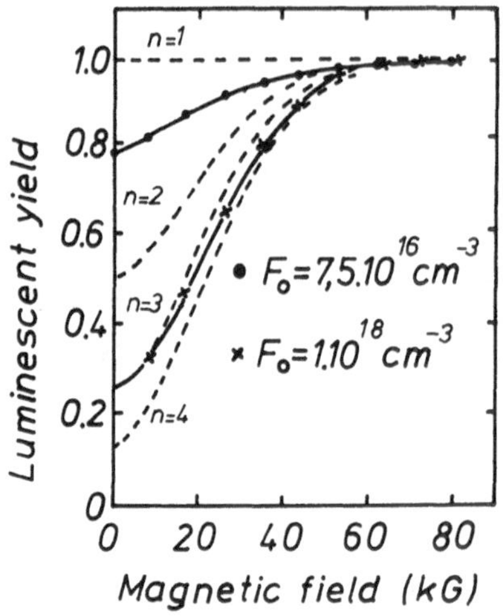

Figure 27: Dependence of the F center luminescence in KCl at 1.7 K on the applied magnetic field for two samples with different F concentrations. The dotted lines represent the expected behavior for clusters of n F centers, as calculated by means of eq.(6.21). (After Porret and Lüty (1971)).

The solid curve for the lightly-doped crystal in Fig.27 has been superposed by 60 % n=1 and 40 % n=2 clusters. A similar analysis involving higher clusters has been made which led to the solid curve for the densely-doped sample. These results suggest the use of F luminescence as a tool of examining the F center distribution within a crystal.

Our present understanding of the nature of the F*-F tunneling process has deepened considerably due to a series of experiments by the Neuchatel group using optical-detection ESR techniques (ODESR). The method is based on the spin-dependency of the radiationless tunneling deexcitation of the F* component of an F*-F pair which leads to a population difference between parallel and antiparallel spin states of the pair, only the latter states deexciting non-radiatively. Consequently, the parallel pairs subjected to radiative deexcitation alone will be more populated (the antiparallel pairs can deexcite in either way). When the resonance condition is met, the microwaves induce spin flips and depopulate partially the parallel pairs. As a result, an ESR can be detected in the form of a resonant dip of the F luminescence intensity. Apparently, the first observation of an optically detected ESR signal from F centers in KCl utilizing the luminescence has been reported by Ruedin and Porret (1968). However, it is clear that monitoring the F luminescence is not the only way to detect the resonance of F*-F pairs: it can also be studied by measuring the F or F' band optical absorptions, a dip or a peak occuring at resonance, respectively, because of the microwave-enhanced reaction (6.a). All these considerations apply to temperatures around LHeT where the spin-lattice relaxation time largely exceeds the F radiative lifetime.

The spin-dependent properties of the F*-F tunneling are well illustrated by a simple model (Jaccard, Ruedin, Aegerter, and Schnegg (1972)). Consider an F*-F pair at LHeT. At this temperature, the nuclear spin orientations can be assumed to be at random. The nuclear spin-lattice relaxation time being sufficiently long, the nuclear field is presumably in a quasi-steady-state. The pair is sufficiently distant so that the nuclear fields $\underline{H}_N^*$ and $\underline{H}_N$ can be assumed to be different. If $\underline{H}_0$ is the applied external field, the fields acting on F(1)* and F(2) are $\underline{H}^* = \underline{H}_0 + \underline{H}_N^*$ and $\underline{H} = \underline{H}_0 + \underline{H}_N$, respectively. Introducing the single-center orbitals φ^* and φ, the two-electron pair can be described by a symmetric and an antisymmetric combination:

$$\Phi_{s,a} = 2^{-\frac{1}{2}} (\varphi^*(1)\varphi(2) \pm \varphi(1)\varphi^*(2)) \quad (6.22)$$

which have to be multiplied by the spin functions

$$\sigma_1 = |++\rangle, \quad \sigma_{-1} = |--\rangle, \quad \sigma_0 = 2^{-\frac{1}{2}} (|+-\rangle + |-+\rangle)$$
$$\alpha = 2^{-\frac{1}{2}} (|+-\rangle - |-+\rangle)$$

to obtain the eigenfunctions of an orbital Hamiltonian in which spin-orbit coupling is taken into account. The interaction with the magnetic fields is described by the following Hamiltonian

$$\mathcal{H} = g^*\beta \underline{H}^*\cdot\underline{S}^* + g\beta \underline{H}\cdot\underline{S} \quad (6.23)$$

Four eigenstates of (6.23) are next found in which the angle ψ between $\underline{H}$ and $\underline{H}^*$ controls the amount of mixing:

$$\begin{aligned}
\Psi_1 &= 2^{-\frac{1}{2}} \sin(\tfrac{\psi}{2})(\Phi_s\alpha + \Phi_a\sigma_0) + \cos(\tfrac{\psi}{2})\Phi_a\sigma_{+1} \\
\Psi_2 &= 2^{-\frac{1}{2}} \cos(\tfrac{\psi}{2})(\Phi_s\alpha + \Phi_a\sigma_0) - \sin(\tfrac{\psi}{2})\Phi_a\sigma_{+1} \\
\Psi_3 &= 2^{-\frac{1}{2}} \cos(\tfrac{\psi}{2})(\Phi_s\alpha - \Phi_a\sigma_0) - \sin(\tfrac{\psi}{2})\Phi_a\sigma_{-1} \\
\Psi_4 &= 2^{-\frac{1}{2}} \sin(\tfrac{\psi}{2})(\Phi_s\alpha - \Phi_a\sigma_0) + \cos(\tfrac{\psi}{2})\Phi_a\sigma_{-1}
\end{aligned} \quad (6.24)$$

The corresponding eigenvalues are

$$\begin{aligned}
E_{1,4} &= \pm\tfrac{1}{2}(g^*\beta H^* + g\beta H) \\
E_{2,3} &= \pm\tfrac{1}{2}(g^*\beta H^* - g\beta H)
\end{aligned} \quad (6.25)$$

This distant-pair model (φ and φ^* assumed orthogonal) is only a first-order approximation to the actual tunneling pair, in which φ and

$\varphi*$ should overlap.

Under the condition of a durable spin-lattice relaxation, the F* electron has no time to thermalize; consequently, the occupation probabilities are those in the ground state which are controlled by the Boltzmann factor:

$$u_1 + u_4 = \cosh(g\beta H_0/k_BT)/2\cosh(g\beta H_0/2k_BT)^2$$
$$u_2 + u_3 = 1/2\cosh^2(g\beta H_0/2k_BT) \tag{6.26}$$

provided H_0 exceeds largely the hyperfine fields H_N , H_N^* . In the final tunneling state (F') both electrons share the same orbital but have an antisymmetric spin function (singlet). Therefore, the tunneling rates should be given by

$$w_{t1} = 2^{-\frac{1}{2}} w_{t0} \sin(\tfrac{\Psi}{2}) = w_{t4}$$
$$w_{t2} = 2^{-\frac{1}{2}} w_{t0} \cos(\tfrac{\Psi}{2}) = w_{t3} \tag{6.27}$$

where w_{t0} depends on the radial parts of the wave functions only and therefrom on the pair separation r , e.g. for hydrogen-like states

$$w_{t0} = w_0 \exp(r/r_0) \tag{6.28}$$

The total tunneling rate will be

$$W_t = (u_1 + u_4)w_{t1} + (u_2 + u_3)w_{t2}$$
$$= 2^{-3/2} \frac{w_{t0}}{\cosh^2(g\beta H_0/2k_BT)}(\sin(\tfrac{\Psi}{2})\cosh(g\beta H_0/k_BT) + \cos(\tfrac{\Psi}{2})) \tag{6.29}$$

At low magnetic fields, $H_0 \ll k_BT/g\beta$ (H_N , H_N^* still neglected):

$$W_t = 2^{-3/2} w_{t0} (\sin(\tfrac{\Psi}{2}) + \cos(\tfrac{\Psi}{2})) \tag{6.30}$$

At high fields, $H_0 \gg k_BT/g\beta$, $\Psi \to 0$, and

$$W_t = 2^{\frac{1}{2}} w_{t0} \exp(-g\beta H_0/k_BT) \to 0 \tag{6.31}$$

Averaging (6.29) over the angles is not that easy since $\Psi = \Psi(H_0)$. At small H_0 we assume $0 \leqslant \Psi \leqslant \pi$ to obtain

$$\langle W_t \rangle_\psi = (2^{\frac{1}{2}}/\pi)\, w_{t0} \quad . \tag{6.32}$$

Inasmuch as drops as H_0 is increased, so does the tunneling rate $W_t(\Psi(H_0),H_0)$. Physically, this implies that both the population and the extension of the asymmetric part of the wave function of the pair decrease in an external magnetic field. While the population drop is effected in sufficiently high fields, the drop in the wave function overlap (resulting from the shape change of the asymmetric part of the initial state) is observed in low fields too (Ecabert, Schnegg, Ruedin, Aegerter, and Jaccard (1972)). The resulting variations of the F luminescent intensity, as well as of the F and F' absorption bands, as H_0 is increased is seen in Fig.28 prior to or at resonance (Ecabert and Jaccard (1978)). The wave function shape variations are rather steep

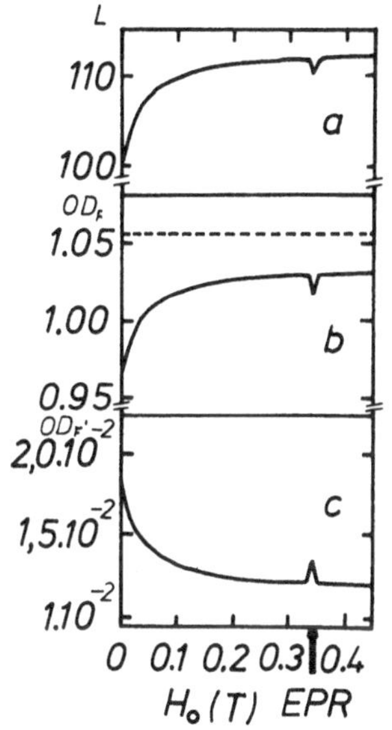

Figure 28: Dependence on the external magnetic field of: (a) the F center luminescence, (b) the optical density in the F band, (c) the optical density in the F' band. The occurrence of resonance under ESR conditions is clearly seen in all the three cases. (After Ecabert and Jaccard (1978)).

and rapidly saturate which is followed by a slower variation of the population of states. The field effect disappears completely at higher temperatures, the low-field condition now being satisfied at any H_0 .

Introducing the radiative deexcitation rate w_R , one can define the tunneling probability of a pair

$$P_t = (u_1 + u_4)\frac{w_{t1}}{w_R + w_{t1}} + (u_2 + u_3)\frac{w_{t2}}{w_R + w_{t2}} \tag{6.33}$$

$$= \frac{v}{2\cosh^2(g\beta H_0/2k_BT)}\left(\frac{\sin(\frac{\Psi}{2})\cosh(g\beta H_0/k_BT)}{1 + v\sin(\frac{\Psi}{2})} + \frac{\cos(\frac{\Psi}{2})}{1 + v\cos(\frac{\Psi}{2})}\right)$$

which holds good for $H_0 \gg H_N$, H_N^* . (At $H_0 = 0$, however, there is an isotropic population distribution $u_i = 1/4$ which is correctly given by eq.(6.26).) Here $v = 2^{-\frac{1}{2}} w_{t0}/w_R$ is the isotropic part of the tunneling probability. P_t should next be averaged over Ψ which has been done analytically for $H_0 = 0$ and $H_0 \to \infty$ only (Jaccard, Ruedin, Aegerter, and Schnegg (1972)), leading to random Ψ and $\Psi = 0$ respectively. It should also be reminded that in general $\Psi = \Psi(H_0)$.

Once P_t has been averaged, one can seek an expression for the luminescent quantum yield η_R . Let dV_i be a small spherical shell between r and $r + dr$ and let an F* center be situated at the origin of the coordinate frame. The probability that there are n F centers in ground state in dV_i is given by Poisson's law:

$$p(n) = A^n e^{-A} / n! \quad , \quad A = F\, dV_i$$

The probability that F* deexcites radiatively, taking into account interactions with F centers within dV_i only will be

$$\sum_{n=0}^{\infty} p(n)(1 - \langle P_t \rangle)^n = \exp(-A\langle P_t \rangle)$$

Now, the luminescent yield will be given by the probability that F* deexcites radiatively in spite of potential interactions with all F centers at any r :

$$\eta_R = \prod_i \exp(-F\langle P_t \rangle dV_i) = \exp(-F \sum_i \langle P_t \rangle dV_i)$$

where the subscript i refers to various shells. Performing the transition to continuous r we obtain

$$\eta_R = \exp(-F \int_0^{\infty} 4\pi r^2 \langle P_t \rangle dr) \tag{6.34}$$

To find $\langle P_t \rangle$ for the particular case of $H_0 = 0$ (neglecting the hyperfine fields) we average eq.(6.33) following

$$\langle P_t \rangle = \frac{1}{\pi} \int_0^{\pi} P_t(\Psi) d\Psi \tag{6.35}$$

which gives for $v < 1$:

$$\langle P_t \rangle = 2(1 - (1 - v^2)^{-\frac{1}{2}} + \frac{1}{\pi} (1 - v^2)^{-\frac{1}{2}} \tan^{-1}(\frac{v^2}{1 - v^2})^{\frac{1}{2}}) \tag{6.36}$$

The latter expression tends to

$$\langle P_t \rangle = \frac{2}{\pi} (w_{t0}/w_R) \tag{6.37}$$

when $v \ll 1$. Assuming eq.(6.9) to hold good for $w_{t0}(r) = p(r)$ we obtain

$$\eta_R = \exp(-\tfrac{4}{3}\pi r_c^3 F) \quad , \quad r_c = r_0(\tfrac{12}{\pi} p_0 \tau_R)^{1/3} \tag{6.38}$$

where $\tau_R = w_R^{-1}$. Equation (6.38) is of the form of Miehlich's equation.

The spectral distribution of the transmitted-light intensity in the F and F'band ranges at microwave resonance during an F-light excitation of the sample is shown in Fig.29 (Ecabert and Jaccard (1978)). This is the spectral distribution of the resonant signal from the optical absorption bands shown in the preceding Figure 28. The ESR-line intensity

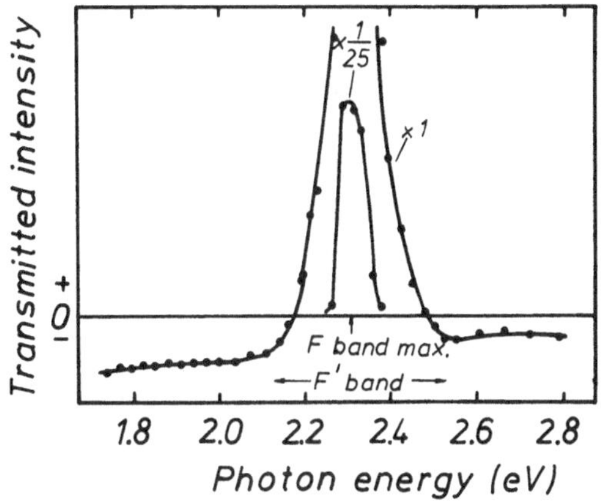

Figure 29: Spectral distribution of the optical transmission in the F and F' band ranges under the conditions of a microwave resonance. (After Ecabert and Jaccard (1978)).

has been obtained by a phase-sensitive detection of the transmitted light. It has been verified that the optical density in the F band is independent of the magnetic field in the absence of an optical excitation of the F centers; consequently, resonance is an inherent property of the excited system. These experiments have revealed some interesting features of the tunneling process. Using eqs.(4.9) and (4.10), the steady-state solutions to the rate equations (6.13) and (6.14) are

$$\begin{aligned} F^* &= q_F \tau_F F \\ F' &= (\tau_{tb}/\tau_{tf}) q_F \tau_F F \\ F &= F_0/(1 + q_F \tau_F + 2(\tau_{tb}/\tau_{tf}) q_F \tau_F) \end{aligned} \tag{6.39}$$

This gives

$$\begin{aligned} \Delta F &= F_0 - F = F_0 \frac{q_F \ {}_F(\tau_{tf} + 2\tau_{tb})\tau_{tf}^{-1}}{1 + q_F \tau_F (\tau_{tf} + 2\tau_{tb})\tau_{tf}^{-1}} \\ \Delta F^{-1} &= F_0^{-1} (1 + q_F^{-1} \tau_{tf} \tau_F^{-1} (\tau_{tf} + 2\tau_{tb})^{-1}) \end{aligned} \tag{6.40}$$

The dependence of ΔF^{-1} on the reciprocal I_F^{-1} of the excitation-light intensity, as measured experimentally by Ecabert and Jaccard (1978), is shown in Fig.30 at two values of the external magnetic field H_0. The result agrees pretty well with the linear dependence predicted by eq.(6.40). Using eq.(6.15), the slope of the straight line is

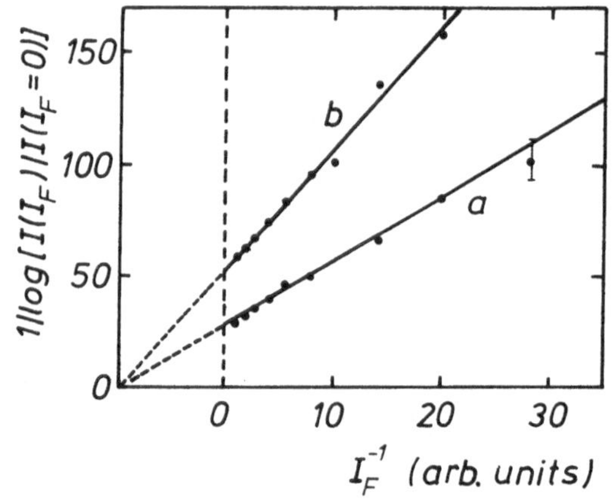

Figure 30: Reciprocal optical density in the F band vs. the reciprocal of the excitation-light intensity at two values of the external magnetic field: 0 (a) and 0.45 T (b). The straight line is indicative of the tunneling process (see eq.(6.40)). (After Ecabert and Jaccard (1978)).

$$s(H_0) = (1 + \tau_{tf}/\tau_R)(\tau_{tf} + 2\tau_{tb})^{-1} \tag{6.41}$$

At $H_0 = 0$ we assume $\tau_{tb} \gg \tau_R \gg \tau_{tf}$ (strong quenching): this yields $s(0) = (2\tau_{tb})^{-1}$. At higher H_0 we have $\tau_{tb} \gg \tau_R \sim \tau_{tf}$ giving $s(H_0) = 2s(0)$, in accordance with the experiment. The F' relaxation times after switching on or off the excitation F-light have been found to depend on the external magnetic field H_0. The latter of these is undoubtedly equal to τ_{tb} and it is difficult to understand just how reaction (6.b) turns out to be spin-dependent. To cope with the situation the authors assume that an additional F' relaxation channel

$$F' + \alpha \longrightarrow F + F^* \tag{6.c}$$

may be operative for close F'-α pairs (e.g. through field-assisted F' ionization followed by electron trapping by the vacancy in excited state). The F*-F pair can deexcite radiatively

$$F^* + F \longrightarrow F + F \qquad (6.d)$$

or it may undergo tunneling via reaction (6.a), depending on the spin state of the pair. This leads to a H_0-dependence of the apparent light-off F' relaxation time.

In a recent paper by Mezger and Jaccard (1981), the virtual F' ionization step in reaction (6.c) is attributed to the probe and the excitation light beams because of the strong F-F' absorption overlap. Both the transient F luminescence and F' population following a pulsed excitation with F light have been measured in KCl at 10 K for F_0 between 5×10^{16} and 1×10^{18} cm^{-3}, with or without any external magnetic field. The luminescent decay, examined over three decades, has been found to be non-exponential. This was attributed to the effect of F*-F tunneling pairs of various separations leading to spatially distributed F* lifetimes (6.15) which results in a superposition of exponentials. The decay curves observed experimentally were analyzed on this basis using a random pair distribution (6.5), equation (6.28) for the tunneling rate, equations (6.27) for the radiative and the radiationless transition rates, respectively, and an angular distribution function

$$P(\Psi, H_0) = \begin{cases} \frac{1}{2}\sin\Psi & , \; H_0 = 0 \\ \frac{1}{2}\,\delta(\Psi) & , \; H_0 \gg H_N^*, H_N \end{cases} \qquad (6.42)$$

The F' decay curves were dealt with likewise assuming a lifetime distribution for the reverse reactions (6.b) and (6.c) based on eq.(6.28). A very good fit to the experimental curves was obtained with parameters: $\tau_R = 890 \pm 5$ ns; $w_{of}^{-1} = 0.4 \pm 0.2$ ns; $r_{of} = 0.8 \pm 0.1$ nm; $w_{ob}^{-1} = 2 \pm 1$ ms; $r_{ob} = 0.8 \pm 0.1$ nm. It should be pointed out that the lattice relaxation, which undoubtedly plays an important role in the tunneling processes considered, has not at all been taken ito account in any of the above models.

Several works by the Neuchatel group have dealt with particular features of the magnetic resonance of distant F*-F pairs in alkali halides using optical detection techniques: the pumping cycle kinetics and the ESR characteristics (Ruedin, Schnegg, Jaccard, and Aegerter (1972)), the spin-lattice relaxation of the F center in its relaxed excited state (Ruedin, Schnegg, Jaccard, and Aegerter (1973)), and the nuclear magnetic resonance (Jaccard and Ecabert (1978)). Inasmuch as the contents of all these goes beyond the scope of the present review, the reader is referred to the original papers for an acquaintance with the details. In a recent paper, Mezger and Jaccard (1980) demonstrate that some of the hyperfine parameters in the spin-Hamiltonian can be obtained from the luminescence of distant pairs in KCl as a function of the static magnetic field only, without applying any rf or microwave field. The luminescent intensity vs. H_0 plot shows three small dips at 790, 1134, and 1516 G, the first one of which corresponding to the hf interaction with the third-sphere K_{III} ions, while the last two arise from the interaction hyperfine with the fourth nearest-neighbor Cl_{IV} ions. These peaks appear superposed on the η_R vs. H_0 plot, the quantum yield nearly saturating at fields as high as a few kG. These "zero-frequency peaks" are conditioned by the efficient optical pumping of the nuclear spin states which is only possible if the spin memory is preserved during an optical cycle. From these measurements the electronic spin-memory losses are found to increase in the 1:4.5:11 ratio in the order KCl-NaCl-KBr and are even higher in KI.

Recently, the ODESR of F center pairs has been measured at 35 GHz in KCl, NaCl, RbCl, KBr, KI, and CsBr (Mezger and Jaccard (1982)). Besides signals from the ground and excited states, a third resonance line is also observed which is attributed to an exchange interaction between nearest-neighbor F centers, one of which is in an excited electronic state (F*), while the other is in its ground state. This exchange interaction has first been considered by Murayama, Morigaki, and Kanzaki (1973)) with regard to ODESR of loose F aggregates (see Section 6.2.2.).

The model of distant F*-F pairs, being based on the random F center distribution (6.5), ignores completely the possible existence of correlations which can considerably reshape the pair distribution function at small r . Such correlations would undoubtedly arise because of the interaction between the components of the pair. Agullo-Lopez and Aguilar (1979) have considered the pair-correlation function $g(r) = \exp(U(r)/k_BT)$ for the distribution of F centers in NaCl, using the Morse potential to approximate the interaction $U(r)$ between a pair of F centers. The pair was treated as a diatomic hydrogen molecule embedded in a dielectric continuum. The conclusion was that appreciable deviations from a random distribution ($g = 1$) occurred for the 1-st (F-F = M) , 2-nd , and 3-rd nearest-neighbor positions, both at 300 and 500 K.

Ishii and Endo (1968) have investigated the photoconductive properties of additively colored KCl crystals at low temperature. The authors aimed at studying the residual photoconductivity below 80 K, attributed usually to colloids (Seitz (1946)). Another point of particular interest is an observation by Wild and Brown (1961) that the photoresponse in the nir is increased on illumination of the crystal with F light at LHeT. The photoresponse in the M band was too small to be observed at

LHeT, even in a sample containing an appreciable amount of M centers. However, when F' centers are formed by F band light, an appreciable photosensitivity is built up in the M band range which reflects the shape and dichroism of the M band. If now F' are completely bleached out optically, the photoresponse vanishes although no change in the M band occurs. It has been concluded that the above photoresponse is due to the F' centers and the following explanation is suggested: Photon absorption in the M band results in emission at 1.08 m which overlaps with the F' band. Consequently, F' will ionize and the resulting photocurrent will reflect the shape and dichroism of the M band. It is also suggested that the F' centers which form presumably on F band illumination at LHeT do so by way of electron tunneling between F*-F pairs. The enhanced ir photoresponse could also be wiped out by thermally annealing the sample within the F' instability range (above 200 K). However, the F' assignment could not be supported by measuring the corresponding optical absorption changes, too small to be detected. Nevertheless, based on the above arguments and on the observed spectral distribution of the enhanced photoresponse, the conclusion was drawn that F' centers were produced by F band light even at LHeT.

Further, the F' yield was measured from the change in photoconductivity at 1000 nm vs. the change collected on the electrode during the F illumination to produce F'. The F band light was found to bring about an order of magnitude higher an F' growth rate than did the 420 nm light, in terms of the F' vs. collected charge plot. Therefore, F' are effectively formed through F band illumination by a process which involves no CB electron trapping and is accompanied by very little free-charge transfer, presumably the F*-F electron tunneling. Similar results were also obtained at LAirT. The F' yield was found independent of the incident-light intensity. These F' centers decayed slowly in the dark at LHeT indicating that tunneling in the opposite direction also occurred. From the F' vs. charge curves, the authors estimated the quantum yield of the tunneling process to be 0.06 for $F_0 = 6 \times 10^{16}$ cm^{-3}. This value corresponds to a tunneling rate of 10^5 s^{-1} assuming $\tau_F = 0.58$ μs and $\eta_t = \tau_F/\tau_t$. The relative F' tunneling quantum η_t was also followed as a function of the F center density F_0: the result is shown in Fig.31. The yield is seen to grow almost proportionally to F_0 below 10^{17} F/cc and less steeply at higher F_0. The F' growth by tunneling was found to saturate on prolonged illumination. F' production by tunneling was also observed in a crystal x-rayed at RT, on illumination with F light at LHeT. Concluding, the authors have worked out a method for investigating the F' centers via the enhanced

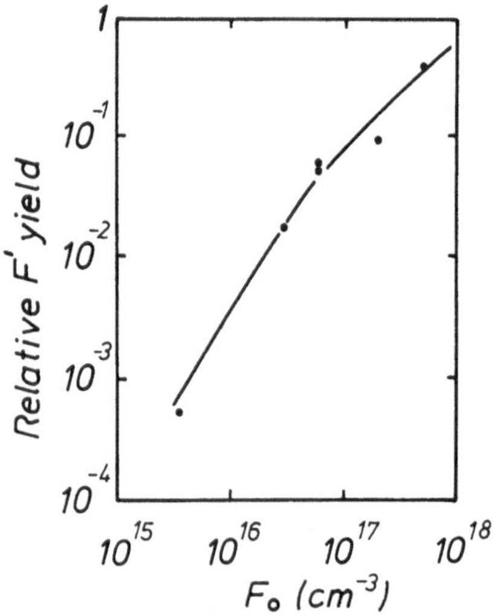

Figure 31: Relative quantum yield of F' formation through tunneling on excitation by light in the F band, vs. the F center concentration. (After Ishii and Endo (1968)).

ir photoresponse they bring about at LHeT, a technique apparently more sensitive than standard optical absorption techniques.

6.2.1.1. Triplet F' center?

Baldacchini, Gallerano, and Lüty (1980) have reported on some early experiments supporting the assignment of the F' band in NaBr and NaI which extends and even peaks to the violet of the F band and is thermally stable up to about 200 K. These materials belong to a group of alkali halides with small cation-to-anion radii ratio ($r_+/r_- < 0.5$). Applying a strong magnetic field to the crystal has led to quenching of the presumed F-F' conversion, as in Porret & Lüty's experiment on KCl (Porret and Luty (1971)). At high F center densities, the F-F' conversion in NaBr and NaI is highly efficient (high $\eta_{F-F'}$) and nearly temperature-independent (Baldacchini, Pan, and Lüty (1981)). This suggests that a significant portion of the F' centers form by an effective forward tunneling of the F* electron to a neighboring F center (reaction (6.a)). However, unlike KCl the tunnel-produced F' centers in NaBr and NaI are not transient, the F' electron apparently not undergoing the reverse tunneling process (6.b) back to the original vacancy in ground state. This implies that the F' state lies lower in energy than the relaxed ground F state (F_0), the lowest possible recipient state in reaction (6.b), in agreement with the observed relative position of the F' and F absorption peaks.

Recently Baldacchini, Gallerano, Grassano, and Lüty (1981), (1983a-d) have again made use of the tunnel effect to initiate an effective F-F' conversion in NaI and NaBr at temperatures as low as 1.9 K. The bell-shaped F' band is believed to be due to optical transitions between the F' ground state and an excited state, which autoionizes during the relaxation of the lattice. Under these circumstances, the optical F-F' conversion has been studied at 1.9 K in strong magnetic fields up to 80 kGauss. A strong reduction was observed of the initial conversion rate $W(H)$ as the field H was increased. However, the reduction at the highest fields, which polarize completely the unpaired electron spins at the F centers, is incomplete. Instead, the experimental points follow well the empirical relation $W(H)/W(0) = 1 - \alpha P^2$, where P is the polarization, while α is a loss factor, found equal to 0.48 (NaI) and to 0.74 (NaBr). This would imply:

(i) some spin memory is lost during the optical cycle to the effect that P is actually lower than the thermal equilibrium value;

(ii) part of F' form in a triplet state.

Incorporating the two leads to $\alpha = (1 - 2\epsilon)(1 - \delta)(1 + \delta)$, where ϵ is the spin-mixing parameter accounting for the loss of spin memory during the optical cycle, and δ is the relative probability for a stable triplet formation in reaction (6.a).

To determine ϵ, the authors have performed MCD measurements monitoring the ground-state spin-polarization P under an intense F light optical pumping. As the pumping light is turned on at $t = 0$, P changes according to $P(t) = P_f + (P_i - P_f)\exp(-t/\tau)$, where $\tau^{-1} = \tau_1^{-1} + \epsilon U$, U being the pumping rate. Performing measurements at various U yields both ϵ and τ_1^{-1}, the spin-lattice relaxation time. The following values are obtained: $\epsilon = 0.03 \pm 0.01$ (NaBr), 0.12 ± 0.04 (NaI). They lead to: $\delta = 0.12$ (NaBr), 0.23 (NaI). This suggests the existence of a triplet F' state which gives an additional contribution to the effective loss factor. The spin-lattice relaxation time was also measured at 1.9 K at various magnetic-field strengths from the rise of the MCD signal as the pumping was ceased. Hyperfine coupling was found to predominate, the relaxation time following

$$\tau_1^{-1} = A\,H^3 \coth(g\beta H/2k_BT)$$

with $A = 1.5 \times 10^{-14}$ (NaBr) and to 15×10^{-14} (NaI), in $s^{-1}G^{-3}$. From the spectral MCD signal the spin-orbit splitting Δ was also determined: -40 ± 4 (NaBr), -50 ± 4 (NaI) in meV (Baldacchini et al. (1981)).

In the quest for a triplet F' absorption, optical absorption measurements during the F-F' conversion have revealed the occurrence of a weak band of large width (about 0.5 eV) centered at about 1.45 eV (850 nm), as opposed to the main (singlet) F' band at 2.45 eV (NaI). Assuming the same oscillator strength for both bands, the fractional area under the nir band (25%) agrees fairly with the above value of δ (Baldacchini et al. (1983b)). Moreover, spectral MCD measurements have revealed a signal for the nir band ($S = 1$) and no signal at all for the main vis band, which is thus definitely assigned to the singlet ground state ($S = 0$)(Baldacchini et al. (1983a)). The nir band is not stable and undergoes decay in the dark which is purely exponential with a time constant of 540 s at 2 K. This time constant is not affected by any external magnetic field, while the band itself is increased in strength by 1.5 times in fields as large as 65 kG. Although the evidence is not conclusive, it is likely that the nir band originates from transitions from the F' triplet level, populated by electron tunneling during the F-F' conversion, to the conduction band (Baldacchini et al. (1983c)). The triplet assignment is supported by:

(i) the nir band's enhancement through spin polarization of the F centered system;

(ii) its spontaneous decay with a long lifetime expected for a forbidden triplet-to-singlet transition;

(iii) its large width suggesting transitions to the ionization continuum.

The observed triplet-to-singlet absorption ratio (25%) suggests that at $H = 0$ the singlet type of reaction channel (6.a) is the more effective one. The nir band increases in strength 1.5 times instead of 2 times in a strong magnetic field which suggests that some spin memory may actually be lost in the pumping cycle. There also are some indications that the F' triplet may decay both to the ground singlet F' state and to the ground F state.

6.2.2. Loose F aggregates

Exposing a freshly-quenched F-centered crystal to F-band light near RT brings about the disruption of the original (random) F center spatial distribution by leading to the formation of clusters of high local

F center density. The physical basis of clustering which results in the formation of close F center pairs (loose F-aggregates) will be discussed in a subsequent Section (6.6).

The effect of clustering on the ground-state ESR signal from F centers has been observed by Schwoerer and Wolf (1963). They reported a continuous narrowing of the F center ESR line in additively colored KCl crystals from 46 to 35 G, as a result of illumination with F band light at RT. The effect was found concentration-dependent, the decrease in line-halfwidth being more rapid with the illumination time as F_0 was increased. It has been explained as an exchange narrowing caused by the loose aggregation of F centers. An effective exchange frequency $\omega_e = E_e/\hbar$ (E_e -the exchange interaction energy) of about 1 GHz has been calculated from the observed narrowing and estimated to correspond to an average pair separation of 4 to 5 lattice spacings. For an account of the early ESR work on close pairs, see Seidel and Wolf (1968).

The ODESR technique has been applied some ten years later by the Neuchatel group to study the photoinduced ESR signal from loose F aggregates in KCl (Schnegg, Ruedin, Aegerter, and Jaccard (1973)). As the amount of RT F-light exposure increased, the positive variation of the F luminescent intensity with the external magnetic field diminished gradually to even become negative at longer exposures. At resonance, a positive peak was observed in contrast to the negative dip of the ESR response of distant F*-F pairs (Fig.32). This has been explained by

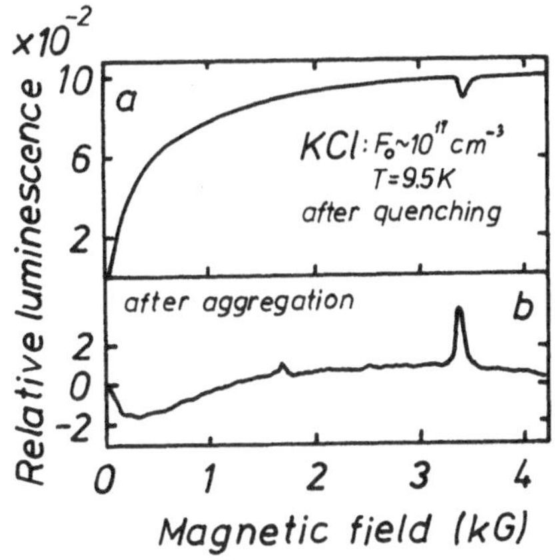

Figure 32: Relative luminescence variation in a KCl crystal as a function of the external magnetic field at 9.5 K for a freshly quenched sample (a), and after a controlled aggregation (b). (After Schnegg, Ruedin, Aegerter, and Jaccard (1973)).

again introducing an exchange term $J\underline{S}_1 \cdot \underline{S}_2$ in the Hamiltonian. In an external magnetic field both the energies of the Zeeman levels and the tunneling probabilities are gradually modified as the pair separation decreases. For a given J^*, only one level will radiate, while the others will deexcite non-radiatively by way of tunneling (Fig.33).

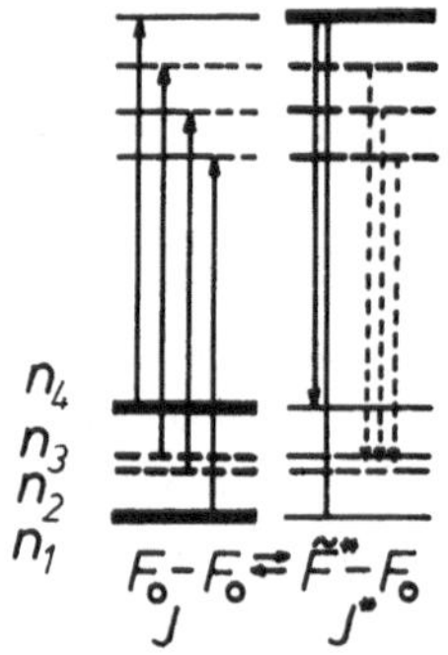

Figure 33: Energy level diagram of a loose F aggregate in ground and excited state. Full lines represent symmetric states, while dashed lines represent antisymmetric states. Full vertical lines label radiative, dotted lines label radiationless transitions. (After Schnegg, Ruedin, Aegerter, and Jaccard (1973)).

(This comes to explain why η_R decreases during the F-aggregation delay-period, to be discussed in greater detail later.) For not too large J^*, antisymmetric states will be favored at the expense of symmetric ones. Consequently, the populations of the ground-state levels will be modified during the optical cycles to attain $n_2 + n_3 > n_1 + n_4$ in steady state. (This causes a transient decrease of η_R which can be observed immediately after an optical excitation with a time-constant of about 0.1 s.) A microwave field at resonance will mix between the spin populations leading thereby to an increase of η_R. Thus the new high-field ESR line, occuring after a sufficient RT exposure to F light, is attributed to transitions in the ground state of loose aggregates, from levels 2 and 3 to level 4 ($\Delta m = 1$).

In this way, the exchange interaction favors the non-radiative deexcitation by changing the spin symmetry of the F*-F pair: whereas in ground state there are two symmetric (luminescent) states Ψ_1 and Ψ_4 and two antisymmetric (non-luminescent) ones Ψ_2 and Ψ_3, in the excited state there are one single symmetric state Ψ_1^* and another one Ψ_4^* assuming appreciable antisymmetry to make it partially non-luminescent. Ψ_2^* and Ψ_3^* remain mainly antisymmetric (Schnegg, Jaccard, and Aegerter (1973)). The exciting light populates Ψ_2 and Ψ_3 at the expense of Ψ_4, thereby decreasing the luminescent efficiency. At resonance, part of the pairs are turned back to the luminous Ψ_1 state and the luminescence rises. The exchange interaction occurs because of the strong wave-function overlap in the excited state of a close F center pair.

In a magnetic field at 10 K, the luminescent intensity also displays ENDOR and NMR in addition to ESR (Schnegg, Jaccard, and Aegerter (1973). ENDOR experiments on the second peak at half field in Fig.32 reveal the spectrum found for the M center: this resonance is therefore due to F-M pairs. A third resonance occuring at about one third of the field after a long bleaching time at RT is shown to arise from F-R pairs.

While a signal has been obtained with the F luminescence, the search for an ODESR from close pairs making use of the F and F' absorption bands has yielded a negative result (Ecabert and Jaccard (1978)). The lack of any F' signal implies either the occurrence of a competitive non-radiative deexcitation mechanism, which has nothing to do with F' formation, or a very short F'lifetime (less than 1 ms). While no preference has been given to any of these, it is worth mentioning that a process of the former type has been considered earlier (Jaccard and Aegerter (1973)) to account for the relatively lower F'-based ESR signal, as compared to the luminescent ESR response, which is at least ten times higher, in distant F*-F pairs. This additional non-radiative process should even be more efficient in close pairs. Provided a suitable spin state is at hand, a covalent bond may be formed within the pair, where the exchange interaction "softens" certain vibrational modes to ultimately make them unstable. As a result, the system shifts from excited F*-F to ground F-F state by tumbling down vibrational modes.

Another interesting investigation employing ODESR in KCl at 1.6 K has been made by Murayama, Morigaki, and Kanzaki (1973). Their result is shown in Fig.34. Part (a) displays the ESR spectrum of a freshly-quenched sample containing between 10^{17} and 10^{18} F/cc, (b) refers to the same sample after a week-long anneal at RT, while (c) has been obtained after a subsequent exposure to F band light at RT. The ESR measurements were made at 34 GHz. The spectrum is resolved into three lines labelled P_1 through P_3. P_1 has a g-value corresponding to the F ground state (1.995), P_3 is attributed to the F relaxed excited state, and P_2 is ascribed to an exchange-coupled F*-F pair. The following

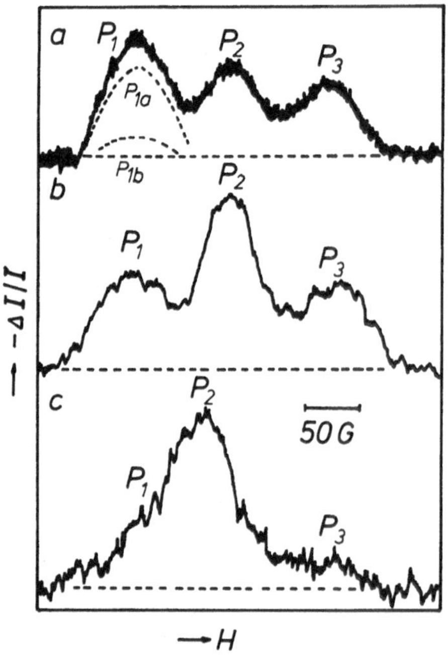

Figure 34: ODESR spectra at 1.6 K of a KCl crystal: (a) freshly quenched; (b) kept subsequently in the dark for a week at RT; (c) subsequently irradiated by F light at RT. The spectrum has been interpreted to be due to an exchange-coupled pair or cluster consisting of an electron in the relaxed excited state of an F center and one in the ground state of a nearby F center. (After Murayama, Morigaki, and Kanzaki (1973)).

spin Hamiltonian has been used:

$$\mathcal{H} = \beta(g^*\underline{H}\cdot\underline{S}^* + g^o\underline{H}\cdot\underline{S}^o) + J\underline{S}^*\cdot\underline{S}^o$$
$$= \beta H(g^*S_z^* + g^oS_z^o) + J\underline{S}^*\cdot\underline{S}^o \quad , \qquad (6.43)$$

$\underline{H}$ being parallel to the Oz axis. J is the exchange-interaction coupling constant. Except for constant terms, eq.(6.43) transforms into

$$\mathcal{H} = \beta g^oHS_z + \tfrac{1}{2}JS^2 - \gamma g^o\beta HS_z^* \qquad (6.44)$$

where $\gamma = 1 - g^*/g^o$, $\underline{S} = \underline{S}^o + \underline{S}^*$. The third term in (6.44) taken as a perturbation, the first-order perturbation theory gives for two limiting cases:

(a) $J \ll \gamma g^o\beta H$ (a weakly-coupled pair)

$|S=1,S_z=1\rangle$ $\qquad E_1 = g^o\beta H - \tfrac{1}{2}\gamma g^o\beta H$

$$
\begin{array}{lll}
2^{-\frac{1}{2}}(|S=1,S_z=0\rangle + |S=0,S_z=0\rangle) & E_2 = \frac{1}{2}\gamma g^{o}\beta H & \\
2^{-\frac{1}{2}}(|S=1,S_z=0\rangle - |S=0,S_z=0\rangle) & E_3 = -\frac{1}{2}\gamma g^{o}\beta H & (6.45) \\
|S=1,S_z=-1\rangle & E_4 = -g^{o}\beta H + \frac{1}{2}\gamma g^{o}\beta H &
\end{array}
$$

The energy level scheme is shown in Fig.35 (a). Two types of ESR transitions ΔE_1^o and ΔE_2^o with the same g-value (g^o) are predicted. Accordingly, the P_1 line should be resolved into two components with different responses to the microwave power which is actually observed. The other ESR transition with g^* occurs as P_3.

(b) $J \gg \gamma g^o\beta H$ (a strongly coupled pair)

$$
\begin{array}{lll}
|S=1,S_z=1\rangle & E_1 = J + g^{o}\beta H - \frac{1}{2}\gamma g^{o}\beta H & \\
(1+\lambda^2)^{-\frac{1}{2}}(|S=1,S_z=0\rangle + \lambda|S=0,S_z=0\rangle) & E_2 = J & \\
|S=1,S_z=-1\rangle & E_3 = J - g^{o}\beta H + \frac{1}{2}\gamma g^{o}\beta H & (6.46) \\
(1+\lambda^2)^{-\frac{1}{2}}(|S=0,S_z=0\rangle - \lambda|S=1,S_z=0\rangle) & E_4 = 0 &
\end{array}
$$

with $\lambda = \gamma g^o\beta H/2J$. In this limit, the ESR transition ΔE_{e1} has a g-value of $(g^o + g^*)/2$. This transition is ascribed to P_2 in Fig.34,

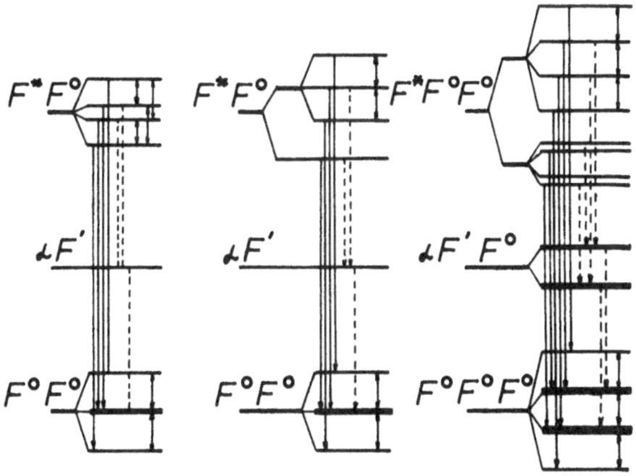

Figure 35: Proposed energy level diagram to account for the spectrum in Fig.34: (a) weakly coupled, (b) strongly coupled pair of F centers; (c) cluster of three F centers coupled strongly. Solid and dashed vertical lines as in Fig.33. (After Murayama, Morigaki, and Kanzaki (1973)).

while the energy level scheme is presented in Fig.35 (b). As expected, the spectra in Fig.34 display an enhancement of P_2 following anneal in the dark, as well as after exposure to F light at RT. However, the P_2

line in Fig.34 (c) is obscured by the occurrence of a fourth line, ascribed to an F*-F-F triple cluster with $J \gg \frac{2}{3}\gamma g^0 \beta H$, as shown in Fig.35 (c).

The essential results obtained by the Neuchatel group on the loose F aggregates are extensively summarized in a paper by Schnegg, Ecabert, Jaccard, and Aegerter (1973): Even a slight optical bleaching of the crystal in the F band at RT leads to drastic changes in the luminescent properties of the F centered system. The luminescent yield now drops rather than increasing in an external magnetic field, while the resonant line appears as an increase rather than decrease of the luminous yield, which it does in the case of more distant pairs. For loose aggregates formed by a short RT exposure to F light, the resonant line has a halfwidth of 64 G and g = 2.001. The interpretation is essentially based on Hamiltonian (6.43), in which H is a resultant magnetic field, which is the sum of the external field H_0 and the nuclear field H_N at the defect site in either electronic state of the pair. The weak coupling limit (a) alone is considered, the coupling constant J assumed non-vanishing in the excited state of the pair only. First-order perturbation theory leads to eigenvalues similar to (6.45). While in ground state (J = 0) there are two symmetrical (luminescent) states and two antisymmetric spin states of the pair, in excited state (J* = 0) only one state remains purely symmetric. Consequently, the effect of the pumping light is to populate preferrably the middle levels at the expense of the lowest-lying level. This makes the quantum yield drop, as observed. ODESR measurements further show that the signal rapidly saturates with the microwave power. Optical detection ENDOR and NMR measurements have also been tried successfully. The latter of these suggest that significant nuclear polarization of the spins surrounding the loose aggregate is created during an optical cycle. Related to this conclusion is the experimental observation that the ground-state spin-lattice relaxation is greatly enhanced for loose aggregates. As a result, the effective magnetic field at the site of the pair will exceed the applied field by the amount of that nuclear polarization. This explains why the ESR line for loose aggregates appears shifted to lower fields relative to the one of isolated F centers. Accordingly, the actual g-value should therefore be lower than 2.001. Depolarizing the nuclei by applying an appropriate rf field would lower the effective magnetic field down to $H_{external}$ and would consequently increase the luminescent yield. As the rf range is swept between 1 and 6 MHz, this would give rise to the observed NMR spectrum.

Further optical-detection NMR measurements are reported by Jaccard,

Schnegg, and Aegerter (1974). Surprisingly, the ENDOR signal of close F center pairs does not vanish as the microwave power is turned off but rather decreases by a factor of 2. The luminescence measured as a function of the rf frequency at a constant magnetic field (or vice versa) results in the spectrum exhibited by ENDOR. An approximate solution is discussed, based on a Hamiltonian containing electronic and nuclear Zeeman, exchange, and hyperfine energies, which explains the observation. The NMR spectrum is predicted to reproduce the standard ENDOR spectrum of single centers, with somewhat broadened lines.

The model assuming an exchange spin-spin interaction leading to a spin-dependent non-radiative deexcitation within close F center pairs is also considered in some greater detail by Schnegg, Jaccard, and Aegerter (1974) and compared with experiments on the variation of the luminescence with an applied magnetic field, optical detection EPR, and population relaxation followed by observing the transient behavior of the luminescence excited under pulsed conditions. All these effects are related to population changes in the ground state caused by optical pumping and are sensitive to the mixing of the states by the different interactions. The paramagnetic resonance has been found to occur in the ground state ($g = 1.998$, linewidth 52 G). It is interesting to note that the average exchange energy ranged between the hyperfine and Zeeman energies depending on the state of aggregation, while the relaxation times ranged between 25 and 250 ms depending on the optical pumping rate. The relations predicted by the model agree generally with the experiment.

6.3. F'-α interactions

Several experimental examples suggesting the role of notable F'-α pair interactions have already been discussed in the preceding Sections. Among these one should recall the presumed effect on the F' thermal lifetime (Schmid and Wolf (1962)), the observed correlated decay-types of F' centers (Cordovilla and Alvarez Rivas (1974)), the collected evidence regarding reaction (6.b) (Mezger and Jaccard (1981)), etc. We will now describe in some detail certain theoretical calculations related to the above interactions.

The F'-α interaction rate constant has been derived in the quasi-classical approximation by Martinez-Negrette and Ruiz-Mejia (1974). In effect, this is a calculation of the F' ionization in the electric field of the vacancy, the field ionization being described as electron tunneling from an excited F' level to the conduction band. The transmission coefficient, derived for tunneling in an applied field $\mathcal{E}$ from a hydrogen-like state of energy E_n to the ionization continuum, is given by

$$t = \exp\left(-2\int_{r_1}^{r_2}\left(\frac{2m}{h^2}(V(r) - E_n)\right)^{\frac{1}{2}}dr\right) \qquad (6.47)$$

where

$$V(r) = -e\mathcal{E}r - e^2/\epsilon_\infty r \qquad (\mathcal{E} = \text{const})$$

$$r_{1,2} = \frac{1}{2e\mathcal{E}}(-E_n \pm (E_n^2 - 4e^3\mathcal{E}\epsilon_\infty^{-1})^{\frac{1}{2}})$$

The tunneling rate τ_t^{-1} is next obtained as t is multiplied by the electron frequency $|E_n|/h$ and by the fractional solid angle subtended by the vacancy at a distance R from the F' center. E_n is taken to be -0.08 eV (Crandall (1965)). τ_t was found to be less than 10^{-4} s which is the measuring time in the experiment of Alvarez Rivas (1970) for a critical field of $\mathcal{E}_c = 2.7 \times 10^5$ V/cm . The field was related to the F' density following a suggestion by Redfield (1963):

$$\mathcal{E} = e/\epsilon\,\bar{r}^2 = (e/\epsilon)(\tfrac{4}{3}\pi\alpha)^{2/3} \qquad (6.48)$$

for the average field created by the empty vacancies to act on the F' center. Using (6.48) the critical F' density corresponding to $\mathcal{E}_c$ was found to be $F_c' = 1.4 \times 10^{16}$ cm^{-3} . Once appearing in the CB, the tunneled electron will be captured by the nearby vacancy in a time

$$\tau_e = \int_0^{\bar{r}}(\epsilon_\infty r^2/\mu_e e)dr = \epsilon_\infty/4\pi e\,\mu_e\alpha \qquad (6.49)$$

which is calculated to be the order of 10^{-11} s for $\alpha = F_c'$. It follows that tunneling is the rate-determining process. An improvement of the model was sought by adding the e-e repulsion energy to $V(r)$. One of the F' electrons was assumed to be in ground (1s) state, the other - in an excited (2p) state. However, this correction to V has proved to be of little significance.

A quantum-mechanical approach to the electron tunneling within an F'-α pair is due to Berezin (1978). He considers the F'-α pair as an excited state of the F-F pair: As a matter of fact, both the optical and thermal ionization energies of the F' center (1.1 eV and 0.53 eV, respectively (Goldberger and Owens (1971))are substantially lower than the optical ionization energy of an F center (about 3 eV (Wild and Brown (1961)) and the F thermal ionization energy (2.05 eV (Markham (1966)) in KCl. The F optical ionization energy can be regarded as a sort of "electron affinity" of the α center. Now, inasmuch as the electronic energy level of the F-F pair lies deeper than the one associated with F'-α, the two states can be regarded as ground and excited states of that same pair, respectively. Under these conditions, a spontaneous deexcitation transition can occur from F'-α to F-F which will materialize through electron tunneling from the F' level to the α center in F ground state. Depending on the pair separation R, the tunneling rate τ_t^{-1} of the deexcitation process can become comparable or even superior to the F' thermal ionization rate $\tau_{F'}^{-1}$. This may account for some experimental observations of the dependence of the F' decay rate on the presence of other nearby defects giving rise to a correlated (exponential) F' decay type (Cordovilla and Alvarez Rivas (1974)).

Berezin's calculation of τ_t^{-1} has been carried out ignoring the electron-phonon interaction. Under these conditions, the tunneling leads to a radiative deexcitation of the pair. The initial-state wave function is taken in the form of a Hulthen orbital

$$|i\rangle = D(\gamma/2)^{\frac{1}{2}} r_A^{-1} \exp(-\gamma r_A)(1 - \exp(-Zr_A)) \tag{6.50}$$

with, in atomic units,

$$D = (1 + 3(\gamma/Z) + 2(\gamma/Z)^2)^{\frac{1}{2}} \tag{6.51}$$

Here $r_A = |\underline{r} - \underline{R}_A|$, $\underline{R}_A$ is the position vector of the center of the F' vacancy; $\gamma = 0.28$ and $Z = 0.47$ (the effective charge of the anion vacancy) are semi-empirical parameters. The final-state wave function is chosen to be in the form of a 1s orbital

$$|f\rangle = (Z^3/\pi)^{\frac{1}{2}} \exp(-Zr_B) \tag{6.52}$$

with $r_B = |\underline{r} - \underline{R}_B|$, B is the point at the center of the α vacancy. The one-electron approximation used is justified insofar as the second electron of the pair does not play any immediate role in the process. Nevertheless, its presence is accounted for by the semi-empirical choice of γ and Z. The tunneling rate is given by

$$\tau_t^{-1} = \frac{4}{3}\alpha_{hf}^3 (\Delta E)^3 \, |\langle i|\underline{r}|f\rangle|^2 \tag{6.53}$$

where $\alpha_{hf} = 1/137$, while ΔE is the transition energy. Inserting (6.50) and (6.52) into (6.53) yields approximately ($Z > \gamma$):

$$\tau_t^{-1} = \alpha_{hf}^3 (\Delta E)^3 \, (32D^2\gamma Z^5/3(Z^2 - \gamma^2)^4) \exp(-2\gamma R) \tag{6.54}$$

$R = |\underline{R}_A - \underline{R}_B|$ is the pair separation. In this model,

$$\Delta E = \frac{1}{2}(Z^2 - \gamma^2) - (\epsilon_0 R)^{-1} \quad , \tag{6.55}$$

the second term giving the Coulomb interaction energy between F' and α.

The origin of the frame is at mid-point between A and B. Further on, Berezin calculated the pair separation R_0 at which $\tau_t^{-1} = \tau_{F'}^{-1} = 1.4 \times 10^{12} \exp(-0.53\text{eV}/k_BT)$ (Goldberger and Owens (1971)): $R_0 = 20$ A at $T = 225$ K .

In this way, the tunneling deexcitation process (TDP) can play an important role at low temperature (where $\tau_{F'}^{-1}$ is small) and can be observed experimentally as a spontaneous decay of F' centers in the F' thermal-stability range. TDP can be particularly strong in clustered F center systems. It can give rise to a slow and weak afterglow in a crystal containing F' and α centers which will be non-monochromatic because of the dependence of ΔE on R . The radiative tunneling transitions (RTT) are also discussed in a subsequent work by Berezin (1980). He considers the possibility of such transitions occuring in more complex systems, e.g. F'-M , F'-M , $F'-F_A$, F'-F'-α, etc.

A luminescence above 700 nm in KCl crystals x-rayed at 4.2 K (afterglow) has been assumed to be the result of electron tunneling within F'-α pairs according to reaction (6.c) with a subsequent radiative deexcitation of the F* center (Grabovskis and Vitol (1979)).

More recently Berezin (1982) considered the possibility for the occurrence of an Anderson transition in an F centered crystal containing F' and α centers. The initial F center concentration F_0 is assumed sufficiently high. At low density of ionized species, the F' electrons will in fact be delocalized undergoing a hopping-like tunnel diffusion over the network of F center sites, because the potential fluctuations due to a random distribution of F centers are weak. However, at higher excitation levels the random fluctuations of the crystalline field will increase, due to the presence of a large number of Coulomb charges, and the tunneling motion of the F' electrons will be more difficult. At sufficiently high magnitudes of the fluctuation potentials, the F' electrons will completely lose their ability to tunnel between neighboring F sites and an Anderson electronic localization of the F' centers will occur. Anderson's transition may manifest itself in the appearance of discontinuities of some physical quantities, such as the high-frequency dielectric constant, etc. Berezin approximates the magnitude of the fluctuation potential, due to a random distribution of F' and centers, by taking the average Coulomb interaction energy of two neighboring ionized centers which is

$$W \sim 2e^2 F'^{1/3}/\epsilon \qquad (6.56)$$

The factor 2 arises from the fact that there are two independent random distributions of oppositely charged defects with the same concentration

$F' = \alpha$. Further, Berezin defines the condition of the occurrence of Anderson's localization as

$$W/V \gtrsim \eta \tag{6.57}$$

where V is the energy-level splitting for a system of an F' center shared between two equivalent F sites. Using a delta-well potential model (see Section 19.3.4) to calculate V , Berezin arrives at

$$V = (2\gamma e^2 a_0/R)\exp(-\gamma R) \tag{6.58}$$

where γ is the reciprocal radius of the F'-electron wave function, a_0 is the Bohr-orbit radius. R is taken to be the average distance between F sites,

$$R = F^{-1/3} \tag{6.59}$$

However, to avoid complications Berezin performed his calculation assuming $F \sim F_0$, that is, $F' < F$. Inserting into (6.57) obtains

$$F'^{1/3} \gtrsim \eta \epsilon \gamma a_0 F_0^{1/3} \exp(-\gamma F^{-1/3}) \quad , \tag{6.60}$$

the condition of Anderson's localization. The following data were used: $a_0 = 0.529$ Å , $\gamma^{-1} = 1.89$ Å , $\epsilon = 4.68$, $\eta = 8$ (KCl). The result is presented in Table 2. Here a = 3.14 Å is the interionic spacing for

Table 2

Estimates of Anderson's localization
a is the interionic spacing in KCl*

R (Å)		F_0 (cm^{-3})	F'_{crit} (cm^{-3})
4a	12.56	5.0×10^{20}	1.3×10^{15}
6a	18.84	1.5×10^{20}	1.8×10^{10}
8a	25.12	6.3×10^{19}	3.5×10^{5}

*After Berezin (1982)

KCl. The above estimates are prescribed for 0 K, provided there are no other instabilities in the (F',α) system. If so, the system of ionized species over the domain of F center sites will be quantum-mechanically stable in Anderson's sense when the density of ionized species exceeds the critical value.

A subsequent paper by Berezin (1983) deals again with the principal instability against radiative tunneling transitions (RTT), this time in triple impurity systems, such as the one composed of F, F', and α centers. If the F-F' separation is much smaller than the F'-α distance, then a fast F'(h) + F(l) — F(h) + F'(l) RTT recharging is possible along with a slow F' + α — 2F transfer. Here h and l stand for the higher and lower energy position of the trapping level for an extra electron at two F sites, which differ by their respective distances from the α center. Such type of RTT can be significant in areas of defect clustering. In a system composed of two F and two F' centers, spontaneous transitions to a lower-lying energy state are conceivable through the simultaneous RTT of both extra electrons (Berezin (1984a)). These will be considered in greater detail later, in Section 6.9.

Some interesting suggestions regarding the deexcitation of crystals containing large concentrations of F' and α centers have also been made by Berezin (1977b). An F'-α pair can deexcite to produce two F centers in ground state in two conceivable ways at low temperature: (i) through a radiative tunneling transition (RTT) or via (ii) nonradiative tunneling, the excess energy being transferred in the form of photons in (i) and phonons in (ii). An additional deexcitation channel may occur in a tri-center system, such as F', F', α : the nonradiative Auger effect. In the particular case, a nearby F' electron tunnels to the α center, while the energy released, as that electron relaxes to the equilibrium ground state of the F center formed, lifts the electron of the second F' component to the conduction band; that is,

$$F' + F' + \alpha \longrightarrow 3F + e \tag{6.e}$$

The rate of the nonradiative Auger process is found by means of the golden rule:

$$k_{Auger} = (2\pi/\hbar) \int \left| \langle \Psi \mid V_{Auger} \mid \Psi_k \rangle \right|^2 \delta(E_k - E) d\underline{k} \tag{6.61}$$

Here $\underline{k}$ is the wave vector of Auger's electron in the conduction band, Ψ, E, and Ψ_k, E_k are the wavefunctions and energies of the initial (F'+F'+α) and final (3F+e) states, respectively. The wavefunctions are chosen of the form

$$\begin{aligned} \Psi(\underline{r}_1,\underline{r}_2) &= \psi(\underline{r}_1-\underline{R}_A)\,\psi(\underline{r}_2-R_C) \\ \Psi_k(\underline{r}_1,\underline{r}_2) &= \varphi(\underline{r}_1-\underline{R}_B)\,\chi_k(\underline{r}_2-R_C) \end{aligned} \tag{6.62}$$

with

$$\begin{aligned} &\psi(r) = (\gamma/2\pi)^{\frac{1}{2}} \exp(-\gamma r)/r\;, \quad \varphi(r) = (a^3/\pi)^{\frac{1}{2}} \exp(-a r) \\ &\chi_k(r) = (2\pi)^{3/2} \exp(i\underline{k}\cdot\underline{r}) \end{aligned} \tag{6.63}$$

The F' electron wavefunctions Ψ are the eigenfunctions of the delta-shaped potential to be considered in detail later, in Section 19.3.4. In the above equations, $\underline{r}_1$ and $\underline{r}_2$ are the position vectors of the two F' electrons, while the color center components are seated at A(F'), B(α), and C(F'). Auger's operator is taken to be the dipole-dipole interaction energy between (F',α) and F':

$$V_{Auger}(\underline{r}_1,\underline{r}_2) = R^{-3}((\underline{r}_1-\underline{R}_A)\cdot(\underline{r}_2-\underline{R}_C) - 3(\underline{r}_1-\underline{R}_A)\cdot\underline{n}\ (\underline{r}_2-R_C)\cdot\underline{n}) \quad (6.64)$$

where $\underline{n} = (\underline{R}_C - R_A)/|\underline{R}_C - \underline{R}_A|$. For A, B, and C forming an isosceles triangle, Berezin obtains

$$E = -\tfrac{1}{2}\gamma^2 - \tfrac{1}{2}\gamma^2 - (\epsilon R_0)^{-1}, \quad E_k = \tfrac{1}{2}k^2 - \tfrac{1}{2}a^2 \quad (6.65)$$

(in atomic units), and

$$k_{Auger} = \tfrac{32}{3}\gamma k_0^3/(\gamma^2 + k_0^2)^4 M^2 R^{-6}(1 + 3\cos^2\vartheta) \quad (6.66)$$

where

$$k_0 = (a^2 - 2\gamma^2 - 2/\epsilon R_0)^{\frac{1}{2}}, \quad M = \langle\Psi(\underline{r}-R_A)|\underline{r} - \underline{R}_A|\varphi(\underline{r} - \underline{R}_B)\rangle$$
$$R_0 = |\underline{R}_A - \underline{R}_B|, \quad (6.67)$$

while ϑ is the angle between AB and AC.

A numerical estimate for $R_0 = 7.7$ Å, $R = |\underline{R}_A - \underline{R}_C| = 30$ Å, $\gamma = 0.28$, $a = 0.47$ yields $1/k_{Auger} = 4 \times 10^{-7}$ s for KCl. Clearly, non-radiative Auger processes can bring about an apparent "thermal ionization" of F' centers in temperature ranges where the thermal stability of an isolated F' center is otherwise guaranteed.

6.4. F*-F* interactions

The first indications of the occurrence of an F*-F* interaction have been obtained in experimental investigations by Park and Hopfield (1965), Park (1965), and Park and Faust (1966) on the optical absorption spectra of F* centers in KI. Additively colored crystals containing F centers between 10^{15} and 10^{17} cm^{-3} were pumped by a pulsed ruby laser (output 5J, pulse duration 500 s) to excite the F centers at low temperature. The induced absorption was monitored using auxiliary-light sources and monochromators, during and after the excitation pulse. From these measurements the F* and F' absorption spectra were obtained. In particular, the F* absorption was monitored by means of a Globar source (Park (1965)) or a Xe gas laser (Park and Faust (1966)), the latter producing ten discrete lines of sufficient intensity between 0.096 and 0.62 eV (12.9 to 2 μm). In a typical case of a crystal, containing initially 7×10^{16} F/cc and excited to about 2×10^{15} F*/cc by a 0.5 J pulse, the F* absorption is composed of an uniform level of $c_{F*} = 10^{-16}$ cm^2 between 0.6 - 0.3 eV (2 to 4 μm) followed by a steep rise to $c_{F*} = 4 \times 10^{-15}$ cm^2 as the photon energy is further decreased to about 0.1 eV (12.4 μm). The steep rise below 0.2 eV was attributed to the optical transition to a higher F bound state, while the level above 0.3 eV was ascribed to photoionization. The F* absorption vanishes for photon energies in excess of 1 eV; nevertheless, the overlap with the F* emission band (peaked at about 0.83 eV, halfwidth 0.19 eV (Fowler (1968)) is appreciable. The two optical bands were found to be correlated closely, apparently because of their common origin. In a particular case, both bands

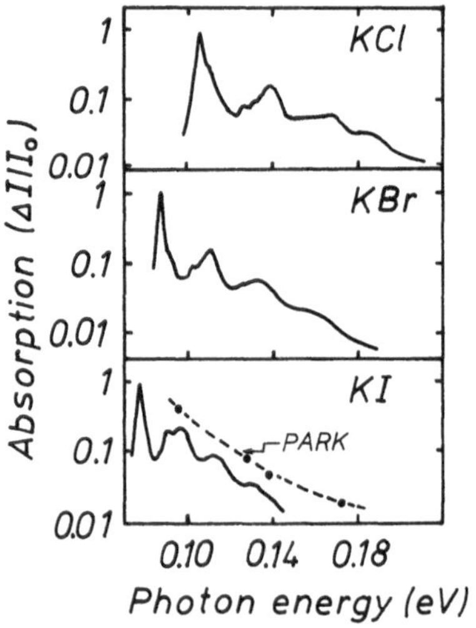

Figure 36: Absorption spectra of F* centers. (After Kondo and Kanzaki (1975)).

have decayed exponentially after a Q-switched laser pulse of 3×10^{-8} s, exhibiting about the same decay time of 2 μs (Park and Faust (1966)). The F* transient absorption bands in KCl, KBr, and KI are seen in Fig.36 (Kondo and Kanzaki (1975)).

The most puzzling observation in Park's work is the occurrence of the F' band itself at temperatures as low as 15 K (Park (1965)). At the F center densities employed, the number of tunnel-produced F' centers cannot be expected to be large (Chiarotti and Grassano (1966b)). Nevertheless, between 10 to 50 % of the initial F centers converted to F' during the laser excitation. A plausible explanation proposed by Park is the energy transfer between two F* centers resulting in the ionization of one of them:

$$F^* + F^* \longrightarrow F + \alpha + e \qquad (6.f)$$

which can materialize either as one of the F* absorbs (and then ionizes) the photon emitted by the other (because the F* absorption and emission bands overlap), or via some interaction within the F*-F* pair. Self-absorption is not likely to be the dominant process in samples of linear dimensions much less than $(c_{F*}F^*)^{-1} = (10^{-16} \times 10^{15})^{-1} = 10$ cm. Park's samples have been as large as 0.3 x 1.0 x 1.5 cm^3. In looking for some alternative, Park examined the dependence of the F* density on F_0 and I to find that $F^* \propto (F_0 I)^{\frac{1}{2}}$, which could be accounted for by introducing a loss term $\propto F^{*2}$ into the rate equations. Such a term can arise from some appropriate pair interaction, which has to be sufficiently long-ranged to produce the observed F' densities in not too heavily colored crystals. Alternatively,it would require large spatial inhomogeneity in the F center distribution or the laser light. In any event, as Park puts it, there must be some mechanism which reduces to a simple pair interaction in the limit of a high-intensity excitation.

In a similar investigation by Fröhlich and Mahr (1966a), the F* decay curves following a Q-switched ruby-laser excitation of a KI crystal were found to be non-exponential. These curves have been obtained by monitoring the induced absorption changes in the F, β, and β* bands (Fröhlich and Mahr (1965)). In a subsequent work (Fröhlich and Mahr (1966b), non-exponential F* decay curves, showing an increase over the exponential luminescent decay rate at low F* densities, are reported for samples containing more than 2 x 10^{16} F*/cc. The non-first-order decay rates are discussed in terms of a resonant energy transfer between neighboring F* centers. In addition to the radiative F* decay, the only F* decay path at low F center densities in KI, excitation may also be transferred to a nearby F*. Energy will be conserved in this process because the F* emission band overlaps with the F* absorption (Park (1965)). As a result, two F* will be destroyed and one electron ejected into the conduction band (reaction (6.f)). The electron can be retrapped by the vacancy or be captured by an F center forming F'.

Several different transfer mechanisms have been discussed in the literature:

(i) Self-absorption (Park (1965)): the photon emitted by one F* is absorbed by another F*. The rate equation is

$$\dot{F}^* = -F^*/\tau_R - P_{SA}F^* \qquad (6.68)$$

where the self-absorption probability per second is assumed to be given by $P_{SA} = (A/\tau_R F_0^*)F^*$ with $A = Gc_{F*}F_0^*$. G is a geometrical factor, the order of the sample dimension. The solution of (6.68) is

$$F^* = F_0^*/((1 + A)\exp(t/\tau_R) - A) \qquad (6.69)$$

(Fröhlich and Mahr (1966b)). This is not a genuine pair interaction.

(ii) Exchange effects: These can take place through the overlap of the wave functions of neighboring F* (Frohlich and Mahr (1966b)).

(iii) Dipole-dipole interaction: It would occur if the dipole fields of neighboring F* overlap. The rate equation for a d-d interaction is solved to give (Frohlich and Mahr (1966b)):

$$B = 1 + (F^*/F_0^* - \ln(F^*/F_0^*) - 1 - t/\tau_F)/(F^*/F_0^*)(\exp(t/\tau_F) - 1)$$
$$B = (16\pi^3/9)R_0^6 F_0^{*2} \qquad (6.70)$$

τ_F is the F* lifetime at low F* density. A comparison with the experiment has been made yielding close fits of eq.(6.70) to the observed decay curves (Fröhlich and Mahr (1966b)), giving $B = 4.2 \times 10^{-35}\ cm^6$ whence $R_0 = 96$ Å .

(iv) Auger's autoionization: Auger's effect could lead to the autoionization of one of the components of an F*-F* pair. Berezin (1969) reports having calculated an interaction distance of about 150 Å , in good agreement with what he has determined from an analysis of Fröhlich & Mahr's decay curves (about 200 Å).

For an account of the early laser-spectroscopy of color centers, see Mahr (1968)).

Fröhlich & Mahr's conclusions on the dipole-dipole interaction as the dominant energy transfer within an F*-F* pair in KI have recently been confirmed for KCl by Schubert and Vogler (1980) using a picosecond laser spectroscopy. A giant ps pulse from a Nd-glass laser was shot onto a sample, colored either additively or x-rayed at RT, and the repopulation of the F ground state was followed by a second monitoring ps pulse. A certain percent of the F* centers were found to return to F in 20 to 150 ps after the excitation pulse at LNT. This rapid initial F* decay was found to depend on both the F concentration and the coloration method. An interaction distance of 106 ± 15 Å was deduced based on Dexter-Forster's theory extended to take into account the non-uniform F center spatial distribution. No F' formation has been observed on the picosecond time scale: apparently, the dipole-dipole interaction leading to energy transfer between F*-F* pairs takes less time than does the F*-F tunneling process.

6.5. F*-F' interactions

An interesting case of an F*-F' interaction has been considered by Berezin (1969a). He points out that the excitation energy of an F*-F' pair (the F* energy) exceeds the ionization energy of the pair (the F' ionization potential). This implies that the F* emission band should overlap with the F' band, as it actually does. As a result, the pair can autoionize spontaneously through Auger's effect leading to an F-F pair and a conduction band electron:

$$F^* + F' \longrightarrow F + F + e \qquad (6.g)$$

In effect, the F* center deexcites non-radiatively, while the F' center is ionized. As a result of Auger's autoionization, the F' lifetime within a pair can be considerably shorter than the isolated F' lifetime. The paired F' lifetime can be expected to depend on the pair separation and, therefore, on the color center concentration and on the excitation light intensity.

The autoionization rate reads (cf. (6.61)):

$$\tau_A^{-1} = 2\pi \int |\langle f | V | i \rangle|^2 \, \delta(E_f - E_i) d\nu \qquad (6.71)$$

in atomic units. $|i\rangle$ and $|f\rangle$ are constructed from the one-electron wave functions in the Heitler-London approximation

$$\begin{aligned} |i\rangle &= 2^{-\frac{1}{2}} \left(\Psi_{F^*}(\underline{r}_1) \Psi_{F'}(\underline{r}_2) \pm \Psi_{F^*}(\underline{r}_2) \Psi_{F'}(\underline{r}_1) \right) \\ |f\rangle &= 2^{-\frac{1}{2}} \left(\Psi_{F}(\underline{r}_1) \chi_{\underline{k}}(\underline{r}_2) \pm \Psi_{F}(\underline{r}_2) \chi_{\underline{k}}(\underline{r}_1) \right) \end{aligned} \qquad (6.72)$$

For an allowed F* — F transition, V is again the dipole-dipole interaction operator

$$V(\underline{r}_1, \underline{r}_2) = \frac{1}{R^3}\left((\underline{r}_1 \cdot \underline{r}_2) - 3(\underline{r}_1 \cdot \underline{e}_z)(\underline{r}_2 \cdot \underline{e}_z) \right) \qquad (6.73)$$

where $\underline{e}_z = \underline{R}/R$, R is the pair separation ($\underline{R}$ is parallel to Oz). Further, using

$$\tau_R^{-1} = \frac{4}{3} \alpha_{hf}^3 (\Delta E)^3 \, |\langle \Psi_{F^*}(\underline{r}) | z | \Psi_F(\underline{r}) \rangle|^2 \qquad (6.74)$$

($\alpha_{hf} = 1/137$, ΔE is the energy of the electronic transition at F, τ_R is the F* radiative lifetime), and choosing

$$\Psi_{F'}(\underline{r}) = (\gamma / 2\pi)^{\frac{1}{2}} \exp(-\gamma r)/r \qquad (6.75)$$

(cf. (6.63)) with $\gamma = (2\omega_o)^{\frac{1}{2}}$ (ω_o is the F' band threshold energy),

$$\chi_{\underline{k}}(\underline{r}) = (2\pi)^{-3/2} \exp(i\underline{k} \cdot \underline{r}) \qquad (6.76)$$

Berezin obtains finally

$$\tau_A = \tau_R(R/R_o)^6 \qquad (6.77)$$

where

$$R_o = \frac{(16\gamma)^{1/6}}{(\gamma^2 + k_o^2)^{2/3}} (k_o/\alpha_{hf}\Delta E)^{\frac{1}{2}} \qquad (6.78)$$

is the interaction distance, while $k_o = (2E_{o_o})^{\frac{1}{2}}$, $E_o = \Delta E - \omega_o$. The following values of R_o were calculated in Å units: 50 (NaCl), 66 (KCl), 75 (KBr), 68 (KI).

6.6. Pair interactions near room temperature

At higher temperatures where the α center is thermally mobile, the photoexcitation and subsequent ionization of an F centered system results in the occurrence of dynamic interactions between color center species described by time-dependent pair separations. Two main interactions involving mobile alpha centers have been considered: the Coulomb attraction between α - F' and the monopole - induced dipole type between α - F. Both interactions would lead to the ultimate formation of the two-F-center aggregate ($M = F_2$) according to the reactions:

$$\alpha + F' \longrightarrow M \qquad \hat{\gamma}_{\alpha 1}\alpha F' \qquad (6.h)$$

(Petroff (1946), Delbecq (1963)), and, respectively,

$$\alpha + F \longrightarrow M^+ \qquad \hat{\gamma}_{\alpha 2}\alpha F \qquad (6.i)$$

$$e + M^+ \longrightarrow M \qquad \gamma_{M^+} n M^+ \qquad (6.j)$$

(Lüty (1961a), Delbecq (1963)). Generally, Petroff has assumed that the n-fold F-center aggregate forms via a sequence of electronic and ionic steps of the form

$$e + F_{n-1} \longrightarrow F'_{n-1}$$

$$\alpha + F'_{n-1} \longrightarrow F_n$$

Lüty has also stressed the role of F' in F-aggregation processes. However, he suggests that the growth of the F-aggregate centers rather occurs via alpha-center precipitation followed by free-electron trapping, through a reaction (6.i)-(6.j)-type sequence. Now F' would merely serve as a reservoir for conduction electrons originating from F centers.

Whatever the particular type of the aggregation mechanism, if the anion vacancy migration were purely random, the M center formation would have been a bulk process with an initial growth rate proportional to the statistical rate constant $\hat{\gamma}_\alpha$ leading to an initial formation stage starting immediately as the exposure to F-band light is switched on. However, early investigations by Petroff (1946), (1950) have revealed that the actual M-band growth is preceded by an incubation or delay period in which the M center production rate is extremely low. He has attributed the delay period to the formation of precursor centers. More detailed investigations within the induction period have later been made by Delbecq (1963). He found that the delay time is generally the longer the lower the F center density F_0 and the lower the temperature T . Although the F band has remained practically unchanged during the induction period, both the radiative quantum yield η_R and the ability

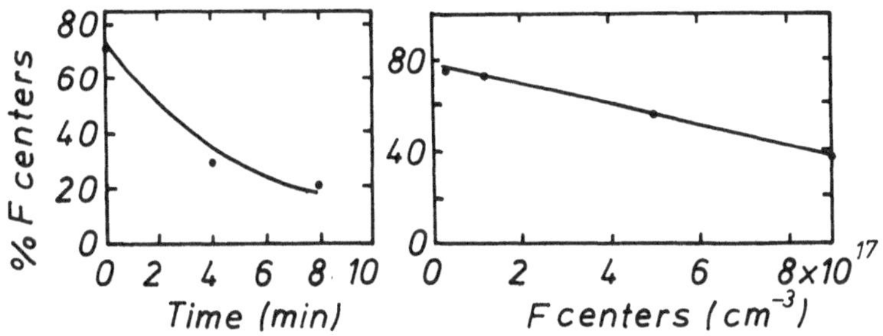

Figure 37: Percent conversion of F centers to F' centers in KCl at 173 K as a function of the time of bleaching with F band light at 273 K (left) compared with the variation of that quantity as the F center density is increased with no bleaching (right). (After Delbecq (1963)).

of F centers to convert into F', $\Delta F/F_0$, have dropped. Fig.37 shows Delbecq's data on the percent photoequilibrium F-F' conversion at 173 K as a function of the time of pre-exposure to F band light at 0 °C in a KCl crystal with $F_0 = 1.2 \times 10^{17}$ cm^{-3} , the pre-exposure producing very little changes in either the F or the M band ranges. Nevertheless, $\Delta F/F_0$ is seen to decrease significantly.

The decrease of the quantum yield $\eta_{F-F'}$ of the F-F' conversion during the delay period has apparently been behind an early observation by Geiger (1955). He investigated the F-F' process in additively colored KCl crystals, subjected to repeated cycles of optical bleaching with F light at 170 K followed by thermal bleaching at RT of the F' centers formed. The quantum efficiency of the F-F' conversion apparently dropped significantly after each cycle. Ultimately, it appears to have decreased down to a point where the crystal is essentially unaffected by the F light at 170 K. Similar results have been obtained by Reinberg and Grossweiner (1961). They studied the optical bleaching at 77 K of F centers in KCl crystals colored either additively or by x-rays at RT. A flash-light bleaching technique has been used to induce conversions between F and F' centers in either direction. Prior RT bleaching with F light was found to suppress both the extent and the rate of the low-temperature optical bleaching.

A temporary bleaching effect (TBE) in KCl at low temperature, following a pre-exposure to F band light at RT within the delay period, has been observed by Schneider and Caspari (1963). It was attributed to the temporary formation of F' centers by electron tunneling from F* to nearby F centers. The TBE spectrum consisted of a prominent F band and a small broad absorption extending to the red and nir region up to about 1 μm, apparently the F' band. The recovery process could not be described by a single exponential. The authors suggest that the process may be a superposition of electron-tunneling events, following reactions (6.a) and (6.b) at various pair separations. Later, Schneider and Patterson (1967) have made further experiments within the delay period by monitoring the F' absorption induced by a ruby-laser excitation in the F band of KI at low temperature. They found that a portion of the F' centers formed are rapidly transient and decay at an appreciable rate without taking part in the establishment of an F-F' photoequilibrium. An additional effect of the RT pre-exposure was observed on the F' optical bleaching rate: F' centers in the pre-exposed state have initially bleached out more rapidly.

Transient absorption changes in the vis and nir produced by F light at 77 K in additively colored KCl, KBr, and KI crystals have all been studied by Schneider and Caspari (1964). The samples contained controllable amounts of M centers produced by F light irradiation at RT. At small F center densities, a small temporary decrease in the F band and a temporary increase in the F' band were observed, the maximum F band change being at the F band peak. The initial F center densities were between 4×10^{16} and 1.2×10^{18} cm^{-3} . The spontaneous decay could not be described by a single time-constant indicating that a superposition of tunnelings to and from F center sites at various intercenter separations were involved. A similar bleaching effect was observed when a major portion of the M band was bleached either thermally or optically. It seems that M centers simply decomposed into two F centers sepa-rated by a few lattice spacings. The transient bleaching effect was found insensible to the application of high external electric fields suggesting that the tunneled F-F' conversion was almost completely saturated by the light intensity alone.

The RT photoconductivity in the F band is also quenched upon bleaching within the pre-aggregation period: A rapid decrease of photosensitivity during the illumination has been found to accompany the relatively less rapid bleaching of F centers in x-rayed KBr (Oberly (1951)) and additively colored KCl (Hardtke, Scott, and Woodley (1960)). Although various explanations have been proposed, there seems to be little doubt that at least in part the observed quenching is related to the deep physical changes occuring during the induction period.

Delbecq (1963) has been the first to suggest that these changes may be due to a photoinduced F center clustering which leads to an enhanced concentration quenching of the observed physical processes. This implies that the aggregate-center formation is preceded by a time-period during which the originally random F center distribution is largely disturbed through the formation of F center clusters, regions of a high local F center density. Were the M center formation a random process, the original F center distribution in a freshly-quenched crystal would not have been so severely disturbed by the buildup of aggregates at random sites. Clustering in a statistical manifold of F centers would only occur if there are some pair interactions between the centers favoring the shorter pair separations, as well as conditions rendering at least one of the components mobile. The role of mobile species has mostly been attributed to the anion vacancy (the alpha center); this assumption has later been confirmed experimentally. The F* center is expected to be even more mobile (Lüty (1968)) but is very short-lived. Now, there are two such pair interactions involving α centers: the F'-α and F-α types, mentioned above. The prerequisite for the occurrence of clustering is that the driving interaction is such that the α center has a much greater chance of pairing with a particular F'(F) partner rather than of having an equal chance of colliding with any F'(F). The latter case, leading to bulk (random) aggregation, would take place if the α center thermal energy was sufficiently high to exceed its potential energy in the field of a given attractive center, thereby preventing an oriented α drift toward that center. The condition for clustering therefore is that there are enough pair separations r within a critical sphere of radius r_c, at which the two energies just equal each other. From eq.(4.22), $r_c = e^2/\varepsilon k_B T = 115$ Å for a Coulomb center in KCl at RT. Now, since $r_c > r_t = 85$ Å, the F*-F electron-tunneling distance for KCl, it follows that all tunnel-produced F'-α pairs at RT would fall into that category. Consequently, the

Markham-Luty tunneling reaction (6.a), leading to concentration-quenching phenomena at low temperature, would also be an efficient clustering mechanism near RT. It would thus lead to the displacement to shorter r of the most probable pair separation as the RT exposure to F band light goes on. Isolated $F'-\alpha$ pairs would not contribute to clustering because of their large pair separations. Aggregation of such isolated centers may be expected to proceed through statistical bulk processes, at later stages of M center formation. These bulk processes are likely to follow reaction (6.i)(the $F-\alpha$ mechanism), which seems to account better for the observed M band growth rates (Delbecq (1963)).

The role of electron traps other than F in M center formation has been studied in experiments by Hirai and Scott (1967)). They first bleached F centers in additively colored KCl crystals by F light at -120 ^{o}C, and then warmed these crystals up to about 0 ^{o}C. M centers did not appear until temperatures well above those of the complete disappearance of F' centers formed by the bleaching. The function of electron traps is concluded to be two-fold: (i) to increase the alpha-center concentration thereby enhancing the rate of M^{+} formation, and (ii) to store electrons until temperatures where a good number of M^{+} centers have been formed. In a photochemical F-M conversion at given temperature, reaction (6.h) is also conceivable.

The F clustering kinetics leading ultimately to aggregation have been considered in detail by Dudek and Grossweiner (1969). Both the condensation of F centers during the delay period and the initial stage of M center formation are believed to result from the same physical phenomenon: the dynamic interaction within the $F'-\alpha$ pair, denoted as F'_{α}. This close pair is assumed to be the precursor of the M center. The F'_{α} entity is engaged in the following model reactions:

$$h\nu_F + F \longrightarrow F'_{\alpha} \qquad \eta_t q_F F \qquad (6.k)$$

$$F'_{\alpha} \longrightarrow M \qquad F'_{\alpha}/\tau_1 \qquad (6.l)$$

$$F'_{\alpha} \longrightarrow 2F \qquad F'_{\alpha}/\tau_2 \qquad (6.m)$$

$$h\nu_F + F'_{\alpha} \longrightarrow 2F \qquad (6.n)$$

$$\longrightarrow F' + \alpha \qquad \eta' q_{F'_{\alpha}} F'_{\alpha} \qquad (6.o)$$

Appropriate experimental conditions have been chosen using pulsed F-light excitation to enhance the role of the dark processes involved. An almost excellent agreement is obtained between model and experiment. Reaction (6.m), the analogue of (6.b) for a close pair, was found to be less important than the optical destruction (6.n) of F'_{α}. Channel (6.o) has been included to account for eventual optical decomposition of F'_α to isolated components. The latter would supply additional material for the statistical bulk processes at the later stages of the M band growth. The temperature dependence of the rate constant in (6.1) yielded the activation energy for correlated migration of the α-center within the pair: 0.60 eV for KCl, in an excellent agreement with earlier findings (Luty (1968)). The rate equation corresponding to reactions (6.k-o) is

$$\dot{F}'_\alpha = \eta_t q_F F - F'_\alpha/\tau_{F'_\alpha} \qquad (6.79)$$

where $\tau_{F'_\alpha}^{-1} = \eta' q_{F'_\alpha} + \tau_1^{-1} + \tau_2^{-1}$. Its solution vanishing at $t = 0$ is (assuming $q =$ const, $F \sim F_0$):

$$F'_\alpha = \eta_t q_F F_0 \tau_{F'_\alpha}(1 - \exp(-t/\tau_{F'_\alpha})) \qquad (6.80)$$

Using a flash-light excitation spectroscopy, Nierzewski, Todorov, and Georgiev (1978) have investigated the F' yield per flash in KCl crystals colored by γ-rays at RT. This yield was found to drop down on repeated flashing at RT within the induction period of M center formation, to become virtually insensitive to the illumination after a total exposure of some tens of flashes. Within the experimental errors, the normalized drop rate proved independent of F . As discussed earlier, the SF F' yield is given by eq.(4.36); consequently, the yield being proportional to the ionization efficiency η_i of the F center, the observed drop can be attributed to a corresponding variation of η_i . The overall behavior was consistent with

$$\eta_i(F_0,N) = \eta_i(F_0)\,\eta_i(N) \qquad (6.81)$$

where N is the exposure in number of flashes, $N = 0$ corresponding to the freshly-quenched state of the sample. The observed drop of $\eta_i(N)$ as N increased was attributed to the corresponding increase of the electron-tunneling probability leading to the production of rapidly transient F' centers which could not be observed with the employed technique of a limited time-resolution. The bimolecular F' decay parameter was found independent of N which was to be expected on grounds that the monitored ms decay range has corresponded to a relaxed system of isolated F' centers.

In a related investigation, Werner, Stradowski, and Sugier (1980) have attributed the observed logarithmic F' decay-type within the μs delay range to an electron tunneling between distributed F'-α pairs (reaction (6.b)). The isoenergetic condition of tunneling requires that the F' electron should tunnel to a ground F level (1s) in an excited vibronic state. As much as about 1.6 eV of excess energy is released during the subsequent lattice relaxation which could promote one or two jumps of the resulting F center to the neighboring F center to form an aggregate. The alternative mechanism is assumed to be the migration of F* centers to nearby F centers. The authors argue that tunnel-produced F' centers following reaction (6.a) are mainly responsible for the F' yield on flash-light excitation, while F' centers produced by free-electron trapping contribute mostly to the F' yield in pulsed-radiolysis experiments.

A solid evidence to support the conclusions on the existence of a dynamic F'-α interaction in alkali halides has been presented by Jaque and Agulló-López (1970). They have investigated the thermally-activated processes occuring in the dark after F-band bleaching of x- or γ-irradiated NaCl at RT. Once the F-light illumination is discontinued, the following effects are observed: (i) The M band exhibits a slight additional increase. (ii) The F band shows an even more pronounced increase. (iii) The spectral range between F and M undergoes an overall decrease. Subtracting the spectra immediately before and after the F-light yields the F' band. After a sufficiently long time in the dark, the spectral analysis gives a pure F band. The F' decay kinetics in the dark were investigated near RT. They were found to be nearly exponential and the decay rates were compatible to those for the trapping of vacancies, suggesting a close correlation. The disappearance of free vacancies in the dark was detected via the initial F-center growth on re-irradiation (restoring the radiation-induced equilibrium), or via the change in F-bleaching rate on subsequent F-light illumination ($\Delta F = \Delta\alpha$). This method enables one to determine changes in the free-vacancy concentration without lowering the temperature to the thermal-stability range of the α-band. All the three bands, F, F', and M, followed a parallel time-dependence suggesting a close relationship between the centers. From the exponential F' decay curves at different temperatures the activation energy of the decay process was found to be 0.48 ± 0.06 eV. Pick's data on $\tau_{F'}^{-1}$ aligned along the same line. These experimental results point to the existence of an F'-α trapping mechanism. To account for the increase of the F and M bands on switching-off the F-light, the authors suggest reactions (6.b) and (6.h), respectively. The observed parallel behavior of F, F', and M indicates that the rates of these reactions are determined by the F' decay process. Both may proceed via a common intermediate (F'_{α}) which rapidly decomposes into an F-F pair or collapses into an M center. The F'-decay efficiency of both reactions is about the same at RT. However, the rate of the latter should be dependent on the temperature, lower-temperature experiments indicating that the M center formation is inhibited below -50 °C. At such temperatures, a relatively fast F' decay is observed leading exclusively to F center production.

Similar conclusions have been drawn from a subsequent work by the same authors (Jaque and Agulló-López (1974)). They studied the kinetics of RT F-band bleaching in γ-irradiated NaCl, colored within the first stage at RT. The F bleaching curve was found composed of two parts: a fast initial drop (a), followed by a slower nearly exponential decrease (b). The F' band was found to rise steeply during (a) and to decrease during (b). The changes in F and F' during (a) obeyed $\Delta F' = \frac{1}{2}\Delta F$, indicating that a genuine F-F' conversion had taken place. The following model was proposed: On switching-on the illumination, an F-F' conversion took place leading to the fast initial drop in F center density until an F-F' photoequilibrium was attained. This equilibrium proved essentially independent of the excitation-light intensity and the temperature. In addition to this purely electronic process, a slow removal of vacancies and F' centers via the thermally-activated reaction (6.h) continuously shifted the F-F' photoequilibrium to account for the slow component at (b). The inclusion of an ionic process at the second F-bleaching stage suggests that the bleaching rate should depend on the repetition rate or on the light and dark periods in case of a sequential illumination. In particular, the rate would grow higher as the dark period increased tending to saturate when that period approached the characteristic time of reaction (6.h). Also, the rate would increase on decreasing the light-on periods at constant dark-pause. A reasonable agreement with the experiment was obtained. From the dependence of the bleaching rate on the temperature, the activation energy of the slow component was found to be 0.50±0.03 eV, in good agreement with the earlier estimate (Jaque and Agulló-López (1970)). One is led to the conclusion that the activation energy of about half an electron-volt for the F' decay in NaCl, as obtained from bleaching experiments near RT, may include a significant tunneling contribution, being characteristic of a correlated rather than random process. The above value is, therefore, not related directly to the one (0.9 eV) found in thermal-glow measurements (Scaramelli (1966)) which may represent the genuine F' thermal-ionization energy. The same interpretation applies to Pick's data on $\tau_{F'}$ in NaCl which agree closely with Jaque & Agulló-López's results, both obtained by similar techniques.

6.6.1. Which species are mobile?

The possibility that the F' center may be the mobile entity during the F-M conversion has been discussed by Crawford (1964). He argues that the 0.6 eV activation energy of the M center formation rate, as measured in KCl, should be ascribed to the migration of F', since the α-center migrational energy from diffusion studies is known to be much larger. Indeed, the enthalpy of anion vacancy migration in KCl has recently been summarized to be 0.9 eV (Fuller (1972)). While most authors have almost ruled out the suggestion of a mobile F' center on grounds of its electronic configuration (Lüty (1968)), the problem of the F' mobility is not yet clear (Sonder and Sibley (1972)). Nevertheless, the apparent migrational energy obtained from aggregation-kinetics studies may be considerably lower than the one obtained from tracer measurements, since it is characteristic of a correlated rather than random motion.

An interaction effect between F' and α centers which can lower the migrational energy of the mobile species has been considered by Brothers and Lynch (1968). The long-range Coulomb force between F' and would not lower the barrier for a jump of either defect sufficiently, because the Coulomb potential did not vary rapidly enough with r. The authors argue that the F' center is almost as compact as an F center. Consequently, as F' and α approach each other, the F' center will polarize. Now, as the F' wave function begins to overlap the α-center, some covalent bond will occur. Both defects now resemble a pair of polarized F centers. A lowering of the jump barrier will result, proportional to dV/dr, not to the binding energy $V(r)$. The idea is similar to the one proposed a little earlier for the interaction within the F-α pair (Farge, Lambert, and Smoluchowski (1967)).

In a cleverly designed experiment, Schneider (1971) has got a convincing evidence that the mobile entity in F-aggregation phenomena is the α center. He used an additively colored and heavily Na-doped KCl crystal, in which F_A centers were produced by a brief pre-exposure to F light. α, α_A, F', and F'_A centers are now subsequently produced by F light at low temperature. By exposure to F' light (outside the F'_A band), the F' are wiped out. F_A formation is then performed by annealing in the dark at a higher temperature: this resulted in the formation of an additional number $\Delta F_A(t)$ where t is the annealing time. These extra F_A centers could have only come at the expense of α centers getting trapped at Na^+ sites. Their amount being $\Delta F_A(t)$, the available α-center density at time t of the anneal is $\alpha(t) = \Delta F_A(\infty) - \Delta F_A(t)$. Next, the latter quantity is inserted in the F_A rate equation to give

$$\exp(-E_\alpha/k_BT) \propto \frac{\Delta F_A(t)}{\Delta F_A(\infty) - \Delta F_A(t)} \frac{Na^{-1}}{t}$$

for small t . An Arrhenius plot of the right-hand side measured at different temperatures has yielded a straight line with $E_\alpha = 0.59$ eV.

A similar technique has been used a bit earlier by Giuliani (1969). He used an x-rayed KCl crystal annealed for 12 h at RT. The sample was cooled down to 150 K and illuminated with F- and then with F'-light. The net result was the decrease of F and the growth of the α band, while no F' was present. The sample was now warmed up to a temperature above 150 K, kept there for 15 min, and then cooled down to LNT for an optical measurement. The data obtained indicate that the α-centers are stable up to 200 K but decay at an increasing rate to totally disappear as the temperature is raised to 250 K. A direct proof has been obtained that the α thermal decay is at least partially due to α-center migration, by observing the growth of an F_A band in a Li-doped sample.

Recently, Rascon and Alvarez Rivas (1983) have observed an optically induced F center aggregation in KCl at temperatures as low as LNT which seems to preclude the participation of any mobile anion vacancies. Moreover, evidence has been collected to support the suggestion of an F* mobile entity. The authors have followed the changes in both the optical absorption spectrum and the thermoluminescence induced by F-light photostimulation in pure KCl samples, pre-irradiated with gamma- or x-rays at RT. At low F center densities (2-3 x 10^{16} cm^{-3}), F bleaching at 110 K destroys most of the F centers without producing either F' centers or F aggregates. At F concentrations near 10^{17} cm^{-3}, however, both F' and F aggregate bands appear in the optical absorption spectrum when the illumination is carried out at 110 K. The aggregate bands alone occur in densely colored samples if the photostimulation is effected at 80 K. When occuring, these aggregate bands (M, N, R) are produced in amounts independent of the presence or the absence of the F' band, at 110 K or 80 K, respectively. The F' thermal annealing curve at 750 nm was obtained by the isochronal method for a sample illuminated at 150 K, and found to correlate well with the main TL glow peak at 226 K attributed to F', a large F' annealing step occuring at the peak temperature. Other, less intense peaks in the PSTL spectrum are also assigned to F', as reported earlier (Cordovilla and Alvarez Rivas (1974)).

The emitted light spectrum was practically composed of the F* emission band alone. F aggregates have generally been found to occur concurrently with the F* luminescence. It is concluded that the F* center can well be the mobile entity in the F aggregation process at temperatures where anion vacancy motion is hindered.

6.7. F'-M and like interactions

The possibility of a radiative tunneling transition occuring between F' and a nearby neutral center, such as the M center, has first been considered by Berezin (1979a). Concrete calculations on several systems of that type are presented in a subsequent paper (Berezin (1980)). A spontaneous radiative tunneling transition (RTT)

$$F' + Y \longrightarrow F + Y' \qquad (6.p)$$

would only occur if the binding energy of the Y' center exceeded that of F'. Data on the binding energy of various primed centers in KCl are presented in Table 3.

Table 3

Binding energies of primed centers*

Center	F'	F'_A	$F'_B(1)$	$F'_B(2)$	$F'_C(1)$	$F'_C(2)$	M'	R'
$E_b=-E_o$(eV)	1.10	1.21	1.29	1.30	1.36	1.37	1.37	1.59
$\gamma=-E_o^2$(a-u)	0.284	0.299	0.308	0.309	0.317	0.318	0.317	0.342

*After Berezin (1980)

E_o is the ground-state energy of an extra electron bound in the short-range field of the neutral core of Y. The RTT rate is given by eq.(6.53), ΔE being the energy difference between the initial and final states of the tunneling electron. In the framework of the one-electron approximation, the RTT of type (6.p) can be represented as an excited-to-ground-state energy transition of the extra electron. The potential field of the two neutral cores (F and Y) is assumed fixed and represented by two delta-well potentials (see Section 19.3.4). The transition is from the shallower to the deeper delta trap. Under these conditions the wave functions used for the calculations are

$$|i\rangle = \varphi(\gamma_i, \underline{r}-\underline{R}_i)$$
$$|f\rangle = N\,(\varphi(\gamma_f, \underline{r}-\underline{R}_f) + A|i\rangle) \qquad (6.82)$$

The $\underline{R}$'s are the position vectors of the two wells, $\underline{r}$ is the electron position vector. N is a normalization factor and the second term in $|f\rangle$ is introduced to orthogonalize it to $|i\rangle$. The delta-potential orbital is given by

$$\varphi(\gamma, \underline{r}) = (\gamma/2\pi)^{\frac{1}{2}}\,\frac{1}{r}\,\exp(-\gamma r) \qquad (6.83)$$

Note that in this model $E = (\gamma_f^2 - \gamma_i^2)/2$. Inserting (6.83) and (6.82) into (6.53) yields τ_{RTT}^{-1}. Results of these calculations are presented in Table 4. Here $\rho = |\underline{R}_f - \underline{R}_i|$. RTT can appear as a weak long-lived

Table 4

Calculated RTT rates*

System	F'+ R	F'+M	M'+R	$F'+F_A$	$F_A'+F_B(1)$
E (eV)	0.49	0.27	0.22	0.11	0.08
τ_{RTT}^{-1} (s^{-1})					
ρ = 10 Å	0.0074	0.035	0.194	0.314	2.47
15 Å	0.95	4.06	39.8	32.7	330
20 Å	150	582	10300	4300	56000

*After Berezin (1980)

deexcitation of a colored crystal following an optical stimulation. The above calculations indicate that the more aggregated a color center, the lower the position of the trapping level for its primed electron. Therefore, it is reasonable to expect a relative increase in the amount of primed aggregates at the expense of F' centers as the crystal relaxes in the dark at low temperature. RTT may be significant in a real crystal, taking into account the inhomogeneous F center distribution.

6.8. α-M and like interactions

Radiative tunnel transitions within positively charged color center systems, such as α-M, M^+-R, R^+-N, etc., have been considered by Berezin (1979a). He presents concrete calculations for the first one of these. Inasmuch as the ionization energies of the F-aggregate centers are inferior to the electron affinity of an anion vacancy (assumed equal to the binding energy of the F center), systems like α-M, α-R, α-N possess an inherent instability against transitions into the configurations M^+-F, etc. In calculating the RTT probability for the reaction

$$\alpha + M \longrightarrow F + M^+ , \tag{6.q}$$

Berezin takes up the simplest case of an equilateral triangle of vacancies, $|\underline{R}_0 - \underline{R}_1| = |\underline{R}_0 - \underline{R}_2|$, where $\underline{R}_0$, $\underline{R}_1$, $\underline{R}_2$ are the position vectors of the center and of the two anion vacancies comprising the M center, respectively. The RTT rate is obtained from eq.(6.53) using

$$\begin{aligned} |i\rangle &= N_i(\varphi(\underline{r}-\underline{R}_1) + \varphi(\underline{r}-\underline{R}_2)) \\ |f\rangle &= N_f(\varphi(\underline{r}-R_0) + A|i\rangle) \end{aligned} \tag{6.84}$$

for the initial and final states of the tunneling electron, where

$$\varphi(\underline{r}) = (Z^3/\pi)^{\frac{1}{2}} \exp(-Zr) \tag{6.85}$$

is an one-electron s-orbital. The A mixing term in (6.84) is introduced to orthogonalize the two electronic states ($A < 1$),

$$\begin{aligned} N_i &= (2(1+S_{12}))^{-\frac{1}{2}} \\ N_f &= (1 - 2S^2/(1+S_{12}))^{-\frac{1}{2}} \\ S_{pq} &= \langle\varphi(\underline{r}-\underline{R}_p)|\varphi(\underline{r}-\underline{R}_q)\rangle \\ S &= S_{01} = S_{02} = (1 + ZR + \tfrac{1}{3}Z^2R^2)\exp(-ZR) \\ A &= -S/N_i(1+S_{12}) \end{aligned} \tag{6.86}$$

The value of Z used is 0.48 a.u., and ΔE is taken to be 0.3 eV. Substituting into (6.53) one obtains

$$\tau_t^{-1} = \frac{1}{6}\alpha_{hf}^3 (\Delta E)^3 S^2 |\underline{R}_1 + \underline{R}_2 - 2\underline{R}_0|^2/(1 + S_{12} - 2S^2) \tag{6.87}$$

A numerical calculation for an M center in the tenth coordination sphere gives $\tau_t = 0.01$ s . This is an order-of-magnitude estimate. It is sensitive to both ΔE , which may be affected by the presence of charged species and by their spatial distribution, and to the asymptotic behavior of the wave functions employed.

6.9. F'-F interchange

Berezin (1984a,b) has recently considered the principal aspects of electron-transfer processes in a system composed of F and F' centers. The simplest case being that of a F-F' pair, he assumes that, due to the interaction with other perturbing centers, the energy level of an F' center at B is raised slightly higher in energy relative to the corresponding vacant level of an F center at A. Under these conditions, the F' electron can be expected to perform a spontaneous tunneling transition (STT) from B to A leading effectively to an F-F' interchange. The excess energy in the process can be emitted in the form of phonons or photons. In the particular case, the initial configuration AB' (the prime standing for the excess electron) is in fact an excited state which undergoes a STT to a lower-lying (ground state) configuration A'B. The lifetime τ of this excited state is composed of phonon- and photon-deexcitation counterparts, and may be expected to increase exponentially with the A to B separation.

Further, polycenter excited states are considered which can only undergo STT to a ground state by way of correlated transitions of several electrons, which jump simultaneously between various lattice sites. These include:

(i) Two-electron correlated tunnel transition in a F-F' dimer A'BC'D in a planar configuration. All the four one-electron transitions: A'-B, A'-D, C'-D, and C'-D would only increase the Coulomb repulsion, if materialized one by one, and are, therefore, forbidden by the conservation of energy. At equal bond-lengths $AB = BC = CD = DA$, the dimer is two-fold degenerate in energy, and in a realistic case an asymmetry occurs, e.g. in the form $AB = BC = a$, $AD = a + \delta$. The situation reminds of the Jahn-Teller distortion in an electron-phonon system but Berezin would not go beyond the formal analogy stressing that the distortion in the F-F' dimer was induced by an external perturbation rather than occuring spontaneously. One way or the other, the AB'CD' configuration of the asymmetric system lies lower in energy than the original A'BC'D arrangement, placing the latter to be an excited state of the former, characterized as a 'genuine ground state'. STT's can now occur between A'BC'D and AB'CD' in the form of correlated double jumps. The excess energy in the process is $\Delta E = 1/AC - 1/BD \simeq \delta/2^{3/2} a^2$. To calculate the tunneling rate, Berezin chooses the F' wave function in the form of a delta-potential orbital (see Section 19.3.4)

$$\varphi(r) = (\gamma/2\pi)^{\frac{1}{2}} \exp(-\gamma r)/r \qquad (6.88)$$

and constructs the initial and final states as

$$\begin{aligned} \Psi_i(r_1,r_2) &= \varphi_A(r_1)\,\varphi_C(r_2) \\ \Psi_f(r_1,r_2) &= N_f(\varphi_D(r_1)\,\varphi_B(r_2) + A\Psi_i(r_1,r_2)) \\ A &= -<\varphi_A(r_1)\varphi_C(r_2)|\,\varphi_D(r_1)\,\varphi_B(r_2)> \end{aligned} \qquad (6.89)$$

The radiative lifetime is next calculated from

$$\tau^{-1} = (4e^2/3hc)^3(\Delta E)^3\,|<\Psi_f|r_1 + r_2|\,\Psi_i>|^2 \qquad (6.90)$$

$$(e^2/3hc)^3(\Delta E)^3|\underline{R}_D-\underline{R}_A+\underline{R}_B-\underline{R}_C|^2\exp(-2\,(|\underline{R}_D-\underline{R}_A|+|\underline{R}_B-\underline{R}_C|))$$

which gives $\tau = 44.1$ s for $\gamma^{-1} = 10$ Å, $a = 25$ Å, $\delta = 5$ Å, $\Delta E = 38.5$ meV in vacuo.

(ii) Three-electron correlated STT in a F-F' trimer in a spatial configuration. Now, all one- and two- electron STT's are energetically prohibited, since leading to the increase of repulsion energy. The only way to reach the genuine ground state is through correlated three-electron STT.

(iii) Four-electron correlated STT in a F-F' quadromer in a spatial configuration. Similarly, a correlated STT of all the four electrons can only occur in the system, STT's involving less than four electrons being forbidden by energy conservation.

The generalization for more-than-four-electrons systems is straightforward. Undoubtedly, similar correlated electron-tunneling transitions, both radiative and nonradiative, may occur in densely colored F centered crystals involving the simultaneous interchange of several electrons between F center sites. The initial excited state of the system will be created on photobleaching in the F band producing F' and α centers. The above estimate of the RTT rate at a dimer indicates that the rate of a correlated tunneling jump may be expected to be similar to the one-electron rate over the combined length of all the individual jumps. The multielectron rate being that small, correlated tunneling processes may affect the late components of the F' lifetime or induce long-term instabilities in color center systems following a photoexcitation. To complement Berezin's investigation, calculations of the nonradiative tunneling rate should next be carried out.

6.10. F'-V interactions

Tunneling processes between V centers and F' centers have been discussed by Teegarden and Maurer (1954). They studied experimentally the optical bleaching of the V_1 band in KCl crystals, x-rayed at 83 K. The initial V_1 bleaching was accompanied by a rapid decrease of the F'band and little if any change of F. No current has been observed to flow across the crystal during the bleaching. A hole is assumed to tunnel from an excited V_1 center to a nearby F' center leading to the formation of an F* center. The F* electron then tunnels on to another V center leaving behind an empty vacancy. This would account for the relative F-band stability during the initial bleaching stage. While the euristic quality of the above suggestions may look uncertain from the viewpoint of our present knowledge of the V bands, they provide an interesting example of a conceivable multiple-tunneling process. Seitz (1955) has further discussed experiments by Maurer's group. He points out that essentially the same result as above is obtained if the crystal is darkened at LNT again but is bleached by V_1 light at 35 K. However, part of V_1 now transform into H. The thermal-annealing behavior of F' on warming from LNT does not appear to be correlated to that of V_1; in this case F rather than F' bleaches preferentially along with V_1. In contrast, F' is bleached in preference to F when V_1 are destroyed by light quanta. It appears that an F'-V_1 tunneling interaction occurs only when V_1 are excited optically.

Evidence for an electron tunneling from F' centers to V_k centers resulting in a recombination luminescence at wavelengths below 460 nm in KCl and KBr crystals is reported by Grabovskis and Vitols (1979). They have studied crystals x-rayed at 4.2 K. During x-raying, KBr exhibited emission bands at 530 and 275 nm, while KCl only emitted one band at 535 nm. After x-raying, the emission spectrum consisted of bands at 550, 410, 350, and 275 nm (KBr) and at 500, 414, 320 nm (KCl). Bleaching with F' light leads to the disappearance of the bands below 460 nm in both crystals. The afterglow has been shown to be essentially the same at 10 K suggesting that it has a tunneling character. The F' bleaching experiment implies that F' is the donor component of the tunneling pair, giving rise to the luminescence below 460 nm. The band at 275 nm in KBr coincides with the luminescent band of the σ-exciton. That band may, therefore, be excited by a tunneling transition within the F'-V_k pair: An F' electron presumably tunnels to the σ-excited state of the V_k-e system which subsequently undergoes a radiative transition to the ground state. In all cases, F band bleaching wipes out the luminescent bands.

Evidence for a dynamic F'-H interaction seems to have been obtained recently by Jimenez de Castro and Alvarez Rivas (1982). They made optical absorption, thermoluminescent, and thermostimulated-current experiments with x-irradiated KCl:Sr (80 ppm) samples between 80 and 300 K. It was found that the glow peaks (eleven of them below 300 K) induced by the irradiation could be regenerated by photostimulation at 80 K with F-band light. F* center emission has occurred in these photostimulated samples at the temperatures at which the glow peaks appeared in the as-irradiated samples. This is explained in terms of the recombination of thermally detrapped interstitials with F' centers, produced by the photostimulation:

$$H + F' \rightarrow 0 + e \quad , \qquad (6.r)$$

where 0 is the perfect lattice, followed by electron trapping by an anion vacancy in the excited F state (see reaction (4.d)). Reaction (6.r) can be paralleled by its counterpart in which free holes emerge:

$$H + \alpha \rightarrow 0 + h \qquad (6.s)$$

However, no TSC spectra could be identified unambiguously. Recently, Bichevin and Käämbre (1987) reported correlated TL and thermoexoemission (TEE) spectra in KCl due presumably to Frenkel-pair anneal by way of reaction (6.r).

Thermal recovery processes following an 80 K x-irradiation have been investigated by thermoluminescent, optical absorption, and EPR techniques in NaCl samples (López, Aguilar, and Aguilló-López (1981)). A second-order kinetics TL peak at 165 K is found to dominate over the TL spectrum of pure samples. It contains a single emission band at 360 nm corresponding to the intrinsic π-emission of the self-trapped exciton (STE). That peak is attributed to mobile V_k - F' center recombination. The V_k migrational energy is found to be 0.37 eV. The authors argue that the V_k-F' emission band should be expected to be the intrinsic STE emission (as it does occur), since the spatial extension of the F' wavefunction makes it probable that the F center would not appreciably perturb the energy of the STE transition.

Radio- and thermo- luminescent studies of F-centered CsI have also been interpreted in terms of F'-V_k interactions (Pellaux, Sidler, Nouailhat, and Aegerter (1973)). The samples, containing Mn and Na impurities acting as electron traps to promote stable V_k center formation, have been colored additively to about 10^{17} F/cc. RL and TL experiments have been made following x- or β-ray irradiation at low temperature. At temperatures below 50 K, V_k are self-trapped and only an intrinsic emission at 3.67 eV due to V_k-F' recombination is observed. Above 50 K, V_k become mobile and another band appears at 2.5 eV resulting from V_k-F radiative recombination. An F-F' conversion brings about the increase of the former band's intensity at the expense of the latter. The F' centers responsible for the 3.67 eV band form on irradiation at low temperature, presumably by way of

$$F + (e\text{-}h\ \text{pair}) \longrightarrow F' + V_k \qquad (6.t)$$

If those F' centers are optically bleached, the 3.67 eV band disappears, while the 2.5 eV emission is enhanced.

Two possibilities are considered: (i) F' autoionization through recombination of one of its electrons with a V_k hole, the energy released bringing about the ionization of the second electron in an Auger process

$$V_k + F' \longrightarrow (e\text{-}V_k)_{nonrad} + F \longrightarrow \alpha + e \quad , \qquad (6.u)$$

followed by electron-hole recombination at another site

$$e + V_k \longrightarrow h\nu\,(3.67\ \text{eV})\ . \qquad (6.v)$$

It should be noted that two V_k holes annihilate the two F' center electrons leaving behind an empty anion vacancy. (ii) F' ionization due to partial absorption of the 2.5 eV emission by the F' band:

$$h\nu\,(2.5\ \text{eV}) + F' \longrightarrow F + e$$

followed by intrinsic recombination radiating at 3.67 eV, as above. TL glow-peak interconversions give preference to (i), at least for a peak at 65 K.

6.11. F'-exciton interactions

Kachlishvily (1962) discusses theoretically the possibility of an exciton interacting with an F' center. Two effects are considered: those of the elastic and inelastic scattering of excitons by F' centers. In the latter case a new defect (F center) and a conduction electron occur as a result of the coupling. The calculation is based on the continuum approach (Section 19.3.1.) taking the interaction energy in the form

$$E_{int} = E_{vac} + E_{el} + e\varphi(r_3) + e\varphi(r_4) \qquad (6.91)$$

where

$$E_{vac} = -eV_1(r_3) + eV_1(r_4) \qquad (6.92)$$

is the interaction energy of the electronic (r_3) and hole (r_4) counterparts of the exciton in the field $V_1(r)$ of the vacancy. The latter is described by

$$V_1(r) = (e/\varepsilon_o)\frac{1}{r}(1 + B(r)) \qquad (6.93)$$

where $B(r) \to 0$ for $r \to \infty$; the electron-electron interaction is

$$E_{el} = (e^2/\varepsilon_\infty)(r_{13}^{-1} + r_{23}^{-1} - r_{14}^{-1} - r_{24}^{-1}) \qquad (6.94)$$

where $r_{ik} = |\underline{r}_i - \underline{r}_k|$, while $\underline{r}_1$ and $\underline{r}_2$ are the position vectors of the F' electrons. Further on,

$$\varphi(\underline{r}) = -ec\left(\int \frac{|\Psi_{F'}(\underline{r}_1,\underline{r}_2)|^2}{\underline{r}_1 - \underline{r}} d\underline{r}_1 d\underline{r}_2 + \int \frac{|\Psi_{F'}(\underline{r}_1,\underline{r}_2)|^2}{\underline{r}_2 - \underline{r}} d\underline{r}_1 d\underline{r}_2\right) \qquad (6.95)$$

is the potential at $\underline{r}$ of the polarization field produced by the F' electrons.

The matrix element of the elastic scattering is

$$E_{if} = \langle f | E_{int} | i \rangle \qquad (6.96)$$

with

$$|i\rangle = \Psi_{exc}(\underline{k}_o)\,\Psi_{F'}(\underline{r}_1,\underline{r}_2) \prod_s \Theta_{n_s}(q_s - q_{s_{F'}}) \qquad (6.97)$$

$$|f\rangle = \Psi_{exc}(\underline{k}_o')\,\Psi_{F'}(\underline{r}_1,\underline{r}_2) \prod_s \Theta_{n_s}(q_s - q_{s_{F'}}) \qquad (6.98)$$

$$\Psi_{exc}(\underline{k}_o) = V^{-\frac{1}{2}}(\pi a_{exc}^3)^{-\frac{1}{2}} \exp(-\rho / a_{exc}) \exp(i\underline{k}_o \cdot \underline{R}) \qquad (6.99)$$

is the free-exciton wave function. Here $\underline{R}$ is the position vector of the center of mass, ρ is the position vector of the electron relative to the hole, $a_{exc} = \pi^2\hbar^2/\mu e^2$ is the exciton radius, $\mu = m_3 m_4/(m_3 + m_4)$. The F' wave function was chosen in the multiplicative form (19.51), each component being given by eq.(19.62) with variational parameter from eq.(19.63). Assuming $q = |\underline{k}_o - \underline{k}_o'| \ll a_o = \pi^2\hbar^2/m_e e^2$ (m_e is the free electron mass), Kachlishvili obtains

$$\sigma_{elastic} = 4\pi e^4 a_{exc}^4 (m_3 + m_4)^2/h^4 \varepsilon_o^2 \qquad (6.100)$$

This yields $\sigma \sim 10^{-15}$ cm^2 and a mean free-path of $\sim 10^{-4}$ cm for $m_3 = m_e$, $m_4 = 2m_e$, $\varepsilon_o = 5.8$, $\varepsilon_\infty = 2.33$ (NaCl), and $F' = 10^{19}$ cm^{-3}. The critical temperature below which F' is the main exciton scatterer is 10 K.

The inelastic scattering was dealt with following Stueckelberg's quasi-classical method. A temperature dependence was obtained for the $\sigma_{inelastic}$ cross-section. However, there does not seem to be any particular experimental data to compare with.

An interesting suggestion had been made at the time by Tomasevich (1957) who considered the Greek bands to result from electronic transitions from the valence band to levels associated with an anion vacancy (α) and an F center (β), with the subsequent formation of F and F', respectively. The hole left behind in both cases was also taken into account. The variational method was employed to calculate the difference between the peak positions of α and β , as well as the halfwidths of both bands. A surprising similarity with the experimental data for KI has reportedly been obtained. For more information on the Greek bands, see Fowler (1968).

7. Electron spin resonance

All theoretical models of the F' center assume a singlet ground state in which the two electrons have antiparallel spins. The existence of triplet states is considered unlikely, at least in KCl (Strozier and Dick (1969)). Accordingly, the F' center has been expected to be diamagnetic in any of its possible bound states. Nevertheless, search has been made for an ESR signal from the F' center yielding a negative result (Schmid and Wolf (1962)). By combined ESR and optical absorption measurements performed during an optical F-F' conversion in KCl at -90 °C, the authors have found that the only changes in the observed ESR spectrum (apparently due to F centers alone in an additively colored sample) are those induced by the variations of the number of F centers as a result of the conversion. These changes affected only the intensity of the F center ESR line but not its lineshape, width, or g-value. No signal due to F' centers could be detected within the g-range from 1.4 to 4.5. The variations of the F center ESR signal were simply correlated (proportional) to those of the F band peak during the conversion. Under these circumstances, the agreement between theoretical expectations and experiment becomes apparent.

However, the above conclusions have recently been questioned, regarding the "odd F' centers" in NaBr, NaI, and possibly LiCl. Some novel experimental evidence for the existence of triplet F' states are discussed in Section 6.2.1.1. Perhaps a decisive ESR check-up is imperative.

8. Production of F' centers by x-rays

Perhaps one of the first observations of the direct production of F' centers by x-rays is due to Dorendorf (1951). He studied the effect of x-raying on the absorption spectrum between 0.2 and 1 μm in KCl and KBr crystals in the temperature range between -180 to +75 °C. The KCl F' band formed at -180, -130, and -80 °C but not at -40 and -20 °C.

Presumably, because of its transient occurrence, the F' band could not be detected at the latter temperatures.

A bit later, Duerig and Markham (1952) have reported on the color center bands produced by x-raying alkali halide crystals at 5 K. They found that just a small F' band occurs in NaCl and KBr and none in KCl. The production and bleaching of F centers in KBr at 5 K and at 78 K have been investigated by Markham, Platt, and Mador (1953) for crystals colored by x-rays at 300 K (x-300), LNT (x-78), and 5 K (x-5), as well as for additively colored samples (A). While A were stable upon bleaching, the F' band occurred to the red of F in x-78 but was absent in x-5. The F band bleached in x-300 at both temperatures and recovered slowly in the dark. To explain the latter observation (and, apparently, the F' formation in x-78 upon F band bleaching), the authors have suggested a tunneling process already discussed in Section 6.2:

$$F^* + F \longrightarrow F' + \alpha \qquad \text{(minutes)}$$

(bleaching under F light) and

$$F' + \alpha \longrightarrow 2F \qquad \text{(hours)}$$

(recovery in the dark). As many as 10 % of the F centers in x-300 could be bleached by F light at 5 K or at 78 K. Over half of the F centers bleached recovered in the dark after several hours. The bleaching - recovery cycle was repeatable: If one shines F' light on a bleached crystal, it recovers in just a few seconds. However, the changes in the F' band itself were too small to be detected. In a discussion over some experimental results by Maurer's group, Seitz (1955) also reports that when KCl crystals are darkened by x-rays a prominent F' band occurs at LNT, while a very low one forms at 35 K.

The production and thermal behavior of the F' band in NaCl on x-irradiation at RT has also been studied by Platt and Markham (1953). NaCl is a peculiar crystal where both F' and aggregate centers can be produced and watched at the same temperature. The F' band was found to be short-lived, disappearing in a matter of minutes following the irradiation. A weak F' band has been monitored in x-rayed KI by means of spectral F' photoconductivity at 78 K (Konitzer and Hersh (1966)). Production of F'centers in NaF on x-raying at 80 K and at 200 K has been reported by Konrad and Neubert (1967). At 200 K an intense F' band formed that bleached to half intensity on warming up to 280 K, even though some of it remained even at 320 K. This band could also be bleached optically by 436 nm light at 80 K, the F band increasing in the process. Otherwise, F' was quite stable at 200 K. A large F' type absorption has been observed in NaF after a prolonged x-irradiation at

N_2- and a smaller amount at CO_2- temperature. Although it was both thermally and optically bleachable, difference spectra suggested that it was not a simple F' band (Amenu-Kpodo and Neubert (1965)).

Color center production on x-raying KCl and KBr crystals at low temperatures has been studied by Faraday and Compton (1965). They found no F' band to form at 5 K in KCl and only a weak one in KBr. At 80 K, the F' production was "normal" in both crystals. The low F'-production efficiency is pretty puzzling in the light of the high yield of F-center formation by x-rays near LHeT (Sonder and Sibley (1972)). Unfortunately, no systematic attempt seems to have been made at that time to correlate the production of F and F' within a wider temperature range, in spite of the fundamental importance of such an investigation. This has been done later by pulsed radiolysis.

Kingsley (1961) has reported the occurrence of an F' band in KBr and KCl crystals upon x-irradiation at 80 K. While the F' concentration produced on bleaching with F light in an additively colored crystal is rather low at LNT (about 2 % of the initial F center density), the number of F' centers created by x-irradiation at 80 K is fairly large (about one-fifth of the F centers produced). If one exposes at 80 K a freshly irradiated crystal to F band light, the F band increases: Apparently, the smaller fraction of F'-absorbed light induces a larger number of F'-F conversion events per quantum absorbed. Also, if one bleaches the F' band immediately after x-raying the crystal, the F band increases, while F' disappears. However, the number of F centers formed is nearly equal to the number of F' destroyed, in contrast to an additively colored crystal where twice as many F centers form. Consequently, the electrons ejected from F' centers in an x-rayed sample do not all reach α centers but are also trapped elsewhere. Measuring the changes in the α band of KBr has shown that about one tenth of the electrons released are trapped by halogen vacancies. Similar measurements in the UV range have also indicated that the major trap for F' photoelectrons are the V_k centers.

To summarize, Kingsley has found that changes in the F', F, V_k, and α bands occur when the F' band is optically bleached in KCl and KBr crystals colored by x-rays at 80 K. The numbers of self-trapped holes and anion vacancies destroyed are 90 % and 10 %, respectively, of the number of F' centers bleached. It has also been concluded that about 2/3rd of all V_k centers are destroyed during the bleaching of the F' band. It follows that nearly as much F' centers have been produced by the x-irradiation.

Kingsley's conclusions have later been utilized by Comins (1971) in seeking an explanation through secondary reactions of the observed increase of the F center production rate in KBr and KCl between 77 and 195 K. In KBr the ratio of F centers to total vacancy concentration was found to be nearly constant (0.7) between 81 to 120 K. The F and α center densities were determined optically from the respective absorption bands, while F' was computed from the increase in F induced by subsequent bleaching with F' light, using $|\Delta F| = |\Delta F'|$. The above constancy was taken to mean that an equilibrium between the above centers was established during the irradiation. This equilibrium could be restored through further irradiation if a deviation was induced, e.g. by F' band bleaching. F' did not form on irradiation at higher temperature, e.g. 162 K in KBr, presumably due to thermal instability. In KCl, F' formed at temperatures as high as 183 K, even though at a reduced rate. Inasmuch as the α center production also dropped, it was concluded that the formation of F' and α was linked in some way. In KCl the α center density was not measured because of inadequacy of the experimental set-up.

A new method of producing F' centers without the simultaneous generation of α centers has been proposed by Kouvalis (1976). It is based on the observation that no F' centers are formed by x-rays at 78 K in a colored KI crystal, which does not contain F aggregates or other excess-electron centers. F' are produced at 78 K by x-rays only in M-center containing samples, presumably by way of:

$$\text{x-ray} + M \longrightarrow M^{+} + e$$

$$e + F \longrightarrow F'$$

The method is similar to the one used sometime later to produce F' in additively colored KCl at RT by exciting F-aggregates with laser light at 1.06 m (Vassilev, Georgiev, Todorov, and Todorov (1978)).

9. F' formation by pulsed radiolysis

F' production in alkali halieds by pulsed electron-beam irradiation has been reported by Ueta (1967). He used the electron beam from a linac (irradiated area 1 cm^2, maximum power of the electron pulse 150 mA at 13.5 MeV), producing pulses of 0.2 μs rise- and fall- times and a half-width of 0.4 μs, and a standard technique to measure transient changes in transmission induced by the pulse. KCl and KBr crystals of various purity have been investigated. A small F' band was produced in KBr at RT which decayed in a microsecond. The F' band in a pure KCl crystal was found to follow the first-order decay type with a time constant of 110±20 μs at RT. However, this lifetime was dependent on the F center density: For a pure crystal the latter was about 1.0 x 10^{16} cm^{-3}, generated by the maximum-power pulse. In a crystal pre-colored additively to about 3 x 10^{17} F/cc, the F' band yield was 16 times larger than in a pure crystal that was not pre-colored. Apparently, secondary electrons generated by the pulse had a larger probability to get trapped by pre-existing F centers than to annihilate with mobile holes. The F' lifetime was found to be 300 μs in a pre-colored KCl crystal. A fraction of the F' bleached in a short time by recombining with holes released from Cl_2^- centers. This was evidenced by establishing a correlation in the decay characteristics of the F' and V_k bands (at 850 and 350 nm, respectively). 25 % of the F' bleached with τ = 3.5 μs and the remaining with τ = 100 μs. At the same time, 60 % of the V_k decayed with τ = 3.5 μs and the remainder with τ = 1.2 μs. Therefore, some 25 % of F' recombined with V_k. The fast F' decay component did not show up in KCl containing U centers (5 x 10^{17} cm^{-3}); presumably mobile holes recombined predominantly with U, due to the overwhelming U center density. This suggests that the V_k-F' recombination occurs via mobile holes rather than by electron tunneling within F'-V_k pairs, the mobile holes being released from V_k centers. Free holes were found to bleach 14 % of F and 25 % of F'. Hence the hole-trapping cross-section is larger for F' than it is for F, apparently as a result of the Coulomb interaction. The slower F' decay component did not occur in U-containing KCl crystals but F' was overlapped by a new band at 730 nm (KCl) and at 780 nm (KBr), ascribed tentatively to an U' center, formed as an electron was trapped by an U center. The U' had a lifetime of 8 μs (KCl) and about 2 μs (KBr) at RT.

Chandratillake, Newton, Robinson, and Rodgers (1977) have investigated the F' thermal decay in pure KCl crystals. The samples were cleaved and annealed under N_2 gas at 650 °C and then cooled down at 1°C/min. The F' centers were produced by pulsed electron-beam irradiation from an electron accelerator of 10 MeV energy. Pulses of 10, 20, and 50 ns were used, the maximum dose being 2 krad per pulse. A standard technique of measuring transient absorption changes was employed. The F' absorption was monitored between 800-900 nm with essentially similar results at all wavelengths except for 825 nm (the M band peak). The F' yield per pulse was found proportional to the F center concentration produced, indicating that F' formed through secondary-electron trapping at F centers already created. A plot of $\log_{10}D_{800nm}$ vs. time was used to represent the F' decay kinetics within the 20-1000 μs delay range following a 10 ns irradiation pulse. The F' kinetics were resolved into 3 exponential components: A, B, and C. Of these, A was long lived with a lifetime of the order of tens of ms. This A component was shown to be brought about through photobleaching by the monitoring beam when it contained components within the F band. Consequently, these F' were presumably in a transient photoequilibrium with F. They have been removed from the time-spectrum by filtering the F band components. At RT, the B component comprised about 75 % of the total F' yield. B is a single exponential component with $\tau_B = 190 \pm 5$ μs at RT. That B was a genuine exponential was checked by varying the pulse duration from 10 to 50 ns: this led to an increased F'_0 but did not affect τ_B. This implies that B does not involve any interaction with other radiation-produced short-lived species, since their concentration would depend on the pulse duration, unless they are very short-lived. The RT τ_B was also independent of the sample history, as well as of the pre-irradiation conditions. It is concluded that longer-lived species have also little effect on B. Component C comprised about 25 % of the F' yield and decayed within about 20 μs. However, C was not a simple exponential and might have involved some reaction with short-lived hole centers. Below RT down to 213 K, B increased as the temperature was lowered, while C decreased rapidly to vanish below 273 K. The genuine B component (corrected for bleaching by the monitoring light) yielded

$$\tau_B^{-1} = 8.1 \times 10^{13} \exp(-(0.60 \pm 0.04)\mathrm{eV}/k_B T) \quad , \quad \mathrm{s}^{-1} \quad \mathrm{(KCl)}$$

No growth of the F band correlated to the F' decay was observed. A search was made for any correlated changes in the M band which was not successful because of the strong F' overlap. It is therefore assumed that a F'-M decay path can contribute only slightly to the overall decay kinetics. Although concluding that τ_B is related to some F' thermal-ionization process, the authors suggest that the F' decay results ultimately in M-center formation and not in F center recovery. Note the similarity between the activation energy of τ_B^{-1} and the M center formation energy measured by a number of authors.

In a subsequent paper, Chandratillake, Newton, Patil, Robinson, and Rodgers (1978) investigated the F' photobleaching properties in KCl between 102-210 K. The photobleaching quantum yield is defined by

$$\eta_{F'} = -\dot{F}'/q_{F'}F' \quad , \qquad (9.1)$$

provided the photoinduced F' decay is exponential at these temperatures. Apparently, this was the case and $\eta_{F'}$ was found to be nearly 1 (or at least constant) up to 160 K and to drop eventually slightly above it. At the same time, Pick's data reveal a sharp decrease of $\eta_{F-F'}$

above 100 K (due to the decrease of γ_α)(Pick (1940)). It is presumed that no F centers are formed as F' are bleached above 100 K, that is

F' — 2F (below 100 K, photobleaching)

F' — M (above 200 K, thermal & photobleaching)

The RT F' (mono) lifetime was found to be 190 ± 10 μs in well-annealed samples and 210 ± 20 μs in samples quenched from the melting point.

The F' decay was also studied in NaF, NaCl, KBr, and KI (Chandratillake, Hamblett, Newton, Patil, Robinson, and Rodgers (1978)). An essentially two-component F' decay was resolved: the shorter-lived B being an intrinsic decay, while the longer-lived A resulting from the transient photoequilibrium due to the monitoring beam; no C component is reported in this work. The following general properties of B are described:

(i) Effect of repetitive pulsing: F' increased from a low value at the first pulse to saturate after 10 to 20 pulses, depending on the repetition rate. This implies that F' is formed through electron capture by pre-existing F centers. The F' growth with N (the pulse number) then reflects the growth of F.

(ii) τ_B was independent of the dose per pulse and of the integrated dose. Consequently, the F' decay does not involve any reaction with other species, either short- or long- lived. The observed F' decay is unimolecular and is assumed to result from thermal ionization.

(iii) The effect of the monitoring light is strongly temperature-dependent. The F' bleaching at any temperature is a combination of photo- and thermo- processes. The former are independent of the temperature, while the latter are strongly temperature-dependent. Thermal bleaching predominates at higher temperatures where the bleaching rate is independent of the light intensity I . At lower temperatures the rate is proportional to I .

(iv) The F' assignment was verified by checking the behavior of τ_B, as the monitoring-light wavelength was varied.

(v) The temperature variation of τ_B , corrected for photobleaching, was found to follow the Arrhenius dependence with parameters listed in Table 2.

There appears to be a reasonable correlation between the thermal activation energy of τ_B^{-1} and the F' absorption band peak for the alkali halides studied. The F' band was obtained by measuring the spectral dependence of the short-lived component B. The F' peaks plotted vs. E_B in Fig.38 are accurate to 10 %. There seems to be a good correlation between F' optical absorption and thermal decay suggesting that the latter may indeed be due to ionization.

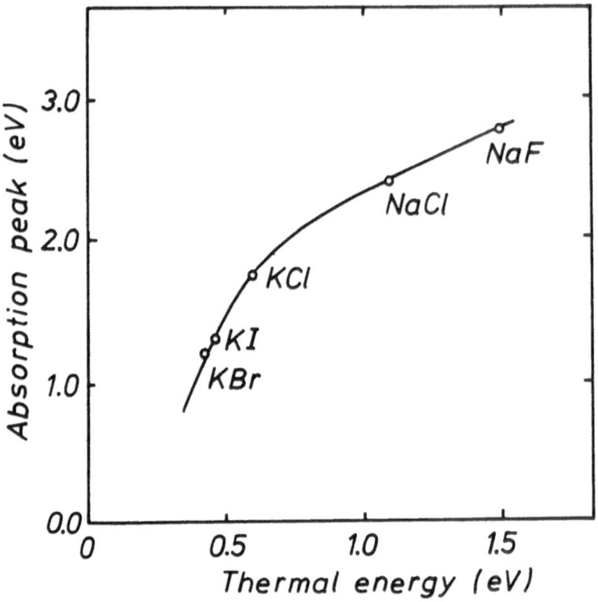

Figure 38: Correlation between the energies for thermal and optical ionization of an F center in several alkali halides. (After Chandratillake, Hamblett, Newton, Patil, Robinson, and Rodgers (1978)).

In a later paper, Chandratillake, Hamblett, Newton, and Robinson (1981) report that both the F' and V_k production in pure virgin KCl crystals is low but increases considerably if the crystals have been F centered prior to the radiolysis pulse. There appears to be a correlation of F' and V_k production processes. The F' yield per pulse increases during repetitive pulsing, presumably because electrons released by a given pulse are captured into F centers formed in preceding pulses. V_k shows a similar behavior: In a single-pulse experiment at RT, the V_k yield increased with the number of F centers already present. It is likely that electrons released as V_k form are trapped by F to form F', thereby securing the stability of the self-trapped hole at low temperature. This is supported by the observed V_k thermal decay between 270-320 K: The V_k decay rate in this temperature range was found the same as the F' thermal-decay rate:

$$\tau_{F'}^{-1} = 10^{14.0\pm0.3} \exp(-(0.61\pm0.02)\mathrm{eV}/k_BT) \quad , \mathrm{s}^{-1} \quad (\mathrm{KCl})$$

$$\tau_{V_k}^{-1} = 10^{13.5\pm0.7} \exp(-(0.59\pm0.04)\mathrm{eV}/k_BT) \quad , \mathrm{s}^{-1} \quad (\mathrm{KCl})$$

It is concluded that the decay mechanism is the F' thermal ionization followed by electron capture and annihilation at V_k. Between 190-260 K, the V_k bleached optically due to F' photoionization, followed by a similar annihilation process. However, as the temperature was lowered down to 140 K, V_k bleached progressively more slowly than F'. The F' optical bleaching rate being temperature-independent, the origin of that difference is not quite clear.

10. F' production by photolysis

Recently, Provoost, Debergh, and Hoebeeck (1982) have examined among other things the occurrence of the F' band in photolytic experiments on several alkali halides. A mode-locked ruby laser followed by a single-pulse selector was used to initiate defect formation through two-photon excitation. Listed below are the basic results of their study:

RbBr: An F' band is produced at 120 K. It is unstable due to spontaneous F'-F conversion at this temperature. The F-F' conversion is also initiated by white light. The F' photolysis growth-time is found to be 100 ms at 140 K.

KBr: F' is steadily observed in the spectrum of photolytic products at 120 K. An F'-F conversion is induced through white-light illumination but some residual F' centers apparently remain. The F' formation time is estimated at less than 30 ps during the photolysis. F' is also found at 160 K where an F-F' conversion is subsequently induced by white light. Apparently, F' appear in the absorption spectra at temperatures as low as LHeT.

RbI: No F' are seen in the resulting spectra at LHeT, 80, and 160 K.

KI: An F' is found at LHeT which is even larger than F. That F' band bleaches on white-light illumination at LHeT, as do the F and H bands.

11. Thermally-stimulated phenomena

In a number of experimental works, valuable information on the F' centers has been collected by studying the thermally-stimulated currents (TSC), thermoluminescence (TL), and thermal bleaching of the optical absorption bands (TB). To begin with, Dutton and Maurer (1953) have investigated both the TSC and the TL in KCl and KBr crystals irradiated with x-rays at -183 ^{o}C. A series of current and glow peaks observed as the samples were slowly warmed up to RT have been assigned to various color centers either by monitoring at pre-selected wavelengths within the absorption spectrum during warm-up, or by studying the entire spectrum after rapid heating to the temperature of a given peak followed by a cool-down to -183 °C for an optical measurement. The TSC and TL peaks ascribed to F' centers were at -68 °C (KCl) and at -130 °C (KBr). These are also the temperatures of largest TB of the F' band. They coincide with the temperatures at which the 'thermally-increased range' of the F photocurrent begins (Seitz (1946)). Consequently, these F' peaks arise from the thermal ionization of the F' centers. Using a simple formula to relate the trap-depth E to the peak temperature T_{max} :

$$E = 1.08\, k_B T_{max} \ln s \tag{11.1}$$

where s is a frequency factor set equal to $10^{9.5}\ s^{-1}$, the F' thermal ionization energies are calculated to be 0.42 eV (KCl) and 0.29 (KBr).

TB experiments have been reported by Halperin, Braner, Schlesinger, and Kristianpoller (1960). KCl crystals were x-rayed at 80 K and subsequently warmed up to successively higher temperatures. After each

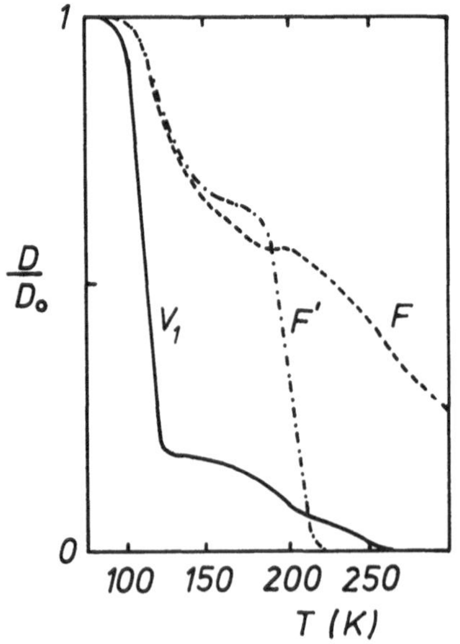

Figure 39: Thermal bleaching curves of F, V_1, and F' centers in KCl normalized to the optical density produced by coloration at 80 K. (After Halperin, Braner, Schlesinger, and Kristianpoller (1960)).

heating step the samples were cooled down to LNT for an optical absorption measurement. Normalized bleaching curves for the F, V, and F' bands are shown in Fig.39. F' bleaches steeply in the vicinity of 200 K, undoubtedly due to the thermal release of F' electrons to the conduction band followed by recombination with trapped-hole centers. This is known to give rise to a green luminescence. Before that, the F' band bleaches partially in the vicinity of 125 K accompanying the steep decrease of the V_1 band which gives rise to a violet luminescence. Now, recombination of mobile holes with trapped electrons is believed to have taken place.

The TB of the F' band in KCl, x-rayed at LNT, and its relation to the M band growth has been studied by Rabin (1963). The optical absorption spectrum was examined between 700-1700 nm on warming up from LNT to 320 K. A small initial M band appeared superimposed on F' at LNT after x-raying. Heating to about 207 K resulted in the disappearance of F', while M remained essentially unchanged. The relative F' and M concentrations were plotted versus the temperature following x-raying at LNT. The F' depleted completely before the onset of M band growth. This seems to have happened at temperatures considerably lower than the ones reported in the previous two works.

Bosacchi, Fieschi, and Scaramelli (1965) have apparently been the first to apply the photostimulated thermoluminescent technique (PSTL) to the alkali halides. The method consists of bleaching a pre-irradiated sample with F-band light at low temperature just before the TL run to create or increase the density of filled-in electron traps (F' centers) and obtain a better resolution. The authors have investigated the PSTL in KCl following x-irradiation at RT and cool-down to LNT. They found two principal glow peaks (labelled I and II). These also appeared in the pure TL spectra after x-raying at LNT but were greatly enhanced by the photostimulation. The stimulation spectra of these peaks followed the absorption spectrum of an additively colored sample with differences arising in the F and K band ranges. Peak I, a doublet at about 120 K, was insensitive to F and K band light, while peak II at 215 K was greatly stimulated by F band light and by K band light as well, in proportion to the absorption spectrum. By stimulating at different temperatures, the intensity of peak II was found to follow closely the temperature dependence of the percentage F-F' conversion, as reported by Lüty (1961b)(Fig.40). The agreement between stimulation and absorption spectra (beyond the F band range) for both peaks suggests that they may arise from conduction-electron capture, the result in Fig.40 implying that trap II is identical with the F center. In addition to capturing conduction electrons (apparently the only mechanism operative for trap I), trap II can be filled with electrons through tunneling from F* centers. This accounts for the strong stimulation of the latter peak by F-band light at LNT. F' bleaching after the stimulation with F band light was found to enhance peak I at the expense of peak II, in agreement with the above identification.

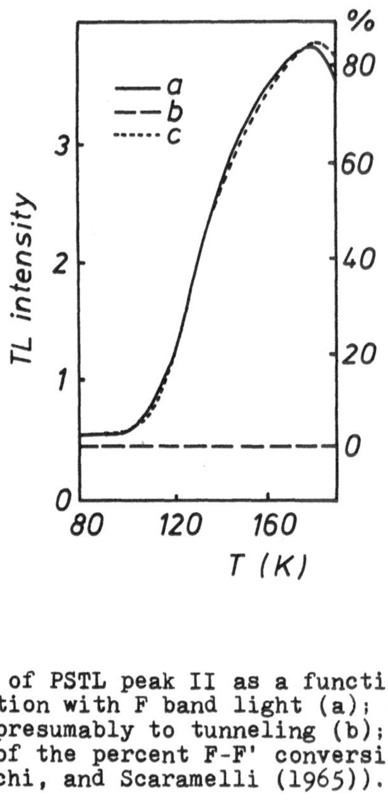

Figure 40: Intensity of PSTL peak II as a function of the temperature of excitation with F band light (a); intensity of the plateau due presumably to tunneling (b); the temperature dependence of the percent F-F' conversion (c). (After Bosacchi, Fieschi, and Scaramelli (1965)).

In a PSTL study of NaCl (Fieschi and Scaramelli (1966)), the photostimulation spectrum proved to compose of excess-electron bands only: L, K, F, R. Measurements of the temperature dependence of the glow-peak intensities in different spectral regions disclosed a 25-fold increase as the temperature was raised from 55 K to 130 K, followed by a slight decrease beyond that range. This temperature behavior agreed generally with the one displayed by the F-photoconductivity (Pohl (1937)) and the F-photobleaching (Pick (1938)).

Careful PSTL measurements have been carried out by Scaramelli (1966) on crystals of KBr and NaCl x-rayed at RT to produce about 5×10^{16} F/cc and 10^{17} F/cc, respectively. Before each TL run, the samples were cooled down and bleached with F light at LNT. Fig.41 shows the glow peak in KBr attributed to F' (the solid line). The dotted curve was obtained in a subsequent run in which the sample was bleached with F' light following the photostimulation with F light, to decrease the number of F' centers. The observed peak positions were characteristic of the salt,

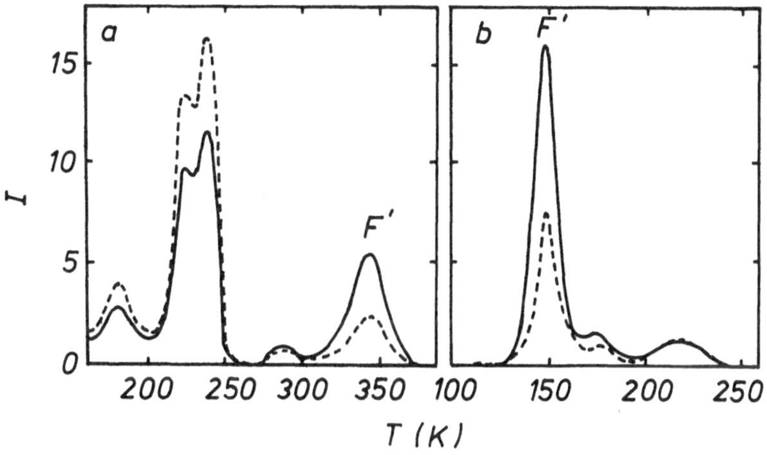

Figure 41: PSTL glow peaks attributed to F' centers in NaCl (left) and KBr (right). (After Scaramelli (1966)).

not the sample. These results have clearly indicated that the glow peaks at 215 K (KCl), 340 K (NaCl), and 149 K (KBr) are indeed due to electrons released thermally from F' centers.

The trap depth E (the F' thermal ionization energy) and the preexponential factor ν_0 (the attempt frequency) have been obtained by studying the TL intensity as a function of the temperature. For a TL glow with unimolecular kinetics the low-temperature tail of the glow peak I(T) obeys

$$\ln(I(T)) = \text{const} - E/k_B T \tag{11.2}$$

On the other hand, both E and ν_0 can be obtained from

$$\tau(T) = \int_{T/\beta}^{\infty} I(t)dt \,/\, I(T) \tag{11.3}$$

where $t = T/\beta$ is the time during a TL run, β being the heating rate, while

$$\tau(T) = \tau_0 \exp(E/k_B T) \qquad (11.4)$$

is the lifetime of the (filled) trap. Now, E and $\nu_0 = \tau_0^{-1}$ can be found from the slope and the intercept of the Arrhenius plot of eq. (11.3), whose right-hand side is an experimentally measurable quantity, provided there are no other TL peaks to overlap with the TL band under examination. The results of Scaramelli's measurements at $\beta = 0.66$ K/s are listed in Table II.

The PSTL in KBr crystals has also been investigated by Crippa, Paracchini, and Felszerfalvi (1968) within the range from 50 to 250 K. The crystals were x-rayed at RT. Photostimulation was performed at 50 K within the spectral range 240-700 nm. The total light emission of the crystal was detected. Six peaks were observed: P_1(77 K), P_2(149 K), P_3(173 K), P_4(214 K), P_5(223 K), and P_6(240 K). P_2 was assigned to the F' thermal ionization. The photoexcitation spectrum of P_2 shows the L, K, and F bands. At 50 K, the filling-up mechanism must be electron-tunneling from F* to F, as well as to other possible traps, which leads to the presence of the F band in the photostimulation spectra of P_{2-5}. That tunneling was the filling mechanism of P_2 was also confirmed by stimulating the crystal at a given wavelength (within the F band) at different temperatures. For $T < 65$ K, the peak intensity remained constant but rose steeply at higher T. A fall for $T > 100$ K was attributed to F' thermal ionization. Generally, the variations of the peak intensity reflected the temperature dependence of the relative photoconductivity of KBr under F-band light. While P_1 and P_6 were found to fill through the conduction band mainly, P_3 through P_5 proved in addition to be closely related to P_2 (F'). All the four peaks of the latter group filled in by way of tunneling from F* centers and decreased when the sample was bleached with F' light following the photostimulation, P_1 and P_6 increasing in the process.

Another important PSTL investigation has been carried out by Fieschi and Paracchini (1969) in additively colored KCl within the 50-300 K range. Freshly-quenched samples are cooled down in the dark to 50 K to be illuminated into one of the F center absorption bands (F, K, L). A fraction of the electrons excited by the light leave the respective vacancies to be captured by traps. During subsequent warming (the TL run) the traps get empty, each kind of trap at its own characteristic temperature, and the electrons released recombine with the activators (α centers) to emit ir photons through the F luminescence. A fraction of the vacancies apparently persists even after the F' centers bleach out thermally, since an F luminescence has been found up to RT. Although the possibility that electrons may also recombine with activators other than α cannot be ruled out, the F luminescence predominates. Two typical glow curves were obtained on photostimulation with: (a)- 410 nm light (the maximum photoconductive response), and (b)- 540 nm light (in the F band). Curve (a) was obviously related to trap-filling via the conduction band, while (b) presumably filled in by way of electron tunneling. In both cases three main glow peaks appeared at 115, 225, and 285 K. The 225 K peak is assigned to F': (i) it falls in the range where F' bleaches thermally; (ii) bleaching the crystal with F' light immediately after photostimulation brings about a decrease of that peak.

Inasmuch as three types of traps have been found to operate between 50-300 K, the assumed value $\eta_{F-F'} = 2$ for the ultimate F-F' conversion quantum efficiency in KCl may be wrong. Deformation by bending was not found to produce any new glow peaks: consequently, no electron traps seem to be associated with fresh dislocations. Estimated trap densities from the glow peaks, taking into account the temperature dependence of the emission efficiency η_R, are: 3×10^{11} (115 K), 4×10^{12} (225 K), and 3×10^{13} cm^{-3} (285 K). The two extreme-temperature traps did not have any measurable effect on the optical-absorption spectrum of the crystal. Inasmuch as the photostimulation spectrum of all the three glow peaks shows the same two maxima, at 410 nm and at 540 nm, all the three traps are filled in both via the conduction band and by electron tunneling. However, tunneling is particularly effective for the intermediate trap, while the extreme ones fill in via the CB mainly.

Studies of F' centers in plastically deformed crystals of KCl and KBr have been made by Hirai and Scott (1966). The crystals, colored additively, were deformed at RT and F' centers were subsequently produced by bleaching with F light at -120 ^{o}C and -140 ^{o}C, respectively. A TB run was then started from -186 ^{o}C up to RT in which the F' absorption was measured at a series of temperatures during warm-up. The TB curves exhibited two temperature regions of rapid F' decrease, corresponding to the thermal bleaching of two types of centers with different thermal stabilities. At the same time, undeformed crystals showed only one such temperature range, in which the "ordinary" F' centers bleached. It has been suggested that two types of F' centers form in deformed crystals whose thermal stability is either decreased or increased by the influence of neighboring dislocations. The TSC during warm-up was also measured. Two new peaks occurred at higher temperatures in addition to the one characteristic of the undeformed crystal. The trap depths were determined by means of the Randall-Wilkins formula

$$E = k_B T_m \ln s \, (1 + \ln(k_B T_m^2/\beta E)/\ln s) \qquad (11.5)$$

using the frequency factor s by Dutton and Maurer (1953). These are shown below along with the data by Dutton and Maurer for x-rayed samples.

Table 5

F'-assigned TL peak data*

T_m ($^\circ$C)	E (eV)	Assignment	Crystal
-112	0.35	F'	KBr undeformed
-120	0.33	F'_d	28% deformed
- 51	0.48	F'_d	,,
-130	0.32	F'	x-rayed (DM)
- 57	0.50	F'	KCl undeformed
- 74	0.46	F'_d	17 or 33% deformed
- 23	0.58	F'_d	,,
- 68	0.47	F'	x-rayed (DM)

*After Hirai and Scott (1966)

The two F' types in deformed samples (F'_d) of lower and higher thermal stability, respectively, are presumably located in regions of dilation and compression around a dislocation line. A positive effect of pressure on the F' thermal-ionization energy is implied. No pressure effects on the F' binding energy have been studied experimentally up to that time.

The possibility that TL from KCl crystals during warm-up can cause optical bleaching leading to both F' destruction and F aggregation has been checked by Sonder, Sibley, and Mallard (1967). They used samples electron (β)-irradiated at 80 K (1.5 MeV electrons producing 2-4 x 10^{18} F/cc). The F' band decayed during warm-up in a similar fashion for all the samples studied: annealing slowly up to 175 K and then disappearing rapidly above 200 K. The anneal was accompanied by TL bands at about 200 K composed spectrally of emission peaks at 680 nm (1.85 eV) and 480 nm (2.65 eV). The 200 K glow peak is double and the rapid F' anneal occurs correlated with the low-temperature component. The high-temperature counterpart is at 209 K. A comparison between the number of F' centers destroyed (about 10^{17} cm^{-3}) and the number of photons emitted at 200 K (about 10^{14} cm^{-3}) reveals that the TL cannot be simply responsible for the observed F' anneal. (The F' concentration was determined from $F' = 10^{16} k_{F'}$, where $k_{F'}$ is the absorption coefficient at the F' band peak in cm^{-1} obtained using $f = 1.3$ and $W = 1.35$ eV in Smakula's equation.) However, it appears that the rapid F' anneal at 192 K occurs at the same time (and temperature) as the light emission does, so that the annealing process and the thermal glow may well have a common origin. The heating rate β seems important for the observed details of the anneal: the ratio of the annealing rates below 175 K and near 200 K depends on β explaining just why Rabin (1963) has found that almost all of his F' centers annealed below 100 K, whereas others have only observed the 192 K anneal, due to a more rapid warming. While little can be said about the slow anneal, the rapid bleaching at 192 K is attributed to electron release from F', in accordance with previous studies (Bosacchi, Fieschi, and Scaramelli (1965); Scaramelli (1966)). The high-temperature component of the twin glow peak (at 209 K) is tentatively attributed to the release of impurity-trapped holes (V_k); indeed it is better resolved in doped samples. Finally, it is concluded that the total TL intensity emitted after irradiation at 80 K is too low, short by several orders of magnitude to be the reason behind the observed F' anneal near 200 K.

In an early work, Ohkura (1961) had observed a current glow peak at 223 K in KBr crystals, colored additively and undeformed, following an F-light irradiation at either 123 K or at 173 K. That peak was attributed to F', as supported by optical absorption measurements. A charge-released-from-F' vs. F-light pre-exposure time at 123 K displayed the same character and was consistent with the F-band growth curve. The F' center density calculated from the thermal peak was nearly equal to the one measured optically. The peak assignment was also confirmed by optical bleaching experiments. However, it was not immediately clear why the peak formed nearly equally well on F-light illumination at both temperatures, in spite of the expected rise in F' production efficiency on increasing the temperature, as found by Pick earlier. The peak position shifted from 223 K to 229 K as a result of plastic deformation. A Randall & Wilkins analysis was applied to deriving the trap depth, density, and frequency factor. The data obtained are listed in Table 6.

Clark and Newman (1969) have examined the PSTL and TL of KBr crystals electron-irradiated at 290 K. Photostimulation was done by F light at 80 K. A visible luminescence was observed peaked at about 450 nm and attributed to electron-hole recombination at some trapped-hole center. A PSTL peak at 173 K was assigned to thermally released F' electrons.

Table 6

F'-assigned TL data (KBr)*

Sample	T_m (K)	E (eV)	N (cm^{-3})	s (s^{-1})
undeformed	223	0.63	4.5×10^{15}	4×10^{11}
18% deformed	229	0.58	3×10^{15}	1×10^{10}

*After Ohkura (1961)

The TL of a crystal plastically deformed at 80 K following an irradiation at 290 K did not show the 173 K peak. A TL peak due to F' centers in a deformed sample was expected on grounds that the movement of dislocations would bring about the thermal ionization of F centers through a change in the F-center potential, as suggested elsewhere. Instead, a peak at 168 K occurred in the PSTL spectrum of such a sample which was assigned to F' in a deformed region.

A TL glow peak at 138 K in KBr crystals heavily doped with Tl was observed after x-raying at LNT and assigned to F', following an earlier unpublished work by Z. Cohen (Roth and Halperin (1982)). The above peak appeared in the PSTL spectrum as well but failed to occur in the TSC spectrum, probably due to low sensitivity of set-up.

Levin, Berggren, and Honnold (1967) have performed PSTL measurements on LiF, NaF, and NaCl crystals x-rayed near RT or above. The glow peaks attributed to the main electron trap (F) were at 395 K (LiF), 283 K (NaF), and 218 K (NaCl). Using Bube's formula (11.1), the authors determined the thermal depth E, the frequency factor s, and the capture cross-section σ_F. Reasonable values have been obtained.

A single TL glowpeak centered at 175 K for a nominally pure and at 184 K for a Na-doped (additively colored) NaCl has been reported following x-irradiation at LNT (Serpi and Serpi Macciotta (1972)). The TL run was made in the 80-300 K range. Inasmuch as the peak largely diminished on bleaching with F' light before warming, it was reasonably attributed to the F' center. Computer analyses yielded 0.50 ± 0.03 eV (NaCl) and 0.48 ± 0.05 eV (NaCl:Na) for the trap depth, presumably the F' ionization energy. However, this assignment does not agree with Scaramelli's identification of the PSTL peak at 340 K in NaCl (Scaramelli (1966)). Recently, a TL and PSTL peak at 315 K observed in NaCl crystals, x-rayed at RT, has been attributed to traps different from F'

(Herreros and Jaque (1974)). This discrepancy again raises the question put forward by Jaque and Agullo-Lopez (1970, 1974) concerning the interpretation of the F' thermal energy in NaCl.

Shibaev, Vasil'ev, Plachenov, and Mochenov (1969) have applied a thermostimulated concentrational emf technique to study the F' centers in KCl. The crystals were β-irradiated at 170 K. The emf occurs during the thermal annealing of the crystal and reflects the temperature dependence of the free-electron concentration. The irradiation leads to F and F' formation, the excess charge presumably accumulating in F'. Subsequent warming initiates TL and emf peaks between 200-230 K. F' bleaching at 80 K following the β-irradiation leads to the decrease of the thermal glow (TL and emf) peaks (at about 210 K), while bleaching at 170 K annihilates the peaks almost completely. It is suggested that the electron-trapping cross-section σ_F may be larger at 80 K than it is at 170 K to explain why the F' centers bleached more easily at the higher temperature.

More recently, Murty and Murty (1974) combined TL and optical absorption studies on alkali halides x-rayed at 80 K. Strong glow peaks were attributed to F' centers based on additional PSTL data and by comparison with available F' thermal energies: 194 K (NaCl), 187 K (KBr), and 219 K (KCl). However, the identification of the NaCl peak is dubious (Murty and Sucheta (1982)).

Undoped NaF crystals have been found to exhibit four glow peaks at 73, 120, 143, and 223 °C (Bhan (1982)). The 73 °C peak is attributed to F' centers, the remaining ones - to F centers. The presumed F' peak was found to only occur after the sample had got certain minimum exposure to RT gamma-rays believed necessary to produce some F center density as the prerequisite for F' formation. The peak enhanced on optical bleaching in the F band; however, prolonged bleaching brought about the destruction of both F and F'. On heating or storage, the F' peak decayed in favor of the F-related peaks. A first-order kinetics is assumed for the 73 °C TL peak leading to parameters listed in Table II. The heating rate has been 0.42 K/s.

A TL glow peak at 60 °C has been observed in undoped NaF crystals following an exposure to 3×10^4 R of gamma rays at room temperature (Bhan and Rao (1982)). The total curve-fitting method (there have also been a number of higher-temperature peaks too) has yielded first-order kinetics parameters for that peak. The peak has been ascribed to F', having been found to enhance on optical bleaching in the F band, in a PSTL experiment. It has also disappeared on prolonged storage in the dark at RT. Persistent optical bleaching made all the peaks lose height.

The above assignment, however, has not been verified by optical absorption measurements. The emission spectrum of undoped samples has consisted of a band at about 420 nm. Earlier Konrad and Neubert (1967) had irradiated NaF at -73 °C to find that F' were stable up to 47 °C on heating at $\beta = 0.75$ °C/min. Bhan and Rao (1982) suggest that the observed higher thermal stability of F' in their experiments may have resulted from the higher irradiation temperature employed. Their heating rate was $\beta = 0.42$ °C/s.

A TL glow peak at 355 K occuring in the PSTL spectrum of RbCl has been reported by Sastry and Sapru (1981) and assigned to F'. The peak obeyed a first-order kinetics. The samples were x- or gamma- irradiated and heated at a rate of 120 K/min. The F' peak was absent in the purely TL spectrum or in the PSTL spectrum of a sample heavily doped with potassium.

A PSTL peak has been found in CsF at 182 K at a heating rate of 8 K/min and ascribed to F' (Cases, Alcala, and Orera (1982)). The peak obeyed a first-order kinetics. The F' thermal-bleaching curve, measured by following the decay of the optical band at the above heating rate, was well-correlated with the TL glow. The spectral TL was also measured: it exhibited an emission band at 830 nm attributed earlier to the CsF F center. F' centers were also found in samples, x-rayed at 75 K.

12. F' exoelectrons

Taft and Apker (1953) have investigated the external photoelectron emission from F-centered RbI at 85 K. Following an F-F' conversion induced by 1.9 eV light, exoelectrons have been detected with photon energies as low as 1.6 eV presumably originating from F' centers. The spectral distribution of the exoelectrons has returned into the one typical of the F centers after the sample has been bleached with F' light (at about 1 eV) or warmed up to RT to destroy the F' centers; this supported the F' assignment. The measured F'-to-F ratio has been as low as 0.01 indicating that F' densities can be recorded which are considerably lower than the ones in usual optical absorption studies. In a later work, Philipp and Taft (1957) have attributed a small photoelectric yield for photon energies between 2.5 and 1.6 eV to F' centers in single crystalline KI at 80 K. The presumed F' exoelectronic spectral distribution exhibited a similar behavior on F'-band optical or thermal bleaching as in RbI.

Invoking two-center Auger processes for describing photoelectric yields goes back to the early fifties. Dexter (1951) has proposed a model for calculating the oscillator strength of the β-band in which Cl^- ion 3p electrons make transitions to bound levels in the field of an F center in close proximity. The final state of the crystal comprises a hole in one of the surrounding Cl^- ions and an extra electron in the F center. However, the latter entity is not a usual F' center because the surrounding ions have not yet moved to new equilibrium positions. Now, Seitz has been the first to point out that one of the F' electrons could recombine with the hole, and that the energy released could ionize the other electron. This Auger process is considered likely to increase the photoelectric yield (Apker and Taft (1951)).

It is also interesting to see just how the above "unrelaxed F' center" has been described electronically. Dexter (1951) argues that both electrons should be assumed bound in 1s orbitals, as in a He atom. The loose F' electron wavefunction is then taken in the form

$$\varphi_{F'*}(\underline{r}) = g(\underline{r}) \exp(-Zr/a_0)$$

where Z is obtained by minimizing the Hamiltonian. Kiselev (1963) has chosen that wavefunction in the normalized form

$$\varphi_{F'*}(\underline{r}) = (\alpha^3/\pi)^{-\frac{1}{2}} \exp(-\alpha r)$$

with $\alpha = 0.3161$, while the F'* electronic potential is chosen in the form of a spherical-box potential

$$V(r) = \alpha_M/d \qquad (r < d)$$
$$= 0 \qquad (r > d)$$

(cf. Section 19.3.2.). A comparison with the bound-polaron wavefunction of Section 19.5 is impending. $g(\underline{r})$ is the CB wavefunction.

Generally, the emission mechanism of exoelectrons from color centers has been assumed to be a two-step process: The first one is lifting the center's electron to the CB, e.g. through photoionization, while the second step is the thermoionic emission from the CB proportional to $\exp(-\chi/k_BT)$. The requirement, therefore, is that χ, the electron affinity at the surface be small. Direct photoionization processes are also considered in which hot electrons are created through direct optical excitation from the ground state of color centers to higher CB states, and are then emitted as exoelectrons. Whether excited optically, thermally, or via Auger processes, the exoelectron emission from color centers seems to be more complicated, being largely influenced by a number of extrinsic factors (Kamada and Tsutsumi (1985), Kamada, Asai and Tsutsumi (1985)).

Bichevin and Kaämbre (1971) seem to have been the first to apply the Auger mechanism to explain thermostimulated exoelectron emission (TSEE) peaks in alkali halides. NaCl and KCl crystals were x-rayed at 90 K, while the subsequent thermal annealing process was followed by TSEE and ESR measurements. A TSEE peak was found to correspond to V_k center annealing which suggested an F'-electron tunneling to approaching mobile V_k leading to a STE (self-trapped exciton) formation near an F center. The energy released as that exciton decays is transferred to the F center to bring about the ejection of its electron. Recently, Kamada, Furikawa, and Tsutsumi (1981) have measured both the TL and the TSEE simultaneously in KCl crystals, undoped or deliberately doped with either Tl or Cu ions. In some cases these were complemented with photostimulated TL and TSEE measurements. For a thermal experiment, the samples have been x-rayed in situ at LNT after being cleaved in vacuo. Undoped KCl exhibited a TSEE glow peak at 207 K which was ascribed to the thermal release of electrons from F' centers. The corresponding TL glow peak was at 211 K. On stimulation with F light at LNT following x-raying at RT, these peaks appeared at 207 K and 203 K, respectively. The decay rates of the exoelectron emission and of the thermoluminescence at 200 K were found to coincide. This seems to show that the exoelectrons did not simply arise as external photoelectrons ejected by TL photons from shallow traps. On the other hand, a TSEE peak at about 210 K in KCl:Tl and KCl:Cu was attributed to Auger ionization involving F' and V_k centers, as in Bichevin & Kaambre's study.

TSEE glow peaks have also been investigated in irradiated NaCl crystals, pure or copper-doped, aimed at checking the Auger F'-V_k annihilation hypothesis (Kamada and Tsutsumi (1981)). In pure NaCl x-rayed at 77 K, a TSEE peak appears at about 160 K which is also observed in NaCl:Cu. This is also the temperature at which V_k centers become unstable thermally in doped samples. The conclusion is that the 160 K peak in both pure and doped samples is due to the Auger process (6.u):

$$V_k + F' \longrightarrow \alpha + e \quad ,$$

where now mobile V_k holes migrate to recombine with F'electrons near the surface. No TL peak has been found to correlate with the 160 K exoelectron emission which lends support to the Auger hypothesis, since the transfer of recombination energy to eject the F electron is essentially a nonradiative process. A small TSEE peak and a TL bump at about 300 K are suspected to result from the F'-α annihilation in NaCl, since both are largely enhanced by F-light stimulation. However, a more recent study by Kamada, Yoshiara, and Tsutsumi (1984) casts doubt on the applicability of a simple Auger mechanism, at least to KCl:Tl. They argue that the anticipated exoelectron kinetic energy is too low, while the energy spectrum observed experimentally is too wide to derive from Auger's model.

Holzapfel (1970) has investigated TSEE of LiF powder samples to find it originating from both bulk and surface centers. A small glow peak at 310 K is observed in the TSEE spectra of samples irradiated at RT and is attributed to F' centers. This same peak appeared more pronounced in samples irradiated at 220 K. The F' assignment is based on the following considerations: While optically detected F' are known to decay slowly at RT (Delbecq and Pringsheim (1953)), a similar behavior is exhibited by the 310 K TSEE centers, which decay by a spontaneous emission in ultrahigh vacuum (Gordan and Scharmann (1968)).

If F' centers near the surface do play a role in exoelectron emission from alkali halides, then the possibility that these may undergo tunnel ionization at relatively low field strengths should be taken into account (Wollbrandt, Bruckner, and Linke (1983)). This appeals for theoretical investigations of surface F' centers.

13. Photomechanical effects

Thermally-stimulated experiments by Hirai and Scott (1966) on plastically-deformed crystals had revealed an effect of neighboring dislocations on the F' lifetime, as mentioned in Section 11. F' centers of both increased and decreased thermal stability relative to ones in undeformed samples had been evidenced. These had been attributed to locations in regions of compression and dilation around a dislocation line, respectively.

It appears that the reverse pinning effect of F' centers on moving dislocations has also been observed somewhat earlier by Nadeau (1964). It occurs as a sudden reversible change (increase) in flow stress when an F-centered alkali halide crystal is illuminated in the F band. This photoplastic effect has been observed in both gamma-irradiated and additively colored crystals, freshly quenched in the latter case to enhance the F-F' conversion. In KCl the photoinduced hardening first increases as the temperature is lowered below 300 K to pass through a maximum at about 175 K, and then decreases to virtually disappear at about 80 K. The temperature ranges above and below 175 K correspond to the ones of thermal instability and the stability of the F' center, respectively. Below 175 K the photohardening decreases due to the decreasing F-F' conversion efficiency, while above 175 K it does so due to the increasing F' thermal instability. With only these qualitative arguments in mind, the observed photoinduced hardening in KCl has been attributed to electrostatic interactions between F' centers and charged dislocations.

Subsequent investigations by Russian workers have further advanced Nadeau's original suggestions. Although Ermakov and Nadgornyi (1968) agree cautiously that the photoinduced pinning could relate to the F-F' conversion, Ermakov, Korovkin, and Soifer (1974a) have later made a substantial contribution. They measured the ultrasonic attenuation, induced by dislocations in plastically deformed crystals, aimed at checking Nadeau's conclusions through experimental means other than the direct flow-stress measurements. NaCl samples irradiated to between 3 and 10 Mrad of Co^{60} gamma rays have been investigated at temperatures between 150 and 300 K. It should be noted that the samples have been deformed plastically to 0.2 % and that the photoplastic measurements have only been performed after the deformation-produced dislocation structure stabilized within 1 to 1.5 hrs following deformation at RT. The photoplastic effect occurred as a light-induced decrease in ultrasonic attenuation, the percent change in attenuation being approximately equal to the relative increase of the pinning centers. The ultrasonic unit has operated at 50 MHz and its time-resolution has been good enough to follow the kinetics of the photoplastic changes on a minute time-scale.

The photoplastic effect has been excited with 500 nm light, within the F band of NaCl. Two kinds of pinning centers have been identified: ones that can and others that cannot be destroyed by a longer-wavelength light (540-750 nm), both occuring as a result of the bleaching of F centers. The photosensitive pinning centers have also proved bleachable thermally, requiring an activation energy of 0.3 0.07 eV. They have been identified as F' centers occuring near dislocation lines. The insensitive pinning centers are assumed to be empty anion vacancies. The number of pinning centers along a dislocation line depends on the excitation-light intensity I : it grows linear in I at low intensities and tends to saturate at high I . The bleaching time of the photosensitive component has proved inversely proportional to the bleaching-light intensity. The observed changes in ultrasonic attenuation, induced by either the excitation light alone or by a combination of excitation- and bleaching- light pulses, are completely reversible within the temperature range studied. While the bleaching-light spectrum coincides with the F' band, the excitation-light spectrum is that of the F band

(Ermakov, Korovkin, and Soifer (1974b)). All the above evidence gives strong support to the F' assignment.

In view of the lack of any specific calculations of the F'-dislocation interaction in alkali halides, the models considered all assume the occurrence of appropriate pinning centers (F', α) along dislocation lines following the bleaching of F centers. This leads to reducing the length of the dislocation segments and, thereby, to the decrease of the ultrasonic attenuation. In any event, the latter decrease, as observed experimentally, reveals the reduction in length of dislocation loops during the illumination, that is, the appearance of new pinning points along dislocations. Inasmuch as F' and α form at F center sites, the photoplastic effect results from the light-induced enhancement of weak obstacles (F) through converting them into strong ones (F', α).

A simple theory has recently been proposed for the photoplastic effect in F center containing alkali halides based on statistical dislocation dynamics (Hagihara, Hayashiuchi, and Okada (1985)). In the particular case under consideration, we introduce P_-, the trapping probability of a dislocation segment by an F' center, v, the velocity of that segment, ξ, an effective interaction length independent of the pinning defect type (whether F' or α), T_p^-, the pinning time of dislocations moving on a glide plane, where the surface F' concentration is F'. Now, since $(\xi F')^{-1}$ is the mean free path of a segment between two successive trappings at F', the following equation obtains:

$$\dot{P}_- = v\xi F' - P_-/T_p^- \qquad (13.1)$$

The F' concentration is found from

$$\dot{F}' = -F'/T' + F/T'' \qquad (13.2)$$

where T' is the F' thermal lifetime, while

$$T'' = F_0 d/I(1 - \exp(-c_F F_0 d)) \qquad (13.3)$$

and

$$F = F(0) - 2F' \qquad (13.4)$$

Comparing with eq.(4.13), eq.(13.2) is seen to be rather approximate. Nevertheless, we retain the latter to follow the author's line of reasoning. Equations similar to (13.1) can be written down for P_+ and P_F, the trapping probabilities at α and F, respectively. The applied flow stress obtains from

$$V_\pi \sigma = U_\pi + k_B T (\ln(v/v_c) + \sum_\beta P_\beta) \qquad (13.5)$$

where U_π and V_π are the magnitude and the activation volume of the Peierls potential, while $\beta = -,+,F$. Solving for $\Delta\sigma$, the following time-dependences obtain

$$\Delta\sigma / \Delta\sigma_{max} = 1 - \exp(-t/T_p^-) \quad \text{for} \quad 0 < t < t_{off} \qquad (13.6)$$

$$= 1 - \exp(-(t-t_{off})/T_p^-) + (1 - T_p^-/T')^{-1} \times$$

$$(\exp(-(t-t_{off})/T') - \exp(-(t-t_{off})/T_p^-)) \quad \text{for} \quad t > t_{off} .$$

The light is assumed to switch on at $t = 0$ and off at $t = t_{off}$. Also

$$\Delta\sigma_{max} = 2k_B TvF(0)\xi T_p^-/V_\pi (2 + T''/T') \qquad (13.7)$$

is the maximum change in flow stress. By measuring the time-kinetics of the flow-stress change as light is turned on and off, one can determine both T' and T_p^- . No comparison with experimentally observed kinetics is reported at this stage.

Another photomechanical phenomenon has been observed which relates closely to the photoplastic effect (PPE): the photoplastic aftereffect (PPAE)(Korovkin (1979)). It occurs in F-centered alkali halides subjected to a short excitation pulse of F light. Korovkin has investigated KCl and NaCl samples, gamma rayed to 5-10 Mrad, by means of the direct flow-stress method. PPAE displays itself in that the flow stress continues to grow after the light pulse, passes through a maximum, and then decreases gradually to its initial level prior to the excitation. It is tempting to regard PPAE as a manifestation of PPE under pulsed excitation, the maximum photomechanical response building up gradually due to the finite accumulation rate of the pinning centers (F', α). The The flow stress restores to its original value as the primed centers decay in the dark to revive the initial F-centered state. The development in time of the flow stress reminds of the evolution of the F' band under a flash-light excitation (Georgiev and Todorov (1976b)). As far as that analogy goes, the dark decay of the flow stress can be used to determine the F' lifetime, the relative change of the flow stress being equal to the one of the pinning centers.

The simple theory developed by the Japanese authors has also been extended to cover PPAE (Hagihara, Hayashiuchi, and Okada (1986)). Using equations (13.1) through (13.5) they arrive at

$$\Delta\sigma = A(\exp(-t/T') - \exp(-t/T_p^-))/(1 - T_p^-/T') \qquad (13.8)$$

with

$$A = 2k_B Tv\xi F(0)T_p^- \Delta t/T'' V_\pi \qquad (13.9)$$

$\Delta\sigma(t)$ first increases with time to reach a peak at

$$T_{aft} = -(\ln T_p^- - \ln T')/(1/T_p^- - 1/T') \quad , \qquad (13.10)$$

following which it gradually decreases to vanish. The above theory assumes that the pulse time Δt is short with respect to all the incorporated time-constants.

The Japanese authors have also carried out experimental measurements on KCl crystals colored to about 5×10^{16} F centers/cm^3 aimed at determining the F' lifetime in the range from 160 K to RT. An optical excitation during plastic deformation has been used. Both PPE and PPAE curves for the $\Delta \sigma$ vs. time dependence have been obtained and found to depend markedly on the temperature. From the PPE growth curves the pinning times T_p^- have been determined using the upper line of eq.(13.6), while T_{aft} have been measured as the PPAE peak times. Now, using eq. (13.10) the F' lifetime vs. temperature $T'(T)$ dependence has been obtained. The resulting curve is shown in Fig.42. The F' thermal energy

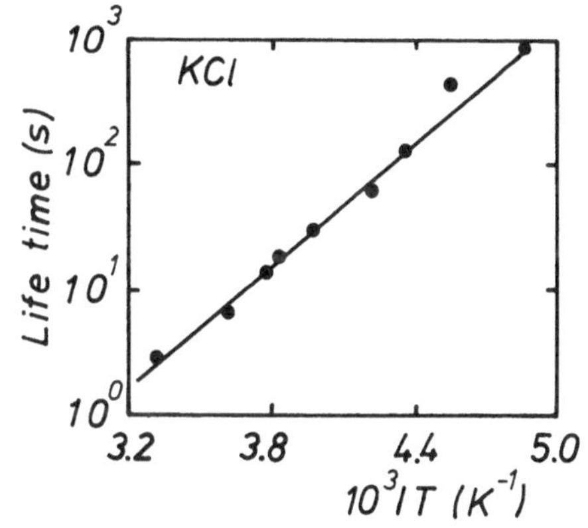

Figure 42: Temperature dependence of the F' lifetime in KCl, as measured by means of the photomechanical effect. (After Hagihara, Hayashiuchi, and Okada (1986)).

is found to be 0.29 eV, while the F' lifetime amounts to about 2 s at RT. It should be noted that the aforementioned analysis makes no use of the flow stress decay characteristics, so that the approximations introduced on using eq.(13.2) are probably less significant. Undoubtedly, such measurements can throw considerable light on the nature of the F'-dislocation pinning interaction.

Already Nadeau's early experiments had revealed that the PPE in some alkali halides changes sign below some critical (low) temperature, e.g. about 80 K in KCl. Thus, as the temperature is lowered to approach the critical point, the photoinduced hardening attributed to the pinning action of F' centers vanishes to give way to a photoinduced softening. Hagihara, Inoue, and Okada (1982) have reported more accurate measurements of the critical point in KCl, KBr, and NaCl crystals: 110 K, 190 K, and about 80 K in that order. The critical point is assumed to be the temperature above which the excited F centers effectively ionize to give rise to F' centers. Below the critical point only F* centers form, which, if in a 2p state, would rotate their principal axes so as to relax the applied stress. While the relax-through-rotation hypothesis has to be substantiated further, the attempt to explain the photoinduced softening in terms of a particular F*-dislocation interaction seems plausible. (See the literature cited therein for references to alternative explanations of the negative PPE). In a subsequent paper, Hagihara, Inoue, and Okada (1983) report on a measurement of the photoconductivity, the 'critical points' of which agree favorably with those of the PPE.

The above simple theory of photomechanical effects due to dislocation pinning by F' and alpha centers has also been exposed at a recent Conference (Hagihara, Hayashiuchi, and Okada (1986b)). The authors claim to have determined the pinning time T_p from PPE $\Delta\sigma(t)$ curves and the remaining time-constants from PPAE giving

$$((T'/T_p)(1 + T_{aft}/T_p) - T_{aft}/T_p)\exp(-T_{aft}/T_p) =$$

$$(T_p/T')(2T'/T_p - 1)\exp(-T_{aft}/T') \tag{13.11}$$

to obtain the F' lifetime T' in KCl from 200 K up to room temperature. An extension of the theory is also reported which separates moving dislocations into two groups: (i) dislocations in the field of a Peierls barrier, and (ii) ones in the field of point defect barriers, liberal state and pinned state, respectively (Hagihara, Hayashiuchi, and Okada (1986c)). If P_i is the probability of the dislocation being in state 'i', the strain rate is taken to be $de/dt = b\rho\sum v_iP_i$, where b is the Burgers vector, ρ is the mobile dislocation density, and v_i is the dislocation sweeping velocity. Assuming $\Delta\sigma$ to be small, one obtains at constant de/dt:

$$\Delta\sigma \doteqdot (k_BT/V_\Pi)\tanh(V_\Pi\,\sigma_o/k_BT)\sum_a \Delta P_a/P_\Pi \tag{13.12}$$

where σ_o is the stress level in the dark, while the sum is over all kinds of point defect barriers. The probability that a dislocation is trapped at a-th kind (P_a) satisfies the master equation

$$dP_a/dt = P_\Pi w_a^+ - P_a w_a^- \tag{13.13}$$

where w_a^+ and w_a^- are the pinning and de-pinning rates at the a-th defect, respectively. Solving the above equations combined leads to a theoretical equation for $\Delta\sigma$ in both PPE and PPAE to explain: (i) the physical meaning of the PPE rise-time; (ii) the dependence of $\Delta\sigma_{max}$ on the excitation-light intensity I, on the F center concentration, and on de/dt; (iii) the relation between T_{aft} in PPAE, T' and T; (iv) the dependence of T_{aft} on de/dt.

J. Stepien-Damm (1981) and Stepien-Damm and Mugenski (1982) have made precise lattice-parameter measurements on several alkali halides aimed at unraveling the changes induced by an F-F' conversion. These can help in understanding the nature of the elastic counterpart of the F'-dislocation interaction. The Bond method was applied with an accuracy of about 3×10^{-6} for $\Delta d/d$ to KCl, KBr, and KI crystals irradiated by gamma rays at RT, or colored additively or electrolytically. The F-F' conversion was initiated by F-band light at 173 K (KCl), 133 K (KBr), and at 127 K (KI). Lattice-parameter measurements were performed at the respective conversion temperatures immediately prior to and just after an F-F' conversion. The initial F center density has been the order of 1.8×10^{17} cm^{-3}. The following percent F' yields were obtained at saturation: F'_{max}/F_0 = 35 % (KCl), 29 % (KBr), 13 % (KI). In all the crystals studied the lattice parameter has been found to increase after an F-F' conversion and to saturate along with the F' density. $\Delta d/d$ values in excess of 10^{-5} were measured at saturation. Both the initial rate and the saturated value of the relative change increased in the order: KCl, KI, KBr. The size factor was also calculated from the experimental data which reflects the relative volume change $3\Delta d/d$ per F'-α pair. The following values were obtained: 12.9 % (KCl), 23.7 % (KBr), and 32.4 % (KI), which increased with increasing the anion radius (decreasing the electrostatic attraction between the lattice ions).

While the lattice around an F' center should undergo compression, the one around an alpha center should expand. The net increase in lattice parameter following an F-F' conversion implies that the vacancy-induced expansion exceeds largely the compression due to F'.

14. Positron annihilation at F' centers

The appearance of positrons in a crystal containing F, F', and centers is expected to lead to the formation of $^{a}A^{+}(F_{e^+})$ (a positron-electron pair trapped at an anion vacancy) and $^{a}A(F'_{e^+})$ (a positron plus two electrons in an anion vacancy) centers. The former result from the capture of positrons by F centers, the latter - from positron capture by F' centers. Due to its positive effective charge, the empty anion vacancy is not expected to trap positrons. There is considerable experimental evidence on the $^{a}A^{+}$ centers (see Dupasquier (1979)). On the other hand, the occurrence of ^{a}A centers can be expected to bring about significant changes in the positron-annihilation characteristics of the crystal. The ^{a}A center would contribute to the short-lived portion of the lifetime spectrum and to the small-momentum region of the momentum distribution. As a matter of fact, annihilation experiments performed after an F-F' conversion have revealed that:

(i) The narrow component of the angular-correlation curves narrows further (Nagarajan, Ramasamy, and Ramasamy (1976)).

(ii) The intensity of the component associated with $^{a}A^{+}$ in the lifetime spectra and in the angular-correlation curves is reduced (Bosi, Dupasquier, and Zappa (1973); Dannefaer, Trumpy, and Cotterill (1974)). The reduction has been found to be at its extreme at the temperature of maximum F-F' conversion.

(iii) The intensity of the short-lived component in the lifetime spectrum is increased (Dannefaer, Trumpy, and Cotterill (1974)). This is believed to result from the negative effective charge of the F' center which not only leads to a larger positron-capture cross-section of F' relative to F but also initiates a higher annihilation rate. Nevertheless, the component due to ^{a}A in the lifetime spectrum has not been resolved.

From combined optical-absorption and positron-lifetime measurements, Bosi, Dupasquier, and Zappa (1973), as well as Dannefaer, Trumpy, and Cotterill (1974) have estimated the ratio $\gamma^{p}_{F'}/\gamma^{p}_{F}$ of the positron-trapping coefficients of F' to F centers. Neglecting the untrapping, the intensity of the long-lived component attributed to the F centers is given by

$$I_3(F) = \frac{3}{4}\,\frac{\gamma^{p}_{F}\,{}_{F}}{\tau_0^{-1} + \gamma^{p}_{F}\,{}_{F} + \gamma^{p}_{F'}\,{}_{F'}} \qquad (14.1)$$

where τ_0^{-1} is the rate of all other positron-trapping processes, while the factor 3/4 accounts for the statistical weight of the ortho-state of the positronium-like system formed at the F center site. The lifetime of the above component is not affected by the F-F' conversion. Inserting the experimentally measured quantities, the following values have been obtained: $\gamma^{p}_{F'}/\gamma^{p}_{F} = 11.5 \pm 5.8$ at -100 °C (Bosi, Dupasquier, and Zappa (1973)), and $= 8.0 \pm 2.0$ between 150-190 K (Dannefaer, Trumpy, and Cotterill (1974)). This result compares favorably with the estimated values of γ_α / γ_F for electron trapping within the same temperature range. However, in a subsequent work by Bosi, Dupasquier, and Zappa (1975) arguments have presented against interpreting γ^p as the positron-trapping coefficient given by Pekar's equations (4.23) or (4.27) for a diffusion-limited capturing process. Instead, the authors suggest the alternative based on eq. (4.30) where the transition from a delocalized Bloch state to a bound state is rate-determining.

In any event, the existence of an F' capturing process which competes with the trapping by F centers seems to have been firmly established in spite of the lack of any resolved ^aA lifetime spectrum (see also Smedskjaer, Dannefaer, Cotterill, and Trumpy (1973)).

Dannefaer, Trumpy, and Cotterill (1974) have also measured the Doppler broadening of the annihilation quanta. Although the broadening was not found to be very sensitive to the F-F' conversion, it increased to nearly that for an uncolored crystal at the maximum conversion. Quantitatively, the increase of broadening seemed to correlate roughly with the decrease of the intensity I_3 of the long-lived component in the lifetime spectrum. Consequently, the narrow momentum distribution appears related to the long-lived component, caused by positronium (Ps) annihilation, suggesting that no Ps is formed at the F' center.

Apparently, more sensitive to the F-F' conversion has been the angular correlation of the positron annihilation (Nagarajan, Ramasamy, and Ramasamy (1977)). The half-width at half-height of the narrow component in the angular distribution dropped from 2.65 mrad to 1.95 mrad, while its intensity dropped from 25 % to 12 %, as a result of an F-F' conversion in additively colored KCl at -100 ^oC. It is concluded that F' should have a narrow momentum distribution and a much larger capture cross-section for positrons than have the F centers. It is now suggested that a positronium ion Ps^- may form at an F' center.

A theory of positron trapping by F' centers has been developed by Berezin (1972). He considers the F' center as a loosely bound electron moving in the weakly-attractive short-ranged field of the neutral core (F center) approximated by the Hulthen potential

$$V(r) = b \exp(-ar)/(1 - \exp(-ar)) \qquad (14.2)$$

where a and b are adjustable parameters. The Schrodinger equation of an electron moving in field (14.2) can be solved exactly giving

$$\psi_{\gamma,a}(r) = N(\gamma,a)\exp(-\gamma r)(1 - \exp(-ar))/r \qquad (14.3)$$

where

$$N(\gamma,a) = ((\gamma/2\pi)(1 + 3(\gamma/a) + 2(\gamma/a)^2))^{\frac{1}{2}} \qquad (14.4)$$

is a normalization factor, and

$$b = a\gamma + \frac{1}{2} a^2 \tag{14.5}$$

The result of attaching a positron to an F' center is regarded as a positronium atom bound in the field of an F center. Berezin chose a trial wave function for the ground state of ^{a}A in the form

$$\psi_{a_A}(\underline{r}_-,\underline{r}_+) = \psi_{\alpha,a}(\underline{r}_-)\,\psi_{\beta,a}(\underline{r}_+) \tag{14.6}$$

where $\psi_{\alpha,a}$ and $\psi_{\beta,a}$ are of the form of Hulthen orbitals (14.3). The energy functional for ^{a}A now is

$$E(\alpha,\beta,a,b) = \int \psi^*_{a_A}(\underline{r}_-,\underline{r}_+)\, H\, \psi_{a_A}(\underline{r}_-,\underline{r}_+)\, d^3r_- d^3r_+ \tag{14.7}$$

where the Hamiltonian of ^{a}A is

$$H = -\frac{h^2}{2m}(\Delta_- + \Delta_+) - V(\underline{r}_-) + V(\underline{r}_+) - e^2/|\underline{r}_- - \underline{r}_+| \tag{14.8}$$

in which $V(r)$ is Hulthen's potential (14.2). Minimizing (14.7) with respect to α and β gives the ground-state energy of ^{a}A. In Berezin's model, $a = (\frac{8}{3}E_{Fabs})^{\frac{1}{2}}$, E_{Fabs} is the energy at the F band peak. The binding energy of the ^{a}A center (the positron affinity of F') is now found as the difference between the ground-state energies of F' and ^{a}A. It has been estimated at 3 eV, which is almost twice the $^{a}A^+$ binding energy. (The F' binding energy in this model is $-\gamma^2/2$.)

Nagarajan, Ramasamy, and Ramasamy (1977) report to have used Berezin's functions (14.6) in computing the positron angular distribution in KCl: this resulted in a rather broad angular distribution with a half-width at half-height of 5.5 mrad, as opposed to the experimental data of 1.95 mrad. The authors conclude that Hulthen's orbitals are not appropriate for describing the positron angular-correlation curves.

More recently, Berezin introduced two essential improvements into his theory (Berezin (1976b)). He took into account: (i) the e^--e^+ correlation, and (ii) the fact that both e^+ and e^- are attracted by the F center to form ^{a}A and F', respectively. (i) was introduced by choosing a "strongly-correlated trial wave function" of the form

$$\psi_{a_A}(\underline{r}_-,\underline{r}_+) = N \exp(-\alpha r_-) \exp(-\beta|\underline{r}_- - \underline{r}_+|) \tag{14.9}$$

while (ii) was accounted for by changing the sign of $V(\underline{r}_+)$ in (14.8), Hulthen's potential (14.2) now acting likewise on both e^- and e^+. Minimizing (14.7) in which ψ_{a_A} is now given by (14.9), results in the ground-state energy E_{a_A}. Next, the binding energy of ^{a}A is computed as the difference of the free-positronium energy (-6.80 eV) and E_{a_A}. Berezin has also calculated the time constant $\tau_{2\gamma}$ of the ^{a}A two-photon annihilation using the formula

$$\tau_{2\gamma}^{-1} = \pi(e^2/hc)^3 < \Psi_{a_A}(\underline{r}_-,\underline{r}_+) | \Psi_{a_A}(\underline{r}_-,\underline{r}_+) > \quad (14.10)$$

The numerical results of Berezin's calculation are listed in Table 7.

Table 7

Calculated F'_{e^+} characteristics*

Crystal	Binding energy (eV)	Time constant (ps)
NaCl	3.59	350
NaF	3.79	330
KCl	3.26	370
KBr	2.95	380
KI	2.64	390

*After Berezin (1976b)

The $\tau_{2\gamma}$ values were found to be well approximated by a Mollwo-Ivey type equation: $\tau_{2\gamma} = 0.226\ d^{0.43}$ ns; A. Finally, Berezin stresses the unique capability of the F center to trap e^- and e^+ which is obviously due to strong polarization effects. This charge invariance of the F center makes it dissimilar to the hydrogen atom, which cannot apparently form any bound state with a positron (Berezin (1979b)). A Mollwo-Ivey type is established for the ${}^aA$ binding energy as well: $E_B = 7.88\ d^{-0.82}$ eV; A. It should also be stressed that $F_{e^+}({}^aA^+)$ is in a way similar to F': both have the same point symmetry (O_h) and form through charged particle trapping at F centers. Consequently, F_{e^+} may be expected to exhibit corresponding optical and thermal properties (Berezin and Evarestov (1971)).

Some time later, Berezin's methods have been utilized in carrying out another calculation of positron's binding energies in nine salts: NaF, NaCl, KCl, KBr, KI, RbCl, RbI, CsBr, and CsI. The trial wave function was chosen of the form (14.6) with

$$\Psi_{i,a}(r_i) = (2\alpha_i(a+\alpha_i)(a+2\alpha_i))^{\frac{1}{2}}\exp(-\alpha_i r_i)(1-\exp(-ar_i)) \quad (14.11)$$

and the energy functional (14.7) was minimized in $\alpha = \alpha_1$, $\beta = \alpha_2$. The F'_{e^+} binding energy was found to be practically independent of the host, amounting to 2.96 eV on the average. This is to be compared with the F_e+ binding energies, also calculated using $V(r) = a/r$ for the F center potential, which ranged between 0.77 eV and 2.31 eV for twenty salts (Lam and Varshni (1978)).

Inasmuch as the $F_e'+$ binding energy exceeds that of F_e+, positron trapping at one of two nearby F centers may bring about tunneling of the electron of the second F center to the F_e+ formed with the subsequent occurrence of an $F_e'+-\alpha$ pair:

$$F_e+ \; + \; F \; \longrightarrow \; F_e'+ \; + \; \alpha \qquad (14.a)$$

The energy released in recharging the pair components may exceed the thermal ionization energy of a neighboring F' center. Now, Auger's autoionization can occur within the tri-component entity, such as

$$F_e+ \; + \; F \; + \; F' \; \longrightarrow \; F_e'+ \; + \; \alpha \; + \; F \; + \; e^- \qquad (14.b)$$

Berezin (1969) has been the first to calculate the rates of Auger's nonradiative processes in colored crystals. In a particular case (Berezin (1977a)), he derives the rate from eq.(6.61) with V_{Auger} in the form of the Coulomb interaction of the tunneling electron at F with Auger's electron at F', to manifest the importance of Auger's processes induced by positron capture at M centers. Accordingly, the initial wave function is chosen as a product of 1s F and delta F' orbitals, while the final state factorizes out of 1s F and a CB electron wave functions. Auger's ionization times of 0.37 ns are obtained for an F center density of the order of 10^{15} cm^{-3}. The reader is referred to Berezin's original article for details.

15. F' muonics

Positive muons have been used for studying inner magnetic fields and defect centers in solids by measuring the muon spin rotation. The residual polarization of the positive muons after thermalization in ionic materials is found to be rather small, presumably due to spin-depolarizing processes such as muonium formation or the formation of radicals. Depolarization results from the superhyperfine interaction with the magnetic moments of the neighboring nuclei. Alternatively, one can study the effect of various color centers on the residual muon polarization. Recently, Jacobs, Orth, zu Putlitz, Schafer, Vetter, Winnacker, and Herlach (1982) have done just that relative to the F' center. The latter has been chosen because (i) the excess negative charge would attract the positive muon to form a bound state, and (ii) the spins of the two electrons of the resulting (μ^+,e^-,e^-) system would be antiparallel to each other with the effect of creating a diamagnetic environment for the muon component, thereby protecting it from the depolarizing action of the neighboring nuclei. In an experiment of measuring the residual muon polarization in a transverse magnetic field, the F' centers were created optically in single crystalline targets of KCl, pre-colored additively to about 10^{17} F centers/cc, through excitation by an argon-ion laser (514.5 nm) at 170 K. Keeping the F' concentration constant, the authors have investigated the residual μ^+ asymmetry A within the 14 to 170 K range. As the temperature was raised, A first increased very slightly (if at all) up to about 50 K, followed by a sharp drop beyond. It is suggested that since A drops with the temperature, the interaction of the muon with the F' centers may not be diffusion-limited. Apparently, more experimental work is needed to clarify the transport, trapping, and depolarization mechanisms for positive muons in alkali halides. Nevertheless, muons "find" F' of only a few ppm at 10 K.

16. F' centers in the presence of F- aggregate and colloid centers

In a series of temporary-bleaching experiments, Hirai, Ikezawa, and Ueta have observed among other things the occurrence of the F' band on F-light illumination of additively colored KCl and KBr crystals containing F-aggregate centers (M, R, N), formed by pre-exposure to F-band light at an elevated temperature. In a KCl crystal colored to 2.7 x 10^{16} F centers/cc, the aggregates were formed by F light at RT, while the F' band was produced by subsequent F-band bleaching at 90 K (Hirai, Ikezawa, and Ueta (1961)). The F' band was induced in KBr at 130 K in a sample containing originally 8 x 10^{17} F/cc and pre-exposed to F light at 45 ^{o}C to form the aggregates (Hirai and Ueta (1962)). In a later study (Hirai, Ikezawa, and Ueta (1962)), the F' band was produced by F light at -130 ^{o}C (KBr) and at -110 ^{o}C (KCl). The F' thermal stability was also monitored, F' becoming unstable in KCl above -70 ^{o}C. The optical bleaching of the F' band was also observed at 90 K. No systematic attempts seem to have been made to see whether the F' centers are in any way affected by the presence of the aggregate bands. However, the bleaching with F light in both crystals was found to lead to the formation of 'primed aggregate-center bands' (M', R', N') in addition to F'. These primed bands form when photoelectrons from F centers become trapped at aggregate centers. The respective species, analogues to molecular ions, differ essentially from the F' centers and are, therefore, beyond the scope of these Notes.

An electric glow peak at 198 K attributed to F' centers has also appeared in colloid-containing KCl crystals (Haupt (1963)). The colloids were formed by illumination at an elevated temperature.

Apparently, F' centers can again form efficiently via the CB at the later stages of aggregate-center production when the spatial distribution of the F centers turns back statistical after the consumption of the F center clusters.

Other authors have also reported the occurrence of F' centers in the presence of F-aggregate centers. Ikezawa (1964) developed a maximum M band through F-band bleaching at RT of a KBr sample, freshly quenched after being colored additively to 2 x 10^{17} F/cc. Subsequent bleaching with F light at 90 K produced the F' band. Prior to forming the M centers, the F-F' conversion was checked at 90 K in the freshly-quenched state. Higher F' concentrations were produced at higher temperatures but F' was found to become unstable thermally above 130 K. F' decay curves were followed between 130-190 K with halflives reportedly longer than the ones measured by Pick. Several years later, Schneider and Bailey (1969) colored NaCl crystals through irradiation by 2 MeV electrons at 77 K. The crystals were then warmed to 273 K to produce a maximum M^+ absorption. The other species identified were F, M, and F'. The latter were found to bleach substantially at 4 K through exposure to light within the F' band (between 500-700 nm). However, exposure to F light at 4 K caused F' to reappear! A similar F' production at 4 K was observed on F-light exposure of a sample colored by x-rays at RT. The role of F' centers in the reorientation of M centers was clarified by Schneider (1970). M centers were shown to reorient indirectly by first absorbing light to convert them into M^+ which then absorbed light themselves and reorientated through rotation. M center electrons are captured by F centers to form F' centers. Reoriented M centers are observed only after rotated M^+ centers capture electrons stored at F' centers. F' thus play an auxiliary role in the process.

Another comprehensive study of the role of F' centers in the production of $M^+(F_2)$ centers in KCl and other alkali halides is due to Aegerter and Luty (1971a,1971b). They propose a model in which M^+ is formed in a F- and M-center containing crystal through optical excitation in the uv L_M absorption bands leading to the ionization of both F and M centers. The electron-excess centers F' and M' play the role of stabilizing antipode-centers to F^+ and M^+. If so, the optical-bleaching spectrum of M^+ should contain the F' and M' bands. It is also suggested that the M^+ centers are themselves stable thermally up to sufficiently high temperatures but that their effective thermal stability is limited by that of F' and M'. To check the model, the M^+ optical-bleaching spectrum was obtained experimentally at 7 K by studying the bleaching curves of the 600 nm luminescence, associated with the M^+ center. This spectrum was found to follow the F' band almost perfectly. Bleaching in the M' band range produced a similar luminescence quenching. In addition, the relative quantum efficiency of the 600 nm luminescence, induced by excitation in any of the M^+ transitions, was strongly reduced at temperatures above 190 K to disappear around 220 K, the temperature range where the thermal ionization of F' occurs. In KCl crystals x-irradiated at LHeT, only a small fraction (about 4 %) of M^+ could be bleached by F' and M' light. This suggested that the main type of stabilizing antipode center was other than F' and M'. The latter center is believed to be the chlorine interstitial ion.

17. F' centers in impure alkali halides

17.1. F' centers in crystals doped with A-type univalent cations

17.1.1. F'_A centers

The optical excitation in the F band near RT of an alkali halide crystal doped with monovalent substitutional alkali cations of radius smaller than that of the host cation (A-type cationic impurity) brings about the conversion of F centers into F_A centers. The F_A center is an F center of which one of the six nearest-neighboring < 100 > cations is replaced by an A-type ion. The F_A are formed through the F-F_A optical conversion which involves migration and capture of ionized F centers by A-type ions, thereby creating α_A centers, followed by photoelectron trapping at α_A :

$$\alpha + A \longrightarrow \alpha_A \tag{17.a}$$

$$e + \alpha_A \longrightarrow F_A \tag{17.b}$$

The presence of a neighboring A-ion lowers the symmetry of the F_A center and lifts partially the degeneracy of the Franck-Condon 2p-like state, which is the final state of the main optical-absorption transition at the center. That lifting leads to the occurrence of two absorption bands F_{A1} and F_{A2} in lieu of the F band. F_{A1} and F_{A2} have optical-transition dipoles respectively along and perpendicular to the center axis, which connects the impurity ion with the center of the vacancy. The corresponding Franck-Condon states are $|2p_z>$ and $|2p_x + 2p_y>$, of which the former is lower-lying, while the latter is such that the F_{A2} band is only slightly shifted towards the higher energies relative to the F band. (For a comprehensive review on the F_A and related centers, see Lüty (1968)). The F_A center has only one relaxed excited state, the initial state of the F_A^* radiative deexcitation transition, which does not show any anisotropy. The F_A^* center reorientates easily via vacancy jumps around the impurity ion at surprisingly low temperatures, the reason being the low jump barrier when the center is in the configuration of the relaxed excited state. Consequently, illuminating an F_A-centered crystal with linearly-polarized F_{A1} or F_{A2} light can lead to the alignment of F_A centers which will remain stable at sufficiently low temperatures: the crystal will exhibit a photoinduced dichroism.

The F_A^* center can ionize thermally at sufficiently high temperature, or in a field-assisted process at lower temperature. As a result, ionized F_A centers (α_A) are formed, while photoelectrons can be trapped at non-ionized F_A to form F_A' centers, the analogue of F'. When field-

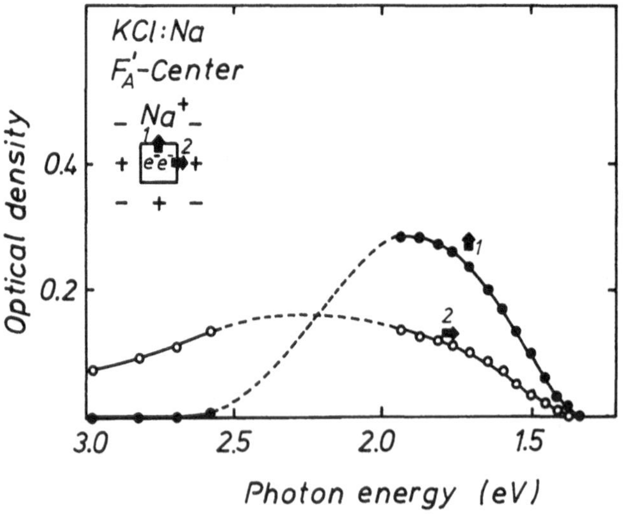

Figure 43: Idealized F_A' bands extrapolated to the case of a completely aligned F_A' center system with an orientation indicated by the center model (see insert). (After Lüty (1968)).

ionization is performed at low temperature of a partially-aligned F_A

center system, correspondingly aligned F'_A occur. By means of such experiments, Link and Luty (1965) have found a dichroism in the F'_A band, the two components, F'_{A1} and F'_{A2}, being polarized along and perpendicular to the center axis, respectively. The shape of these bands, extrapolated to the case of a complete F'_A alignment, is shown in Fig.43.

While the isotropy of the F^*_A state implies a more spread wave function, the observed dichroism of F'_A suggests that either the F'_A ground state (possibly not s-like) or the final Franck-Condon state of the F'_A absorption is in fact strongly influenced by the impurity. This raises again the question of the existence of an excited 2p-like F'^*_A state which either remains stable or becomes autoionized following the lattice relaxation. While the two F'_A bands are seen to have the same long wavelength edge, they are quite different in peak position and halfwidth (Fig.43). Of these, F'_{A2} resembles the F' band, while F'_{A1} is more symmetric and is likely to result from a transition to a bound excited state.

The only theory of the F'_A center available so far seems to be the one proposed by Berezin (1976a) based on his δ-potential semiempirical approach to the color centers. He argued that F'_A should have bound excited states because of its lower symmetry. In particular, he calculated the ionization edge of the F'_A(Na) band in KCl. The field acting on the F'_A electron was approximated by two δ-potentials, one of power λ_1 corresponding to the F center component, the other of power λ_2 due to the Na^+ ion. The F'_A wave function is of the form

$$\Psi(r) = \sum_{i=1,2} C_i \exp(-\alpha r_i)/r_i \; , \qquad r_i = |\underline{r} - \underline{R}_i| \tag{17.1}$$

while the energy is (in atomic units)

$$E = -\alpha^2/2 \tag{17.2}$$

where α is the solution to the secular equation corresponding to the system of linear equations for C_i :

$$(\lambda_i - \alpha)C_i + \sum_{k\neq i} C_k R_{ik}^{-1} \exp(-\alpha R_{ik}) = 0 \; , \; R_{ik} = |\underline{R}_i - \underline{R}_k| \tag{17.3}$$

$\underline{R}_{i,k}$ are the position vectors of the δ-wells (i,k = 1,2), $R_{i,k}$ = d/2 (half the lattice parameter), $\lambda_1 = (2h\omega_0)^{\frac{1}{2}}$, $\lambda_2^2 = (2\Delta I)$, ω_0 is the optical threshold of the F' band, ΔI = 5.14 - 4.34 eV is the difference in ionization potentials between Na and K ; $\lambda_1 = 0.284$, $\lambda_2 = 0.242$ in atomic units. The results of Berezin's calculation are presented in Table 8.

In addition to the F_A centers, F_B and F_C centers are formed, apparently by similar processes. These centers comprise an F center next to

Table 8

Calculated long-wavelength edge of primed bands*

Center	Symmetry	Optical threshold energy (eV)
F'	O_h	1.10
F'_A	C_{4v}	1.21 (0.60)
F'_B(I)	D_{4h}	1.29 (0.77, 0.48)
F'_B(II)	C_{2v}	1.30 (0.68, 0.55)
F'_C(I)	C_{2v}	1.36 (0.77, 0.66, 0.46)
F'_C(II)	C_{3v}	1.37 (0.68, 0.51)

*After Berezin (1976a)

two A-ions (F_B): I- located along a line with F in the midst, II- all three components placed in the vertexes of a triangle; and an F center next to three A-ions (F_C): I- in a planar configuration, II- in a spatial configuration. Berezin's results on F'_B and F'_C are also shown. The

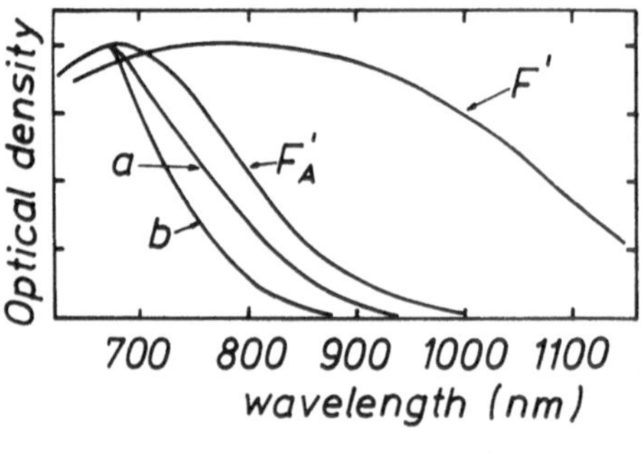

Figure 44: Normalized absorption spectra at LHeT of F' and F'_A centers, of an overlap of F'_A and F'_B centers (curve a) and F'_C centers (curve b). (After Schneider (1969)).

lower energies in parenthesis refer to optical transitions to bound states, still not observed experimentally. The photoionization threshold calculated compares favorably with the experimental data: 1.24 eV (F'_A), 1.32 eV (F'_B), and 1.41 eV (F'_C) (Schneider (1969)), see Fig.44.

Schneider observed the formation of all the primed bands on F-light illumination of KCl heavily doped with NaCl. These F'-type centers also bleached optically. They were found to absorb light at increasingly shorter wavelengths in the order: F', F'_A, F'_B, F'_C, as shown above, in accordance with the theoretical conclusions.

Apparently, the F'_A band has first been observed by Kojima and Nishimaki (1961) and by Kojima, Nishimaki, and Kojima (1961). Irradiation with F light of a KCl crystal containing F and F_A centers at -70 °C has initially given rise to the F' band and then to a new band (F'_A) to the red of F with the following properties: F'_A is relatively stable up to -40 °C. When a specimen containing $F_A$centers in large amounts is irradiated with F_{A1}-light between -50 and -130 °C, F'_A grows rapidly as F_{A1} decreases. F'_A is reported to be slightly narrower than F' and not to absorb light beyond 950 nm. On warming up to RT, F'_A disappears and F_{A1} recovers. The F'_A halflife at -50 °C has been estimated at about 500 s. The lowest temperature at which F'_A could be produced by F_{A1} light was -135 °C. The general conclusion drawn is that F'_A is an F_A having trapped an extra electron. It is also considered likely that $\sigma_F > \sigma_{F_A}$ because F' develops more rapidly than does F'_A.

Reasonably, no principal defference has been expected to occur between the F-F' and the F_A-F'_A conversion kinetics. Terzijski and Popov (1974) have investigated both conversions in KCl:Na within the temperature range -160 to -30 °C. From the temperature dependences of F'_{max} and F'_{Amax} between -160 and -110 °C, the F* and F^*_A thermal-ionization energies have been deduced in accordance with eq.(4.16) at $\beta_{F'} = q_{F'}$: 0.17±0.03 eV and 0.11±0.03 eV, respectively. At the same time, from the data between -70 to -30 °C the thermal-ionization energies of F' and F'_A have been found to be 0.51±0.08 eV and 0.6±0.1 eV, respectively, assuming $\beta_{F'} = \tau_{F'}$. The steady-state concentrations F'_{max} and F'_{Amax} have been independent of the excitation-light intensity below -120 °C for F' and -100 °C for F'_A. In a preceding work, Popov and Terzijski (1974) have examined the F'_A formation kinetics in KCl:Na. From the initial F'_A formation rates at various temperatures, the authors deduce using eq.(4.13) the relative F^*_A thermal-ionization efficiency:

$$\eta_{F_A} = (1 + 2.6 \times 10^{-7} \exp((0.11 \pm 0.03)\mathrm{eV}/k_BT))^{-1} \quad \text{(KCl:Na)}$$

However, later the preexponential factor has been specified to be 2.6 x 10^{-5} between 80-200 K (Popov (1979)). Popov's results on the F and F_A ionization quantum yields are shown in Fig.45 compared with the data by Fedders, Hunger, and Lüty (1961) on the F center. The F_A centers have been produced in an almost complete optical F-F_A conversion at 233 K.

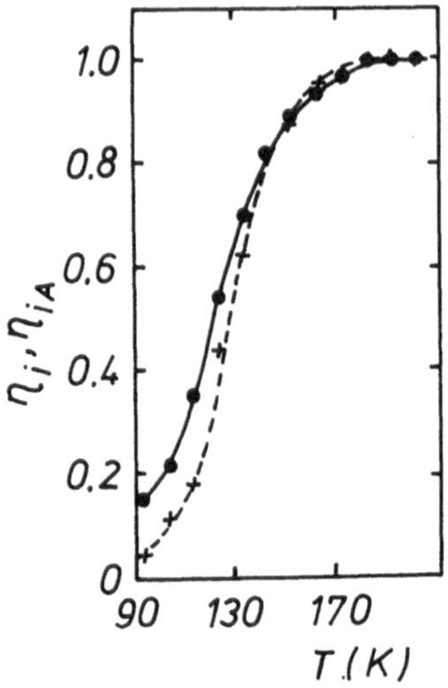

Figure 45: Temperature dependences of the ionization quantum yields of F centers (crosses) and F_A(Na) centers (filled circles) in KCl. The dashed line represents data by Fedders, Hunger, and Lüty (1961). (After Popov (1979)).

In this same work, the F'_A thermal-ionization energy is also specified to be 0.62±0.1 eV. Popov also reports having estimated the height of the cross-over barrier for radiationless F* deexcitation by examining the F_A(Na) formation kinetics between 250-280 K: 0.47 eV (KCl). The initial F'_A formation rate has also been found proportional to the excitation-light intensity (Popov and Terzijski (1974)), again in accordance with the conversion equations (4.13), at -140 °C.

The ratio $\sigma_{\alpha_A}/\sigma_{F_A}$ of the electron-trapping cross-sections has also been found experimentally between 110-170 K, within the F'_A thermal-stability range, from an analysis of the solution of eq.(4.13). The result is shown in Fig.46 where the temperature dependence is described fairly well by (Popov and Terzijski (1979), Popov (1979)):

$$\sigma_{\alpha_A}/\sigma_{F_A} = 10.2(1 + 4.2 \times 10^4 \exp(-0.12\text{eV}/k_BT))^{-1} \quad \text{(KCl:Na)}$$

According to eq.(4.8) this gives $\sigma^*_{\alpha_A}/\sigma_{F_A} = 10.2$, $E_i = 0.12$ eV. Consequently, trapping of a conduction electron by α_A occurs in the excited state F^*_A as a pre-capturing step before the final recombination with the perturbed anion vacancy in ground F_A state. The obtained depth of the pre-trapping level (0.12 eV) agrees well with the thermal ionization energy of F^*_A (0.11 eV) found in F_A-ionization quantum yield

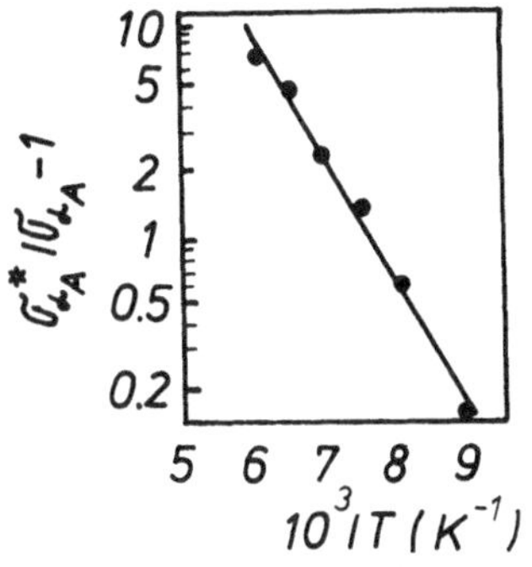

Figure 46: Temperature dependence of $\sigma_\alpha^*/\sigma_\alpha - 1$ for the KCl F_A(Na) center from the kinetics of the F_A-F'_A conversion in the F'_A thermal stability range. (After Popov and Terzijski (1979)

measurements (Popov and Terzijski (1974)). Terzijski and Popov have republished their 1974 data on the temperature and intensity dependences of the photoequilibrium F' and F'_A densities (Terzijski and Popov (1975)). The primed centers were produced between 80-240 K by illuminating F or F_A containing KCl or KCl:Na crystals with 500 nm light until a photoequilibrium was attained. The temperatures of ultimate F-F' (F_A-F'_A) conversions were found to be 183 (193) K. The quantities $b_{F'} = F_0/F'_{max} - 2$ and $b_{F'_A}$ were found to vary linearly with the reciprocal

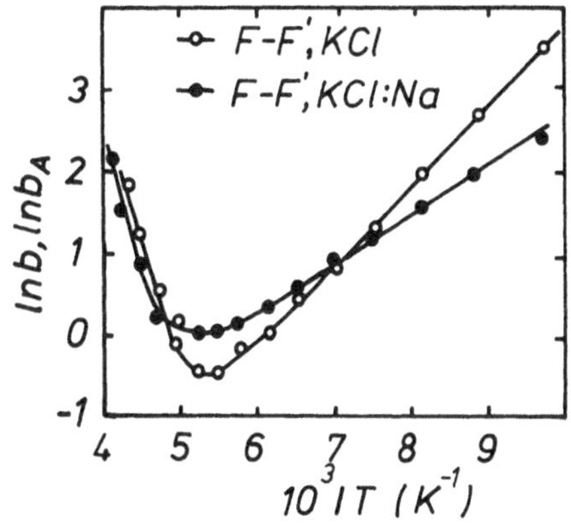

Figure 47: Temperature dependence of $F_0/F'_{max} - 2$ for F and F_A(Na) centers in KCl (hollow and filled circles, respectively). The reversal of slope at higher temperatures indicates the

inclusion of thermally-activated cross-over deexcitation of the excited centers. (After Popov (1979)).

temperature between 110-160 K and 200-240 K, exhibiting opposite slopes in the two ranges (Fig.47). Below 150 (170) K, F'_{max} and F'_{Amax} proved independent of the excitation-light intensity, as required by eq.(4.16) for $\beta_{F'} = q_{F'}$. The F-F' conversion model has been used as a basis for interpreting the b's. In the lower temperature range, electron trapping takes place presumably in the excited state and F' is assumed to be thermally stable. This leads to $b \propto I^0 \exp(E_{F*}/2k_BT)$. Within the higher-temperature range, the electron is captured by the vacancy in ground state and F' is thermally unstable. Now, $b \propto I^{\frac{1}{2}} \exp(-E_{F'}/2k_BT)$. From the respective experimental branches the authors obtained: $E_{F*} = 0.17 \pm 0.03$ (0.11 ± 0.03) eV between 110-160 K and $E_{F'} = 0.51 \pm 0.08$ (0.6 ± 0.1) eV between 200-240 K.

While Terzijski and Popov's assertion concerning the direct ground-state trapping of a conduction electron cannot be qualified categorically, their estimates of $E_{F'}$ seems fairly realistic. It also seems likely that the required constancy of γ_α / γ_F results from the inclusion of the thermally-activated cross-over F* deexcitation rather than from the presumed existence of any nonvanishing γ_α^0 term, as predicted by Georgiev and Todorov (1976b) for that same temperature range.

The F'_A band thermal decay is similar to that of the F' band. Inasmuch as the decay type has been found to differ from the single-exponential law, Tomura, Kitada, and Takebayashi (1965) plotted the F'_A band halflife, measured as the time-period during which the optical density at 719 nm decreased to half its initial value, versus the reciprocal temperature between -60 and 0 °C to obtain an Arrhenius type

$$t_{F'_A \text{ halflife}} = 4.2 \times 10^7 \exp(0.41\text{eV}/k_BT) \text{ , s} \quad (\text{KCl:Na})$$

The F'_A were produced by a flash of F_A light shone on a sample containing about 10^{17} F_A/cc, produced by optical F-F_A conversion at -30 °C. From eq. (4.33) assuming a constant F' lifetime we have

$$t_{F' \text{ halflife}} = \tau_{F'}\left(\frac{\gamma_F}{\gamma_\alpha}(F_0/F'_0 - 2\ln 2) + \ln 2\right) \qquad (17.4)$$

It is now clearly seen that halflife measurements do not give the F'_A thermal-ionization energy directly, since $t_{F'_A}$ depends also on the cross-sections' ratio $\gamma_{F_A}/\gamma_{\alpha_A}$ and on the F'_A yield F'_{A0} which are both temperature-dependent in addition to $\tau_{F'_A}$. In a subsequent work, Tomura and Kitada (1967) report having obtained a two-slope halflife vs. $(\text{temperature})^{-1}$ dependence, yielding 0.28 eV and 0.60 eV from the lower and the higher temperature branches, respectively. The entire temperature dependence was interpreted by means of Kubo & Toyozawa's tunneling theory (Kubo and Toyozawa (1955)), the lower energy implying thermal excitation to a higher excited F'_A state rather than to the conduction band. This state was assumed to be of a s-like type, since no optical absorption originating from transitions to such a state was found in the vicinity of 4 μm (0.3 eV). However, it was suggested that an excited F'_A state might occur due to the deepening of the F_A-core potential caused by the Na ion.

The F'_A band of non-aligned F'_A centers in KCl:Na has been examined at RT by Todorov, Dechev, and Georgiev (1972) using flash-light excitation spectroscopy. In a subsequent paper, the authors report to have investigated the decay at RT of the F'_A band and found it to follow eq. (4.33) with a constant F'_A lifetime (Todorov, Dechev, Tomova, and Georgiev (1973)). The F'_A centers were produced at the temperature of measurement by shining a flash of F_A light onto a sample containing F_A centers, produced by a preliminary optical F-F_A conversion at that same temperature. This made it possible to also watch the yield of the flash-produced bands which provided additional information, described in a separate paper (Todorov, Baltova, and Georgiev (1975)). In a complete qualitative analogy with the F' data (see Section 4.2), the F'_A yield was found to increase as the 0.5th power of the excitation-light intensity before saturating at the highest intensities, and to decrease with the temperature between 290-350 K. For a sufficiently durable flash (LF), assumed to bring about a photoequilibrium between the excitation products, the F'_A yield was found to decrease as $\exp(0.3\text{eV}/k_BT)$, while for a short flash (SF) of the same energy the yield (presumably far from equilibrium) was proportional to $\exp(0.15\text{eV}/k_BT)$. At the same time, $\gamma_{F_A}/\gamma_{\alpha_A}$ obtained by fitting eq.(4.33) to the F'_A decay kinetics, decreased as $\exp(0.33\text{eV}/k_BT)$, in spite of considerable scatter, apparently due to uncontrollable flash-light intensity variations (Todorov, Tomova, and Georgiev (1974)). Another important result from the F'_A decay measurements was the F'_A thermal-ionization energy estimated at 0.75 0.05 eV, as shown in Fig.48. As in the F' case, the obtained quantitative data have been explained assuming the inclusion between 290-350 K of thermally-activated radiationless F^*_A deexcitation, related to a cross-over barrier of about 0.45 eV (Georgiev and Todorov (1975)). Reasonably, the obtained larger thermal-ionization energy corresponds to the larger optical depth of the F'_A level relative to F'. The dependence of both $E_{F'_A}$ and the optical depth on the impurity suggests that the F'_A ground state may be more compact than the F' state, since the

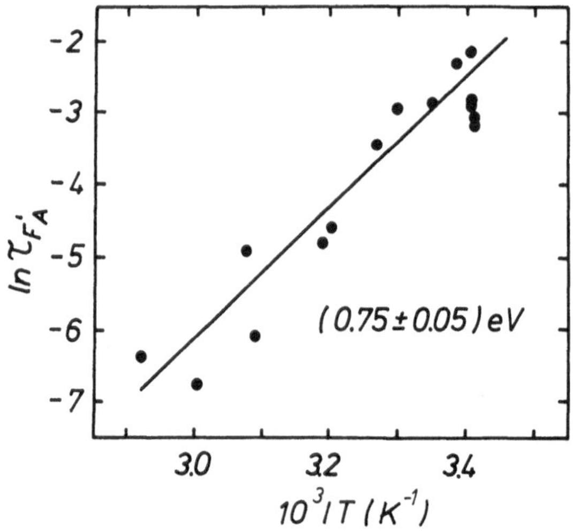

Figure 48: Temperature dependence of the F_A' thermal lifetime in KCl:Na. (After Todorov, Tomova, and Georgiev (1974)).

final states in both ionization processes lie in the conduction band. Apparently, this also leads to an increased stability of F_A' relative to F'.

The observed intensity and temperature dependences of the F_A' yield have once again been interpreted assuming the occurrence of non-linearities in the capture cross-sections and the ionization F_A quantum yield similar to those introduced for the F center (see Section 4.2.). The reason again being the presumed inclusion of some interaction which switches on under a high-intensity excitation, the observed close parallelism btween F and F_A lends strong support to the authors' hypothesis (Georgiev and Todorov (1976b)).

Fast reversible photoprocesses (presumably F_A reorientation and F_A-F_A' conversions) in F_A-containing KCl:Na have been investigated by Nikolova, Todorov, Popov, and Terzijski (1979) between 130-360 K. The maximum F_A' yield (about 33 %) at 800 nm due to exposure to 500 nm light was obtained at about 200 K, the yield exhibiting a drop as the temperature was moved in either direction. The F_A reorientation kinetics on illumination with a linearly-polarized He-Ne laser light (632.8 nm) followed a bimolecular (second-order) equation of the form

$$F_A^{-1}(t) = kt + F_{A0}^{-1} \qquad \text{along } [100] \tag{17.5}$$

where k has been derived based on a simple model:

$$k = \frac{2}{3}(\gamma_{\alpha_A}/\gamma_{F_A})(\eta_{i_A}q_{F_A})^2/\beta_{F'_A F_A} \qquad (17.6)$$

The temperature dependence of kF_A was explained assuming $\beta_{F'_A} = q_{F'_A}$ which agreed qualitatively with the predicted temperature behavior of η_{i_A} and $\gamma_{\alpha_A}/\gamma_{F_A}$ for the F center (Georgiev and Todorov (1976b).

17.1.2. Intermediate perturbation

Apart from the existence of F'_A centers, which are in a sense F' centers strongly perturbed by an A-type impurity in a nearest-neighbor position, there is no solid evidence for the occurrence of F' states intermediate between F'_A and the pure isolated F' center which states should be weakly perturbed by the impurity in a more distant position. Among the specific F' properties expected to be influenced by such a perturbation are the F'-band shape and the ionization rates $q_{F'}$ and $\tau_{F'}$. Because of the detailed balance, changes in the latter could well result from changes of the trapping cross-section σ_F, or of the thermal ionization energy $E_{F'}$ as well. While no evidence seems to have been found for any variation of $q_{F'}$ with the impurity concentration in a freshly-quenched additively colored sample, there are some indications that $\tau_{F'}$ may be longer in heavily-doped KCl:Na relative to that in a high-purity gamma-irradiated crystal (Nierzewski, Todorov, and Georgiev (1978)). The difference, if any, may result from a proportionally smaller σ_F in the doped sample, as if an A-type impurity in proximity would lower through screening the trapping capability of an F center. It has also been suggested that the crossover barrier may gradually decrease as the distance between the F center and the Na^+ ion shortens (Georgiev and Todorov (1975)): this would lead to an increased σ_α and to a related drop of F'_{max}. However, no detailed measurements have been made so far to substantiate this view. Still longer F' lifetimes have been reported for a gamma-irradiated Na-containing KCl sample (Nierzewski, Todorov, and Georgiev (1978)). Another quantity which is likely to be impurity-dependent is the F* lifetime τ_F. The above suggestions would imply a gradual transition from F' to F'_A affecting nearly all the intrinsic properties of the F' center.

In this respect, some early observations by Ohkura, Awane, and Miyamoto (1960) are worth mentioning. They found a substantial decrease of both the initial F' formation rate (measured as $\Delta \dot{F}/F_0$) and the photo-equilibrium F' density (measured as $\Delta F_S/F_0$) in additively colored KCl at -100 °C, when F_A and/or M, R, and Z_1 centers were present. F_0 was between 1.7-2.7 x 10^{16} cm^{-3} . An explanation was proposed by assuming an increased number of electron traps. Apparently, the F_A-F_A' conversion has not been known at that time.

Later, Härtel and Lüty (1964a) have investigated the F-F' conversion in KCl:Na to find that it is not primarily affected by the impurity. They came to the conclusion that Na^+ did not effectively trap electrons between -60 °C and RT. The F' thermal decay was found to closely correlate with the F_A growth following photoexcitation at -50 °C. The initial F_A- formation rate was found proportional to the F' (α) concentration which led to suggesting reaction (17.a,b) for the F-F_A conversion. The F_A' absorption appeared modifying the primed band in an F_A- containing crystal. More recently, the correlation between the F' decay and the F_A growth has been confirmed in a RT study using flash-light spectroscopy (Dechev, Georgiev, Koralov, Gochev, Todorov, and Georgiev (1971)).

Following unsuccessful first attempts, Bosi and co-workers have finally managed to initiate an F-F' conversion type in additively colored samples of NaF:Li subjected to repeated quenching from 560 °C (Bosi, Gagliardelli, and Nimis (1981,1982)). It was found to bear nearly all the basic features of a genuine F-F' interconversion. However, the precise shape of the F' band could not be established because of the presence of a large colloid band peaked at about 500 nm. Nevertheless, nearly 13 % of the F centers have been estimated to convert into "F'" at 228 K. The same authors have also reported accurate data on the F* center lifetime $\tau_F(T)$ and the radiative quantum efficiency $\eta_{rad}(T)$ for additively colored NaF:Li crystals (Bosi, Gagliardelli, and Nimis (1983)). These are presented as a function of the temperature starting from 7.6 K. An anomalous behavior of the F* decay, observed earlier, is now confirmed: It suggests the availability of some additional decay channel of the F* electron, other than jumping into the conduction band. Various possibilities are considered, including a polaron ionization of the F* center, as proposed recently for KCl (Georgiev, Gochev, Christov, and Kyuldjiev (1982))(see Section 19.1.1.).

17.2. F' centers in crystals containing divalent cationic impurities

17.2.1. Intermediate perturbation

The F-F' conversion in a KCl crystal containing divalent Ca^{++} cations has been investigated by Hartel and Luty (1964b). Prior to that, nearly half of the F centers were converted to Z_1 centers (the Z_1 center is the association of an F center with an impurity-cation vacancy dipole) by F band bleaching at -30 °C. By subsequent F-band bleaching at -90 °C, F' centers were formed, the F' band that occurred being by all means identical with the one in pure crystals. It has been concluded that the F-F' conversion is not affected by the presence of divalent ions and Z_1- centers. The latter centers, unlike F_A, are not electron traps.

However, Härtel and Lüty have not followed, or at least reported, just how the F-F' conversion proceeded but rather stressed the identity of the F' band produced and its relationship to the F band bleached. This has been attempted later by Nierzewski, Todorov, and Georgiev (1978) by studying the reverse F'-F process at RT in a Sr^{++}- containing gamma-irradiated KCl sample. The F' lifetime was unusually long in this sample presumably due to a drop in σ_F again, probably as a result of an impurity-induced screening of the F-center potential. However, the σ_F/σ_α ratio was not proportionally smaller in KCl:Sr^{++} than it was in pure KCl.

In a TL study aimed at establishing the role of copper in the luminescence of x-irradiated NaCl, Inabe and Takeuchi (1978) have concluded that the luminescent center responsible for the main TL peak at 327 K is the substitutional Cu^{2+} ion. The following radiative process was assumed:

$$e + Cu^{2+} \longrightarrow Cu^{1+*} \longrightarrow Cu^{1+} + h\nu \quad .$$

The electrons were presumed to originate from thermally decomposing F' centers, formed as F centers trapped electrons released from Cu^0 during the irradiation. The experimental heating rate was $\beta = 0.33$ K/s . The F' decay kinetics was found to be first-order by studying the isothermal annealing of the luminescence associated with the 327 K TL peak. Therefore, eq.(11.5) was applied to derive the F' parameters. Heavy Cu-doping (240 ppm) has reduced the F' thermal energy by 10 % and the frequency factor by 97 %.

Recently, Vassilev, Karamikhailova, Mladenova, and Georgiev (1980) have reported failing to produce F' centers by a powerful pulsed-laser excitation at 530 nm in a gamma-irradiated KCl:Eu^{2+} crystal at RT. Similar results have been reported by Murty and Murthy (1974) where copper doping in NaCl has been found to largely suppress a TL glow peak, attributed to F' in pure crystals, following x-irradiation at 80 K. Two new peaks appeared instead which were ascribed to copper atoms. Presumably, in both cases the impurity incorporation introduces competing electron traps (Eu^{2+}, Cu^+) which overpower the F centers to form Eu^{1+} and Cu^0, respectively.

However, Nedashkovskii, Avdonin, Dugarova, Plachenov, and Savel'ev (1974) have found a remarkable glow peak in the PSTL spectrum of a KCl: Eu^{2+} crystal at about 200 K which they attribute to F'. A corresponding knee appeared in the temperature dependence of the photostimulated afterglow-emission intensity. The crystals have been β-irradiated at RT, and bleached with F light at LNT before warming them up for the PSTL run, or bleached at increasingly lower temperatures in the afterglow-yield measurements.

It can be expected that any intermediate perturbations will be felt more efficiently in divalent-ion containing crystals as a result of the electric field of the impurity-vacancy complex. As a matter of fact, Kojima, Ebata, Tamura, and Kojima (1976) have reported resolving the optical properties of a center, intermediate between F and Z_1, perturbed by the electric field of the IV dipole in KCl doped with divalent cations, such as Sr^{++}, Ca^{++}, Ba^{++}, Eu^{++}, and Cd^{++}. The perturbed F center was found to have a radiative lifetime of 230 ns, much shorter than the isolated F center. It would be very interesting to know the extent to which the perturbation would affect the corresponding F' center. Unfortunately, no specific measurements of this kind seem to have been made.

Bhan and Rao (1982) have observed an orange-red afterglow at RT following an irradiation with gamma rays of Mn^{2+} doped samples of NaF. The afterglow, lasting to within one hour after the irradiation, was accompanied by a reduction in absorbance on the long-wavelength side of the F band. Inasmuch as a TL glow peak has been found at 60 °C and attributed to F' in undoped samples, the afterglow in NaF:Mn^{2+} was assigned to F' too. If justified, this would imply a somewhat reduced thermal stability of F' in doped samples.

Recently, the occurrence of a broad F' band has been reported for NaCl crystals, gamma rayed (10^7 to 10^8 R of Co^{60}) to 5×10^{17} F/cc, mainly on illumination with F band light (Nishimaki and Shimanuki (1982). The crystals were deliberately doped with divalent impurities (Ca^{2+}, Sr^{2+}, and Ba^{2+}) or with a monovalent ion (Li^{+}) in a search for a nir luminescence attributable to perturbed M and N centers, as observed previously in KCl. Both types of ion were found to stimulate the perturbed emission. No data on any possible perturbing effect on the F' centers are reported.

17.2.2. Strong perturbation

Presumably, as a result of the presence of a repelling cation vacancy in the Z_1- center configuration, the F-center component cannot trap another electron to form an F'-type center nearest-neighboring an impurity-vacancy dipole. However, this can eventually materialize once the cation vacancy has been removed from the complex Z_1 center. The resulting entity would be an F center in the vicinity of a divalent cation which can now turn into an impurity-perturbed F' center on trapping a conduction-band electron. This is exactly what Ohkura has suggested as a model for the Z_2 center (see Radhakrishna and Chowdari (1972)). However, in view of the still speculative nature of this model, the Z_2 centers will not be considered here any more.

17.3. Quenching of the F-F' conversion in crystals doped with anionic impurities

17.3.1. OH^- doping

Preliminary experiments by Gomes and Lüty (1983) are reported to have disclosed a drastic effect of high (10^{-3} to 10^{-2}) concentrations of OH^- molecular impurities on the electronic processes originating from the relaxed excited state of the F center in KCl, KBr, etc. As a matter of fact, the intrinsic F luminescence, the F-F' conversion, and the F aggregation all change to virtually disappear following a short light-induced migration of (additively colored) F centers. At the same time, the F band itself transforms into a (slightly) red-shifted and broadened band of a higher optical stability. Presumably, the relaxed excited state of F centers that have come closer to OH^- defects undergoes a significant perturbation. More systematic investigations reported later (Gomes and Lüty (1984)) have revealed a gradual degradation on OH^- doping of all the above processes, including photoconductivity. In quenched crystals with statistical defect distributions, the observed effect can be fitted to a model that involves total degradation below a critical distance of about 5 lattice spacings between the F center and the OH^- defect. Light-induced migration leads to the complete association of all F centers with OH^- defects resulting in the formation of $F_H(OH^-)$ centers, characterized by the new optical absorption band.

Measurements of the time-dependent ground-state repopulation following a pulsed-laser excitation, have shown OH^--perturbed F centers to undergo nonradiative deexcitation at rates shorter than 10 ns. Curiously, the relaxed excited state M center does not experience any luminescence quenching on OH^- doping. This may be due to the compact M center relaxed-excited-state wave function, as opposed to the diffuse character of the F relaxed excited state. The extended character of the latter is by all means favorable for an interaction with a nearby OH^- center to produce a perturbed excited state. That perturbed state will stretch towards the OH^- defect (or vice versa) depending on whether the positive (or negative) end of the impurity dipole points to the F center. Assuming that the perturbed F center undergoes a low nonradiative rate in the former case and a high rate in the latter, the effective nonradiative deexcitation process at the $F_H(OH^-)$ center will depend on the rate at which the OH^- dipole rotates. The latter being as high as 10^{10} s^{-1} at temperatures near 10 K, the authors argue it would lead to a high nonradiative rate at the F center in KCl whose intrinsic radiative deexcitation rate is only 10^6 s^{-1}. Clearly, experiments on materials with lower reorientational rates (RbBr) could prove crucial for checking Lüty's idea.

Lüty's line of reasoning as regards the F-OH^- interaction distance is based on arguments similar to the ones leading to eq.(6.7): The fraction of F centers that will have an OH^- neighbor at a distance between R and R + dR is

$$\frac{dn_F(R)}{F} = \pi\frac{4R^2}{R_o^3}\,\frac{n_{OH^-}}{N}\left(1 - \frac{n_{OH^-}}{N}\right)^{\left(\frac{4}{3}\pi R^3/R_o^3 - 2\right)} dR \qquad (17.7)$$

for a statistical defect distribution. Further, given an OH^- neighbor at R , the luminescent quantum yield of the F center counterpart in a pair will be

$$\eta(R) = \tau_{rad}^{-1}/(\tau_{rad}^{-1} + \tau_{nrad}^{-1}(R)) \qquad (17.8)$$

Here $\tau_{nrad}^{-1}(R)$ is the nonradiative rate at F due to the F-OH^- interaction; the radiative rate is τ_{rad}^{-1} , while $N = R_o^{-3}$ is the density of lattice-site pairs of the alkali halide. The overall quantum efficiency due to all pairs will now be

$$\eta_{rad} = \int \eta(R)\, dn_F(R)/F \qquad (17.9)$$

Assuming $\eta(R) = 1$ for $R > r_c$ and $\eta(R) = 0$ for $R < r_c$ leads to eq.(6.7), with F therein now replaced by n_{OH^-} . Gomes and Luty demonstrate that equation to work surprisingly well for the F-OH^- system. A more refined approach would invoke physically meaningful assumptions for $\tau_{nrad}^{-1}(R)$. The authors check the applicability of two of these:

$$\tau_{nrad}^{-1}(R) = AR^{-6} \quad \text{(induced-dipole interaction)} \qquad (17.10)$$

$$\tau_{nrad}^{-1}(R) = \tau_o^{-1}\exp(-AR) \quad \text{(electron-exchange interaction)}$$

The latter is shown to work much better, on a proper choice of the incorporated parameters.

Another remarkable result of Luty's work is the observation that the luminescence degradation is virtually the same in OH^- and OD^- systems. This seems to rule out energy transfer from F* to OH^-, such as to excite vibrational modes (e.g. librational or stretching) of the latter, which transfer is expected to be strongly isotope-dependent. In any event, the authors did not find any measurable vibrational fluorescence excited by F band pumping which could be expected to come out of the hydroxil group by virtue of energy transfer.

18. Practical applications of the F-F' conversions

18.1. Calibration of light sources

The quantum yield $\eta_{F-F'}$ having the ultimate value of 2 at sufficiently high temperatures (about -100 °C for KCl), the F-F' conversion in alkali halides has been used for absolute intensity calibration of incandescent light sources (see for example Tomiki (1960)). However, the accuracy of such calibrations depends essentially on the F center density: these must be performed at low F_0 because of the concentration quenching of $\eta_{F-F'}$.

Apparently, similar possibilities are provided by the quantum yield $\eta_{F'-F}$ of the reverse F'-F reaction. However, this would require the knowledge of η_i , γ_F/γ_α^* , and F/F' to check the experimental value of $\eta_{F'-F}$. From measurements by Fedders, Hunger, and Luty (1961) we get, for instance, $\gamma_F/\gamma_\alpha^* = 1/6$, F/F' = 1/2 , $\eta_i = 0.05$ for $F_0 = 4 \times 10^{16}$ cm^{-3} at -180 °C (KCl), wherefrom eq.(4.15) yields $\eta_{F'-F} = 1.84$. The experimental value is $\eta_{F'-F} = 1.9$ for $F_0 = 10^{17}$ cm^{-3}. Care must also be taken to use a correct value for γ_F/γ_α^* , since γ_α^* is known to fall as the F' center density is increased at low temperature (Crandall (1965)).

For other examples of using $\eta_{F-F'}$ for intensity calibration, see Schmid and Zimmerman (1968).

18.2. Optical information storage

The use of the F-F' conversions for an optical information storage has been proposed in the early 70's (Kiss (1970), Busse and Weiss (1971).

However, Tubbs and Wright (1971) have been the first to examine the practical performance of the F-F' system. They did it in two ways: the photochromic and the electro-photochromic operational modes. These will be described in some detail now.

In the photochromic mode, a freshly-quenched additively-colored KCl crystal is kept at 135 K, where the $\eta_{F-F'}$ and $\eta_{F'-F}$ temperature curves intersect, and an F-F' conversion is effected. One can write with F' (red) light, erase with F (green) light, and read by monitoring the optical absorption in the F band. The operability of this system has been checked up to 500 write-erase cycles. The readout was non-destructive, since the photoequilibrium F' density proved dependent on the excitation-light intensity, in agreement with earlier observations (Costicas and Grossweiner (1962)). Consequently, information written onto a crystal with a high-intensity light beam is not destroyed even on prolonged reading with a low-intensity interrogating beam, because the final photoequilibrium F' density is lower. However, the signal-to-noise ratio worsens. The disadvantages of using the photochromic mode are: (i) the necessity of maintaining a constant temperature for the crystal (because the equilibrium F' density is temperature-dependent); (ii) the finite F' lifetime (which can be an unfavorable factor for prolonged storage times); (iii) possible defferences in the initial bleaching rates of F and F' leading to a drift of the F band level until the amount of bleaching in the F and F' bands becomes the same at the combination of F (readout) and F' (write) intensities used. (iii) can be overcome by carefully balancing the F- and F'- light intensities.

Most of the difficulties inherent to the photochromic mode have been removed by introducing the 'electro-photochromic mode'. Information is now written by F light in the presence of a strong electric field. The crystal temperature is kept low (about 90 K). F light is supplied by the 5145 A argon laser line for KCl. Under these conditions, the F band is bleached in 'bit 1' regions because of field-assisted ionization at low temperature (Luty (1958)). Erasure is effected by F' light (the 6328 A He-Ne laser line for KCl) in the absence of a field. In this way, the written page is a matrix of locations of normal (0) and reduced (1) F band optical density. Readout is practically non-destructive if carried out by the same write-beam of F light in the absence of a field. The operability of the electro-photochromic F-F' system has been checked up to 10^5 write-erase cycles and no fatigue found.

It is to be noted that the F-F' system exhibits nearly the ultimate recording characteristics of an iueal photochromic material, because of the: (i) high oscillator strength of the F band (providing a good contrast), (ii) high conversion quantum yields (sensitivity), and (iii) intrinsicly short switching times of the F-F' conversions.

In addition to the above-described application to a bit-by-bit recording, the F-F' system can also be used to store volume holograms in a "thick crystal", as reported by Tubbs (1973). He also argues that "color center materials based on the reversible F-F' conversion in alkali halides at low temperature are the only practical material with sensitivity close to the ideal photochromic". The following experimentally obtained recording characteristics of the F-F' system are listed:

Maximum resolution: better than 1500 l.p./mm

Maximum diffraction efficiency η : better than 3 %

Exposure to $\eta = 0.2$ % : 4 mJ/cm^2

Exposure to erase: 10 mJ/cm^2 .

In a subsequent paper, Tubbs and Scrivener (1974) argue that the photochromic F-F' system exhibits two significant disadvantages: a destructive readout (because both F-F' and F'-F have about the same efficiencies at about 140 K) and a limited contrast (because only as much as 40 percent of the F centers can be converted to F' at this temperature). In the electro-photochromic process, as many as 80 % of F can be converted to F' and α with $\eta_{F-F'} \sim 1$ at 80 K. Here too, some problems remain with a destructive readout due to the overlap of the F and F' bands. Another problem is maintaining a low temperature during storage although the high recording sensitivity warrants this. The sen-

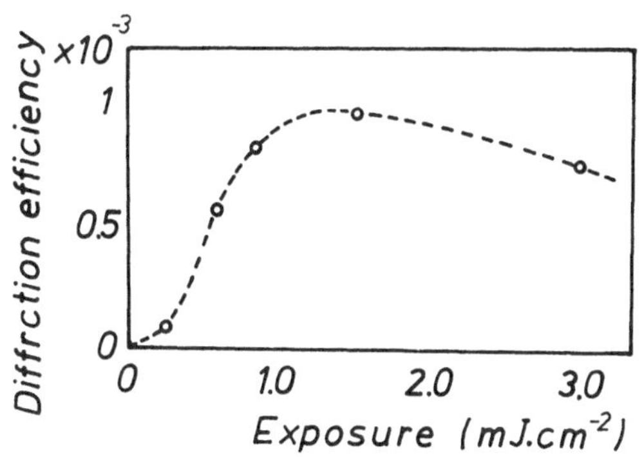

Figure 49: Variation of the holographic efficiency in a photochromic KCl crystal with the total incident light energy. (After Tubbs and Scrivener (1974)).

sitivity obtained is close to the maximum theoretical prediction: optical-density changes of 0.1 on exposure of 0.3 to 1.0 mJ/cm^2 have been produced in KCl samples as thick as 100-500 m. Applications are advised in image recording, holographic recording, data storage, etc. In the holographic recording the advantage lies in the high recording sensitivity (low write-energy). For instance, maximum diffraction efficiencies of 0.1 % have been obtained in the photochromic mode on exposure to only 1.5 mJ/cm^2, as shown in Fig.49. Although the authors complain of problems with maintaining the crystal temperature, they report having obtained diffraction efficiencies as high as 0.18 %.

As the temperature is raised to fall into the F' thermal instability range (above -80 °C for KCl), the F-F' conversion can no longer be used for permanent storage. However, as suggested by Georgiev and Todorov (1976), the F-F' system can now be employed in making rapidly transient (self-erasing) recordings with a variety of potential applications to super-operative optical memories, time-transformers, etc. However, a fatigue problem arises limiting the recyclability of the F-F' system because of F center clustering near RT which lowers the F' yield and thereby the recording contrast and the diffraction efficiency. Impurity-perturbed F-F' conversions, such as F_A-F'_A in KCl:Na, exhibited an improved recyclability due to the suppression of clustering by the F-F_A conversion. Unfortunately, insufficient experimental material collected for just one crystal host - impurity ion system (KCl:Na) made it difficult to substantiate the utilitary value of these observations. It would be fair to mention that the use of F_A-F'_A conversion in optical recording has first been suggested by Kiss (1970).

Although no holographic recording has been reported using the F-F' conversion near RT, the system provides potential possibilities for applications in dynamical holography. However, other authors have expressed generally pessimistic views on the prospects of employing the F-F' system in technology (Shvarts, Gotlib, and Kristapson (1976)).

An experimental investigation of the F-F' switch-over times defined by the F-bleaching times in which a given contrast F'/F is achieved, has been made by Popov and Terzijski (1974) for both F-F' and F_A-F'_A. In both cases, the maximum recording speeds for F'/F = 1 % were found to be between -110 and -70 °C at a bleaching-light intensity of 10^{15} cm^{-2}s^{-1} . It can be argued that the rise of the switch-over time on both sides of the above range is due to the drop of η_i below -100 °C and of $\tau_{F'}$ above -70 °C, respectively. Indeed, very high F' yields have been obtained at RT within microseconds when using a flash-light excitation and fast readout techniques (Georgiev and Todorov (1976a), as opposed to F-F' times of the order of 10 s achieved by conventional light sources (Popov and Terzijski (1974)). Popov also reports data on the reciprocity-law failure for both F-F' and F_A-F'_A (Popov (1979)). The non-observance of reciprocity seems convincing in both cases, since it follows even within the frameworks of the conventional conversion model.

19. Theory

19.1. F-F' conversion quantum efficiency

A radiationless F-F' conversion has first been considered by Rampacher (1965) based on Stumpf's approach. However, no comparison with the experiment was possible, since the model precluded the radiative lifetime. Later, Wille and Wahl (1966) have carried out a quantum-mechanical calculation of the temperature dependence of the F-F' quantum yield $\eta_{F-F'}$, essentially the F photoionization efficiency η_i. The authors argue that the lowest-lying conduction-band states in the vicinity of an anion vacancy must have an s-type character to account for the low oscillator strength of the direct optical-ionization transition of the F center electron. At the same time, the observed temperature dependence of $\eta_{F-F'}$ suggest F* thermal ionization through some direct non-radiative process. However, a quantum-mechanical analysis using the non-radiative operators has shown that no such process is allowed between F* and the conduction band. This follows from the symmetry of the states involved in accordance with group theory: along with the radial symmetry of the operators, it leads to vanishing matrix elements of the thermal transition. At the same time, a radiation-induced transition from the 2p-type F* state to the s-type CB states is permissible. The authors postulate that the ir radiation needed arises from the black radiation of the crystal between 80-220 K resulting from the transversal optical (TO) phonons, strongly coupled to the thermal radiation field. Physically, the black radiation is absorbed leading to F* ionization which is temperature-dependent, since due to thermal broadening the F* absorption (ionization) band overlaps increasingly with the TO emission as the temperature is raised. The electron coupling to the radiation field is assumed small.

Stumpf's quantum theory of ionic crystals has been employed to calculate τ_i^{-1} and τ_R^{-1} (Stumpf (1961)):

$$\tau_R^{-1} \propto \frac{\omega_{max}^2 e^2 2^{12}}{3\hbar(c/n)^3} \frac{\alpha^3 \beta^5}{(\alpha + \beta)^{10}} \tag{19.1}$$

$$\tau_i^{-1} = \frac{4}{3}\int \rho(y)\bar{b}(y)H_{AL}^2 S_{AL}(\mu)\frac{y\omega_0^2}{\hbar\omega_1}\,dy \qquad (19.2)$$

where α and β are the variational parameters of the F and F* wave functions:

$$\psi_F = \frac{\alpha^{3/2}}{\pi^{1/2}}\exp(-\alpha r)$$
$$\psi_{F*} = \frac{\beta^{5/2}}{\pi^{1/2}}\,r\cos\vartheta\,\exp(-\beta r) \qquad (19.3)$$

which are

$$\alpha = \frac{m_g e^2}{\hbar^2}\left(\frac{1}{\varepsilon} + \frac{5}{16}\left(\frac{1}{n^2} - \frac{1}{\varepsilon}\right)\right)$$
$$\beta = \frac{m_e e^2}{\hbar^2}\left(\frac{1}{2} + \frac{1}{5}\left(\frac{1}{n^2} - \frac{1}{\varepsilon}\right)\right) \qquad (19.4)$$

m_g and m_e are the corresponding effective masses, different for F and F*, $y = \omega/\omega_0$, ω_0 is the limiting TO frequency

$$\rho(y) = \frac{\omega_0^2}{c^3\pi^2}\,y^2\left(\varepsilon_\infty + \frac{\varepsilon_0 - \varepsilon_\infty}{(1 - y^2)^2}\right)\left(\varepsilon_\infty + \frac{\varepsilon_0 - \varepsilon_\infty}{1 - y^2}\right)^{\frac{1}{2}} \qquad (19.5)$$

is the radiation density. Note that due to the strong TO phonon - ir photon coupling (19.5) differs from the density of free radiation

$$\rho(\omega) = \omega^2/\pi^2(c/n)^3$$

Also

$$\bar{b}(y) = \frac{\bar{\omega}(\bar{\omega}^2 - \omega_0^2)}{\omega(\bar{\omega}^2 + \omega^2)}\left(\exp(\hbar\bar{\omega}/k_B T) - 1\right)^{-1} \qquad (19.6)$$

$\bar{\omega}$ is an effective phonon frequency

$$\bar{\omega} = \omega_0\left(2 - y + \frac{2\varepsilon_\infty}{\varepsilon_0 - \varepsilon_\infty}(1 - y^2)^2\right)^{\frac{1}{2}} \qquad (19.7)$$

$$H_{AL}^2 = 2^{10}\pi e^2/V\beta^5$$
$$S_{AL} = \exp\left(-\frac{1-\lambda}{1+\lambda}X_{AL} + \frac{1}{2}\lambda\mu\right)I_\mu(z_{AL}) \qquad (19.8)$$

where $z_{AL} = 2X_{AL}\lambda^{\frac{1}{2}}/(1 - \lambda)$, $\lambda = \hbar\omega_1/k_B T$,

$$I_\mu(z) = \sum_{m=0}^{\infty}(z/2)^{2m+\mu}\frac{1}{m!\,\Gamma(m+\mu+1)} \qquad (19.9)$$

is the modified Bessel function, while

$$X_{AL} = (e^2\beta/5\hbar\omega_1)\left(\frac{1}{n^2} - \frac{1}{\varepsilon}\right) \qquad (19.10)$$

ω_1 is the limiting frequency of the LO phonon, μ is a parameter. The F-F' quantum yield is calculated from (4.14), (4.7), and (4.3) for

$\tau_r = \tau_R$, using eq's (19.1) and (19.2). Free parameters of the model are m_g and m_e , which are adapted so that the calculated yield should coincide with the experimental value at 130 K (Fedders, Hunger, and Lüty (1961)): this resulted in $m_g = 1.62\ m$ and $m_e = 0.238\ m$. The calculated $\eta_{F-F'}$ vs. temperature dependence reminds the experimental curve, as shown in Fig.50. An apparent F* "thermal-ionization energy" of 0.13 eV has been deduced from the theoretical curve. However, the agreement between theory and experiment falls short at both

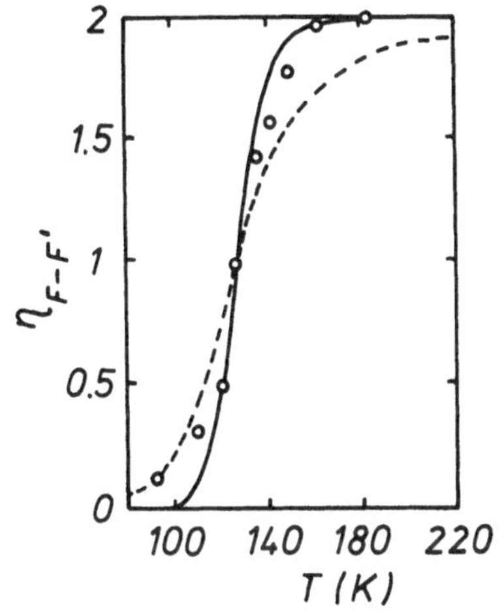

Figure 50: Calculated temperature dependence of the F-F' quantum yield after Wille and Wahl (1966) (---) and Stocker (1968) (——). The experimental points (circles) are from Fedders, Hunger, and Lüty (1961). (After Stocker (1968)).

the higher and the lower temperatures.

Another model suggested by Wille and Wahl (1966) has been investigated by Stocker (1968). While Wille & Wahl's conclusion on the vanishing matrix element of the thermal transition to s-type CB states is assumed decisive, it is now proposed that the F* ionizes through a nonradiative transition from a 2p-like relaxed excited state to a 2p-like polaron band. The F center electron is presumably coupled strongly to the LO lattice phonons. The F center Hamiltonian used reads

$$\mathcal{H} = \frac{\underline{p}_e^2}{2m^*} - \frac{e^2}{\varepsilon|\underline{r}_e - \underline{r}_o|} + \hbar\omega \sum_{\underline{k};i=1,2} \left(q_{i\underline{k}}^2 - \frac{\partial^2}{\partial q_{i\underline{k}}^2}\right) +$$

$$4\pi e(\hbar\omega e/8\pi\bar{\varepsilon}V)^{\frac{1}{2}} \sum_{\underline{k}\neq o} \frac{1}{k}(q_{1\underline{k}}\cos(\underline{k}\cdot\underline{r}_e) + q_{2\underline{k}}\sin(\underline{k}\cdot\underline{r}_e)) \qquad (19.11)$$

The field of the anion vacancy is assumed purely Coulombic. The third term in (19.11) is the free Hamiltonian of the normal lattice vibrations, while the fourth is the electron-phonon interaction term. p-type wave functions of the form (19.3) are used for both the F* center and the polaron, with variational parameters and radial coordinates β_1, $r = |\underline{r}_e - \underline{r}_o|$ and β_2, $r = |\underline{r}_e - \underline{R}|$, respectively, where $\underline{r}_e$, $\underline{r}_o$, and $\underline{R}$ are the position vectors of the electron, the anion vacancy, and the polaron in that order. β_1 is found by minimizing the F center functional

$$E(\underline{P}) = \int \psi^*_{F*}\mathcal{H}_e\psi_{F*}d\underline{r}_e + 2\pi\bar{\varepsilon}\int \underline{P}^2 d\underline{r} + \int d\underline{r}\underline{P}\int d\underline{r}_e\psi^*_{F*}(\nabla\frac{e}{|\underline{r}-\underline{r}_e|})\psi_{F*} \qquad (19.12)$$

where $\mathcal{H}_e$ is given by the first two terms in eq.(19.11). Subsequent variations in $\underline{P}$ and β_1 lead to

$$\underline{P}_{min} = \frac{1}{4\pi\bar{\varepsilon}}\int\psi^*_{F*}(-\mathrm{grad}\frac{e}{|\underline{r}-\underline{r}_e|})\psi_{F*}d\underline{r}_e$$

$$\beta_{1min} = \frac{m^*e^2}{\hbar^2}(\frac{1}{2\varepsilon}+\frac{1}{5\bar{\varepsilon}}) \qquad (19.13)$$

Inserting back into (19.12) one obtains

$$E_F = -\frac{\hbar^2}{2m^*}\beta^2_{1min} \qquad (19.14)$$

The polaron functional (in the strong coupling limit) is minimized by

$$\underline{P}_{min} = -\frac{e}{4\pi\bar{\varepsilon}}\int|\psi_P(\underline{r}_e-\underline{R})|^2\,\mathrm{grad}\frac{1}{|\underline{r}-\underline{r}_o|}\,d\underline{r}_e$$

$$\beta_{2min} = 0.196\, e^2m^*/\hbar\bar{\varepsilon} \qquad (19.15)$$

which leads to

$$E_P = -0.09\, m^*/m \quad (\mathrm{eV}) \qquad (19.16)$$

These results indicate that the 2p-polaron is energetically possible. However, it would relax into the 1s-polaron ground state, a short while after the ionization transition.

The following operator of the nonradiative transition is postulated:

$$\mathcal{H}'(q) = \hbar\omega\sum\int\psi_n(\frac{\partial\psi_m}{\partial q_{i\underline{k}}}\frac{\partial}{\partial q_{i\underline{k}}} + \frac{1}{2}\frac{\partial^2\psi_m}{\partial q^2_{i\underline{k}}})\,d\underline{r}_e \qquad (19.17)$$

$\mathcal{H}'$ is regarded as a small perturbation. The kinetic problem is further dealt with by means of Pauli's master equation. The following result is obtained

$$\tau_i^{-1} = \frac{2\pi}{\hbar} \sum_{l^P, l^F} |\langle P,l^P|\mathcal{H}'|F,l^F\rangle|^2 \, P_{l^F} \, \rho(\Delta E,T) \tag{19.18}$$

where the density function is given by

$$\rho(\Delta E,T) = \sum_{l^{\underline{K}},\underline{K}} \delta(\Delta E/\hbar\omega) \, |\langle \underline{K}|l^{\underline{K}}\rangle|^2 \, P_{l^{\underline{K}}} \tag{19.19}$$

in which

$$\hbar\omega\Delta E = E_F - E_P + \hbar\omega(l_1^F + l_2^F - l_1^P - l_2^P) + \frac{3}{2}\hbar\omega \tag{19.20}$$

Here l^F and l^P are the F- and P- coupled oscillators, respectively, $l^{\underline{K}}$ are the free oscillators. P_l are the occupation probabilities of the vibrational states. In thermal equilibrium,

$$P_l = (1 - \exp(-\hbar\omega/k_BT))\exp(-l\hbar\omega/k_BT) \tag{19.21}$$

$\underline{K}$ is the free-polaron momentum

$$|K_s\rangle = V^{-\frac{1}{2}} \exp(iK_sR_s)$$

$$|l_s\underline{K}\rangle = \pi^{-1/4} \frac{H_{l_s}(\zeta_s/\zeta_o)}{(2^{l_s})^{\frac{1}{2}} \, l_s! \, \zeta_o} \exp(-\zeta_s^2/2\zeta_o^2) \tag{19.22}$$

in normal coordinates, $|P,l^P\rangle$ and $|F,l^F\rangle$ are the polaron and F center wave functions, respectively.

The F-F' quantum yield was again calculated assuming an isolated F center undergoing optical excitation, thermal ionization of the relaxed excited state, and radiative deexcitation transitions. The electron effective mass m^* (the only free parameter of the theory) was adapted so as to fit the experimental $\eta_{F-F'}$ value at 130 K (Fedders, Hunger, and Lüty (1961)): this gave $m^* = m$, the free electron mass. The computed $\eta_{F-F'}$ vs. temperature dependence shows an improved agreement with the experimental curve, except for the lower temperature range (Fig.50). The calculations predict an energy difference of 0.21 eV between the 2p-polaron band and the 2p- relaxed excited F state. However, both Wille & Wahl's and Stocker's results are very sensitive to the choice of m^* , the quantum yield dropping quickly as m^* is increased. Nevertheless, the polaron model seems to work generally better suggesting that a more refined calculation based on the nonradiative processes can possibly be developed (see Section 19.5 for a recent example).

19.1.1. F* thermal ionization rate

Perhaps the first attempt to treat the thermal ionization of a trapped electron on a rigorous quantum-mechanical basis has been made by Kubo (1952) using the adiabatic approximation. The multiphonon transition is attributed to the dependence of the lattice vibrations on the electronic state. The transition probability from an initial vibronic state $|\,l,n\,\rangle$ to a final state $|\,l',n'\,\rangle$ is given by

$$w(ln \to l'n') = \frac{2\pi}{\hbar}\,|\langle l,n|\mathcal{H}'|l',n'\rangle|^2\,\rho_f\,\delta(E_{l'n'}-E_{ln}) \tag{19.23}$$

where the density of states in the final state is approximated by

$$\rho_f = 4\pi V m^* hk/h^3 \tag{19.24}$$

if the final state is one of a free electron, while $\mathcal{H}'$ is the non-adiabatic operator. The total thermal ionization rate is further found by summing (19.23) over all the initial states weighed by their occupation factors, and over all final states which satisfy the energy-conservation condition. The trapped electron wave function was assumed to be of the eq.(19.3) 1s-type. A simplified model of weak-coupling adiabatic potential was adopted in an Einstein one-frequency approach with different frequencies for the bound and ionized electron states. The following result is obtained for the low-temperature range

$$\tau_i^{-1} = 256\,(m^*/M)\,\omega\left(\tfrac{1}{2}(1-\omega'/\omega)\right)^{E_0/\hbar\omega - 1}\exp(-E_0/k_BT) \tag{19.25}$$

where M is the mass of the donor atom and E_0 is the thermal-ionization energy (the energy difference between the minima on the ionized- and bound- state potential curves). Note that a vibrational-frequency difference between the two electronic states is essential for obtaining a nonvanishing ionization rate. However, eq.(19.25) leads numerically to too small rates when applied formally to F- center-like donors, suggesting that some approximations made may have been quite unappropriate. For the high-temperature range, Kubo's model predicts a thermal-ionization rate similar to that based on the activated states theory:

$$\tau_i^{-1} = 256\,(m^*/M)(k_BT/\hbar)\exp(-E^*/k_BT) \tag{19.26}$$

where

$$E^* = E_0\,\omega^2/(\omega^2 - \omega'^2) \tag{19.27}$$

In a subsequent paper by Kubo and Toyozawa (1955), Kubo's methods have been extended to cover more general cases of vibronic situations which include the possibility of tunneling of the lattice equilibrium coordinates. For the single-frequency weak-coupling case depicted in Fig.51 the authors obtain

$$\tau_i^{-1} = (2\pi)^{\frac{1}{2}}\,\gamma\cdot\gamma^{-\frac{1}{2}}\,(\hbar\omega^2/E_0)(k_BT/E_0)^{\frac{1}{2}}\exp(-E^*/k_BT) \tag{19.28}$$

at high temperatures, while at low temperatures

$$\tau_i^{-1} = (\pi/8)^{\frac{1}{2}} (\hbar\omega/E_o)^{\frac{1}{2}} (\gamma/2\exp(1 - \gamma/2))^{E_o/\hbar\omega} ((2/\gamma - 1)^2 \gamma''^2 + 4(\gamma\gamma' - \gamma''^2)(\hbar\omega/E_o)/\gamma^2) \exp(-E_o/k_B T) \qquad (19.29)$$

where γ is the interaction coupling constant, γ' and γ'' are also constants. E_o is the energy separation between the minima of the two potential curves, at q" and q', respectively,

$$E^* = E_o(1 + \gamma/2)^2/2\gamma, \qquad E_r = (1 + \gamma/2)E_o \qquad (19.30)$$

$E^* - E_o$ is the height of the crossover point over the minimum of the ionized-state curve (see Fig.51), E_r is the reorganization energy.

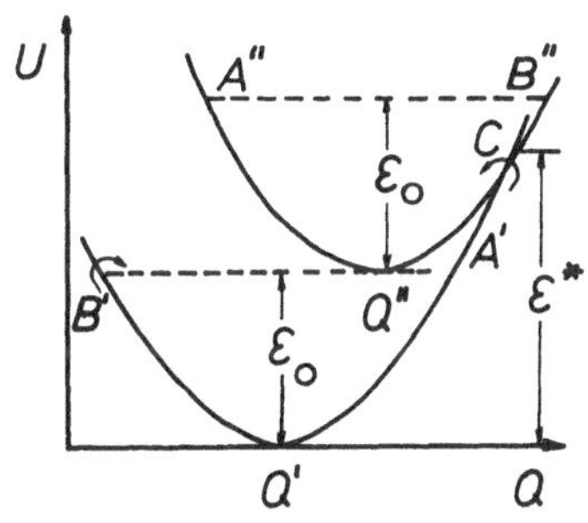

Figure 51: Configurational-coordinate diagram depicting the weak-coupling situation. (After Kubo and Toyozawa (1955)).

Kubo's earlier calculation (Kubo (1952)) has been made for $\gamma = 0$ (zero coupling), that is, q" = q'. Note that now the different-frequencies-condition is no more required to achieve nonvanishing ionization rates if $\gamma \neq 0$: this results because of the inclusion of the lattice-tunneling effect by allowing

$$q'' - q' = (E_o/\hbar\omega)^{\frac{1}{2}} \gamma^{\frac{1}{2}} \neq 0$$

Equations (19.28) and (19.29) predict a two-slope τ_i^{-1} vs. reciprocal temperature curve. Of particular interest is the low-temperature portion. Inasmuch as usually $E_o \gg \hbar\omega$, the third factor in eq.(19.29) is very sensitive to the coupling constant γ and attains its maximum value of 1 at $\gamma = 2$; now $E^* = E_o$ so that the crossing point of the two potential curves coincides with the equilibrium position of the upper curve. In this case, the tunneling distance of the lattice coordinate will be minimal, as the system undergoes the transition from the lower to the upper electronic state. $\gamma = 2$ is also a middle-of-the-road case which separates the weak and strong coupling limits. Now the reorganization energy is just twice the energy separation between the two minima: $E_r = 2E_0$.

Equation (19.29) predicts a thermal ionization rate that vanishes exponentially as the temperature approaches 0 K. This behavior is inherent to weak-coupling situations. However, Georgiev, Gochev, Christov, and Kyuldjiev (1982) have recently calculated the temperature dependence of the F* thermal ionization rate using literature experimental data on the ionization efficiency η_i and the lifetime τ_F in KCl, as well as eq.(4.7). The resulting temperature curve points to: (i) considerable deviations from the Arrhenius law below 100 K, and (ii) the existence of a nonvanishing residual ionization rate, about 40 s^{-1}, at

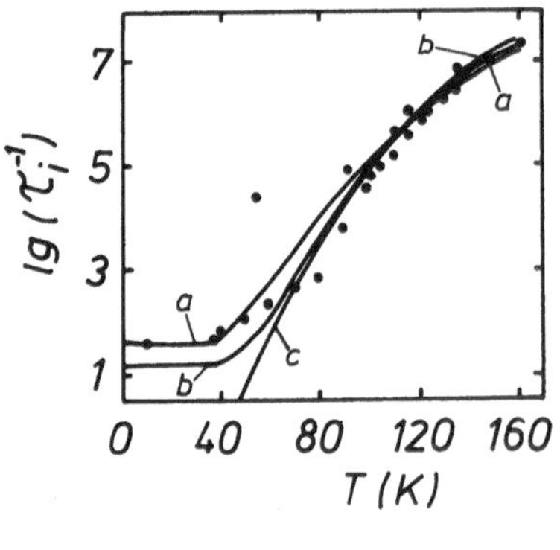

Figure 52: Temperature dependence of the F* thermal ionization rate in KCl. The experimental points from various sources are marked by black dots. The solid curves are calculated by means of a reaction-rate approach assuming three types of processes: (a) isothermic, (b) exothermic, and (c) endothermic. (After Georgiev, Gochev, Christov, and Kyuldjiev (1982)).

low temperature, as shown by the dots in Fig.52. Both observations being serious obstacles to a weak-coupling interpretation, accepted usually by F center theories (Stoneham (1975)), the authors constructed a configurational-coordinate diagram of the KCl F center using optical absorption and emission data. This diagram includes the parabolae of the ground F_0, first-excited F*, and low-lying ionized state F^+(Fig.53).

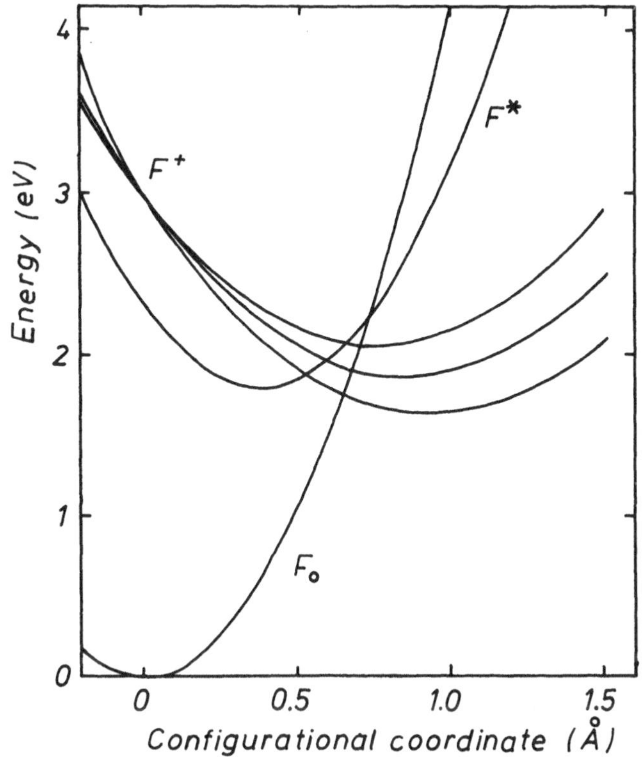

Figure 53: Configurational-coordinate diagram of the KCl F center calculated empirically from optical absorption, emission, and photoconductivity data. The three parabolae for the ionized state (F^+) are at three different values of the F* optical ionization energy: 0.3, 0.4, and 0.5 eV. The result suggests a strong-coupling situation between the F* and F curves. (After Georgiev, Gochev, Christov, and Kyuldjiev (1982)).

The result suggests strongly a strong-coupling situation between F* and F^+.

A reaction-rate approach (Christov (1982)) has been used to analyse the experimental rate vs. temperature dependence in Fig.52. For a single-frequency one-dimensional coordinate model in the harmonic approximation, the theory predicts

$$\tau_i^{-1} = 2\nu \sinh(h\nu/2k_BT) \sum_n W_1(E_n)W_e(E_n)\exp(-E_n/k_BT) \qquad (19.31)$$

applicable to the whole temperature range, comprising both a range of predominating lattice tunneling and one of overwhelming classical jumps. The strong-coupling situation is depicted in Fig.54 (idealized). In

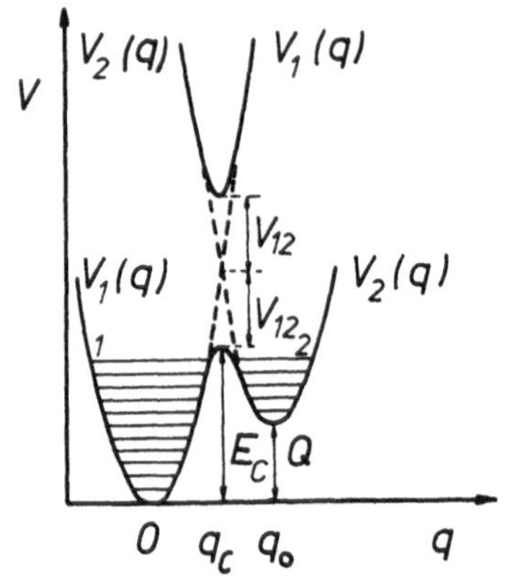

Figure 54: Configurational-coordinate diagram depicting the strong-coupling situation. V_{12} is the resonance splitting. (After Georgiev, Gochev, Christov, and Kyuldjiev (1982)).

eq.(19.31):

$$W_1(E_n) = \frac{\pi F^2(\xi_o,\xi_c)}{2^{n+m}n!m!} \exp(-(n-m)^2 h\nu/E_r - E_r/h\nu) \qquad (19.32)$$

is the probability for lattice tunneling, where

$$F(\xi_o,\xi_c) = \xi_o H_n(\xi_c)H_m(\xi_c-\xi_o) - 2nH_{n-1}(\xi_c)H_{m-1}(\xi_c-\xi_o) + 2mH_n(\xi_c)H_{m-1}(\xi_c-\xi_o)$$

$H(\ )$ are Hermite polynomials, $\xi = 2\pi(M\nu/h)^{\frac{1}{2}}q$ is a dimensionless- and q is a Cartesian- vibrational coordinate, M is the mass and ν is the frequency of the lattice oscillator, assumed to be the same in both electronic states; ξ_c and ξ_o are the coordinates of the crossover point and of the minimum of the final-state parabola, respectively, the minimum of the initial-state potential being at $\xi = 0$. $E_n = (n + 1/2)h\nu$ is the vibrational energy, n and m are the vibronic quantum numbers in the initial and final electronic state, respectively, $E_r = h\nu\xi_o^2/2$ is the reorganization energy of the system.

$$W_e(E_n) = \frac{2\pi}{\gamma_n\Gamma^2(\gamma_n)} \exp(-2\gamma_n(1 - \ln\gamma_n)) \qquad (19.33)$$

with

$$\gamma_n = (V_{12}^2/2h\nu)(E_r|E_c' - E_n|)^{-\frac{1}{2}} \qquad (19.34)$$

is the electron-transition probability, V_{12} is the resonance splitting which arises because of the coupling of the two electronic states. Finally, $E_c' = (E_r + Q)/4E_r$ is the height of the crossover point, where Q is the reaction heat at 0 K, $E_c = E_c' - V_{12}$ is the crossover barrier, as shown in Fig.54. The rate of the reverse reaction obtains as the forward rate is multiplied by $\exp(Q/k_BT)$:

$$k_{21}(T) = k_{12}(T)\exp(Q/k_BT) \qquad (19.35)$$

The theory thus involves four parameters: ν, E_r, V_{12}, and Q.

Three types of reaction were checked while attempting to fit the experimental data in Fig.52: endothermic ($Q = +h\nu$), isothermic ($Q = 0$), and exothermic ($Q = -h\nu$). The following values were obtained of the four parameters after machine processing of the data: $Q = 0$, $\nu = \nu_{LO}$ ($\omega_{LO} = 4.02 \times 10^{13}$ Hz (Fowler (1968))), $E_r = 0.73$ eV, $V_{12} = 0.019$ eV. These values can be considered quite reasonable from the viewpoint of the expectations 'a priori' (Georgiev, Gochev, Christov, and Kyuldjiev (1982)). The apparent F* thermal ionization energy has been estimated at 0.15 eV in excellent agreement with the experimental value (Fowler (1968)).

The adopted model presumes F* ionization through strong lattice tunneling from the relaxed excited state to a virtual polaron state (cf. Stocker (1968)). The conclusion that the polaron equilibrium energy is about equal or even slightly lower than the RES electronic energy leads to far-reaching consequences, since now the F* center would ionize and F' would form at very low temperature even if the F centered system was fairly dilute. However, it should be kept in mind that the electron transfer from F* to the polaron states will not be an autoionization process, being initiated by tunneling of the surrounding lattice through an energy barrier. The total ionization rate, as given by eq's (19.31) through (19.34), leads to residual low-temperature effects observed experimentally in the F-F' conversion (Luty (1961b)) and in the F-band photoconductivity (see Bosi, Podini, and Spinolo (1968)).

19.1.2. F* radiationless deexcitation rate

The F* nonradiative deexcitation rate τ_{nR}^{-1} inherent to an isolated F center has been discussed and compared with experiment by Honda and Tomura (1972) based on Kubo & Toyozawa's theory (Kubo and Toyozawa (1955)). With w_{1n} standing for the probability that the system initially is in the n-th vibrational state, the general expression used is

$$\tau_{nR}^{-1} = \frac{2\pi}{\hbar\Delta E} \sum_{n,n'}^{0<E_{1'n'}-E_{1n}<\Delta E} |\langle 1'n'|\mathcal{H}'|1n\rangle|^2 w_{1n} \qquad (19.36)$$

where again $\mathcal{H}'$ is the nonadiabatic operator. Following Kubo and Toyozawa, eq.(19.36) is approximated by

$$\tau_{nR}^{-1} = \frac{2\pi}{\hbar} f_1(T)f_2(T) \qquad (19.37)$$

Honda and Tomura argue that $f_1(T)$ = const , while

$$f_2(T) = \frac{h^3\nu\,|\underline{v}_{1'1}|^2}{8\pi^2(W_1 - W_{1'})^2} \coth(h\nu/2k_BT) \qquad (19.38)$$

at sufficiently low temperatures, e.g. below 60 K. Inserting into (19.37) one obtains

$$\tau_{nR}^{-1} = f_1 \frac{\nu h^2|\underline{v}_{1'1}|^2}{2(W_1 - W_{1'})^2} \coth(h\nu/2k_BT) \qquad (19.39)$$

Here $\underline{v}_{1'1}$ is the matrix element of the electronic transition, while W_1 are the expectation values of the electronic Hamiltonian $\mathcal{H}_e$. At low temperatures the contribution to the deexcitation rate, as described by eq.(19.39), comes from the lattice tunneling effect exclusively. The configurational-coordinate diagram is similar to the one depicted in Fig.51 with the upper curve now corresponding to F* and the lower to F. Note that quite different vibrational frequencies can be involved in the different transitions undergone by the F electron.

Honda and Tomura have also checked eq.(19.39) experimentally in a variety of alkali halides. From measurements of the radiative quantum yield η_R and the F* lifetime τ_F , $\tau_i^{-1} + \tau_{nR}^{-1} = (1 - \eta_R)/\tau_F$ was calculated and its temperature dependence compared with eq.(19.39) assuming a classical ionization rate

$$\tau_i^{-1} = \nu_i \exp(-E_i/k_BT) \qquad (19.40)$$

Experimental data are shown in Fig.55: it is remarkable that even at temperatures as low as 10 K the total nonradiative rate $(1 - \eta_R)/\tau_F$ is nonvanishing. The best fit to the theory is presented by the solid lines. From such fits, vibrational frequencies characteristic of the nonradiative deexcitation process of the order of $2 \times 10^{12}\ s^{-1}$ were found for NaCl, RbF, KF, and NaF. The obtained order of magnitude for the pre-exponential factor in (19.39) was $10^6\ s^{-1}$. (For RbF both estimates were generally lower.) However, it seems rather difficult to draw any quantitative conclusions on the applicability of the theory, due to the mathematical complexity of the pre-coth factor in (19.39).

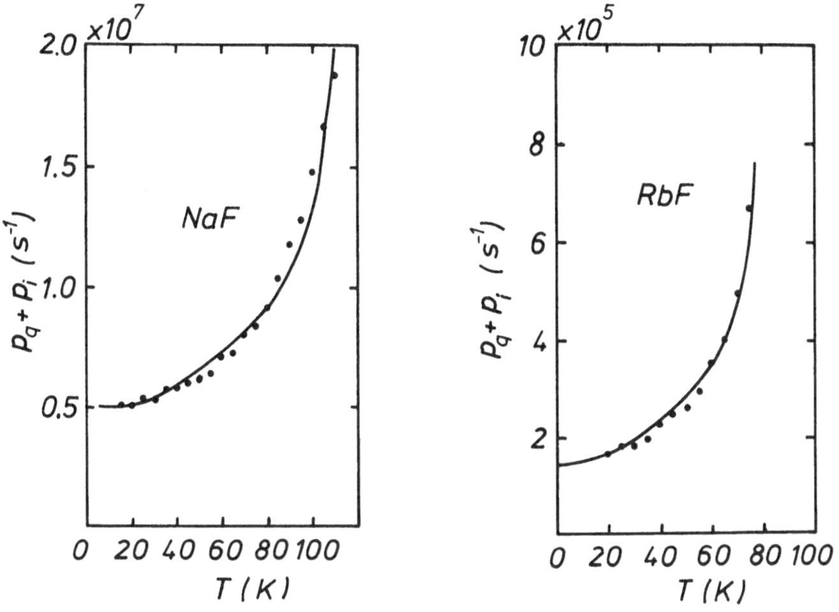

Figure 55: Temperature dependence of the total nonradiative rate at F* (thermal ionization + radiationless deexcitation). The circles are experimental points, while the solid lines have been calculated by means of Kubo & Toyozawa's theory assuming a classical thermal-ionization rate. (After Honda and Tomura (1972)).

19.1.3. F* radiative deexcitation rate

To complete Section 19.1. on the F-F' quantum yield, we shall briefly discuss the F* radiative rate, the last component of the lifetime equation (4.3) for an isolated F center. Following Honda and Tomura (1972), the adopted model of a relaxed excited state active in emission is based on three physical assumptions:

(i) The RES is composed of hydrogen-like 2s- and 2p- states, the 2s lying lower in energy than the 2p states.

(ii) 2s and 2p couple strongly through Γ_4^- optic phonons, giving rise to new mixed states, to be labelled 2s' and 2p'.

(iii) The radiative deexcitation transition starts from the 2s'-state.

The resulting radiative lifetime will be temperature-dependent, due to the temperature variation of the mixing ratio of 2s and 2p in 2s'.

The Hamiltonian of the RES electron is assumed to be

$$\mathcal{H} = \mathcal{H}_0 + \mathcal{H}' \qquad (19.41)$$

where $\mathcal{H}_0$ is the hydrogenic Hamiltonian, while the electron-phonon interaction potential is (see Henry and Slichter (1968)):

$$\mathcal{H}' = V_xQ_x + V_yQ_y + V_zQ_z \qquad (19.42)$$

in terms of the electronic operators V_j and the generalized nuclear coordinates Q_j, both transforming as Γ_4^-. The hydrogen-like states 2s and 2p are eigenstates of $\mathcal{H}_0$, while the mixed-up states are eigenstates of $\mathcal{H}$, and are assumed to be linear combinations of 2s and 2p:

$$\begin{aligned}
|2s'\rangle &= \beta\,|2s\rangle - \alpha\,|2p\rangle \\
|2p_1'\rangle &= \beta\,|2p\rangle + \alpha\,|2s\rangle \\
|2p_2'\rangle &= a_{2x}|2p_x\rangle + a_{2y}|2p_y\rangle + a_{2z}|2p_z\rangle \\
|2p_3'\rangle &= a_{3x}|2p_x\rangle + a_{3y}|2p_y\rangle + a_{3z}|2p_z\rangle \\
|2p\rangle &= a_{1x}|2p_x\rangle + a_{1y}|2p_y\rangle + a_{1z}|2p_z\rangle
\end{aligned} \qquad (19.43)$$

with

$$a_{ix}^2 + a_{iy}^2 + a_{iz}^2 = 1 \quad (i = 1,2,3); \quad \alpha^2 + \beta^2 = 1 .$$

The radiative transition rate is given by

$$\tau_R^{-1} = \tfrac{4}{3}e^2nE_m^3\hbar c^{-3} \sum \alpha_j^2\, |\langle 1s|\,j\,|2p_j\rangle|^2 \quad (j = x,y,z) \qquad (19.44)$$

where E_m is the emission band peak, while n is the refractive index. Introducing $\alpha_j^2 = (\alpha a_{1j})^2$, we obtain from symmetry considerations $\alpha_j^2 = \frac{1}{3}\alpha^2$. The following expression is obtained:

$$\alpha^2 = \frac{\gamma^2\coth(h\nu/2k_BT)}{2(1 + \gamma^2\coth(h\nu/2k_BT) + (1 + \gamma^2\coth(h\nu/k_BT))^{\frac{1}{2}})} \qquad (19.45)$$

where

$$\gamma^2 = h(2\pi^2\delta^2M\nu)^{-1}\,|\langle 2s|V_j|2p_j\rangle|^2 \qquad (19.46)$$

in which M and ν are the effective mass and frequency associated with the promoting optic mode. Inserting into (19.44), we obtain

$$\tau_R^{-1} = \tfrac{4}{3}e^2n\hbar^{-4}c^{-3}E_m^3 \sum |\langle 1s|\,j\,|2p_j\rangle|^2\alpha^2 \qquad (19.47)$$

The radiative transition rate (19.47) is indeed dependent on the temperature. Eq.(19.47) has been compared with experimental data on η_R/τ_F in several alkali halide hosts. Some results are depicted in Fig.56.

Radiative lifetimes of the usual F center emission starting from

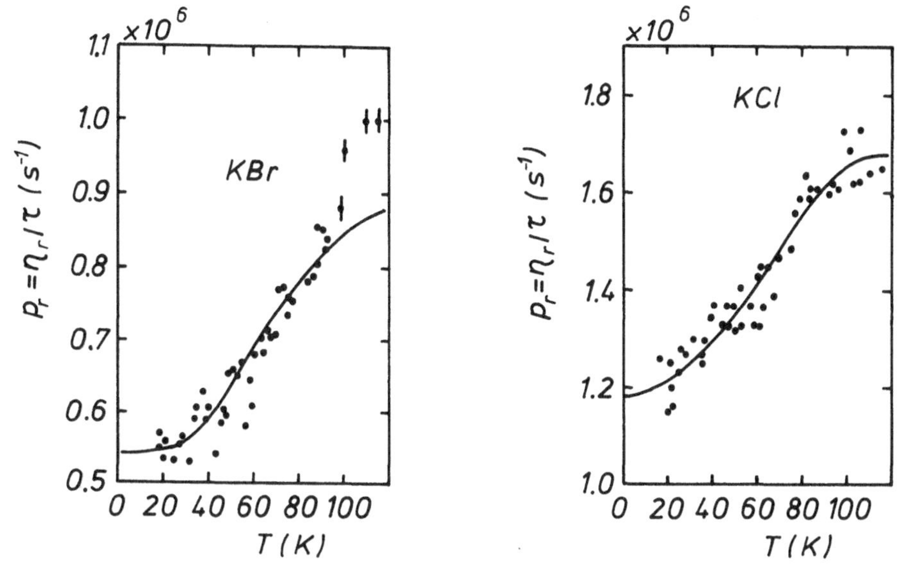

Figure 56: Temperature dependence of the F* radiative deexcitation rate. The circles are experimental points, while the solid lines have been calculated by means of eq.(19.47). (After Honda and Tomura (1972)).

the 2s' state are rather long, typically the order of 10^{-6} s (see Fowler (1968)). At the same time, emission lifetimes starting from the 2p'

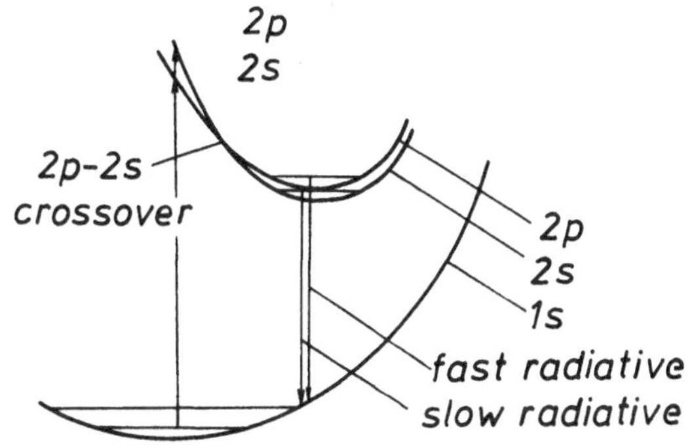

Figure 57: Mixing of 2s- and 2p- like excited states of the F center (schematic). These states are reached through 2- or 1-photon absorptions, respectively. (After Georgiev (1985a)).

state should be considerably shorter, due to the allowed optical transition. It is interesting to note that such lifetimes, the order of 10^{-8} s, have actually been observed, following a two-photon optical pumping of the F center (De Martini, Giuliani, and Mataloni (1975)), as sketched in Fig.57. One-photon optical absorption of the F center ends up in the 2s' state.

19.2. Electron trapping rate by a halide vacancy

The nonradiative electron-capture rate by a halogen-ion vacancy in alkali halides has been calculated quantum-mechanically by Fraser (1974) using Stumpf's methods. An important problem which arises in the treatment of nonradiative transitions between discrete electronic levels is energy conservation. If energy is to be strictly conserved, the discrete electronic states must couple to some form of continuous spectrum. This requirement is easily met for radiative transitions where the radiation field itself provides such a spectrum. Otherwise, appropriate quasi-continuous spectra are provided by the acoustic phonons. However, from optical absorption, emission, and thermal-ionization measurements, the F center electron is known to couple with phonon modes that lie closer to the optical rather than the acoustical branches. To overcome this difficulty, Stumpf (1961) has introduced "dressed phonons" which secure the indirect coupling of the F center electron to the acoustic phonons of the crystal. These dressed phonons couple the local mode to the acoustic phonons by the anharmonic phonon-phonon interaction. The use of dressed instead of purely harmonic phonons satisfies the energy conservation exactly. Stumpf's method is also applicable to transitions from a band of levels to a discrete level; now the coupling of an additional continuous spectrum greatly increases the range of electronic energies from which an energy-conserving transition is possible.

Introducing the nonadiabatic supplement $\mathcal{H}'$ in the expression for the total Hamiltonian of the defect-electron system $\mathcal{H} = \mathcal{H}_{ad} + \mathcal{H}'$, the differential nonradiative transition probability is defined by

$$W_{l'n',ln} = \frac{2\pi}{\hbar} \left| \langle l',n' | \mathcal{H}' | l,n \rangle \right|^2 \delta(E_{ln} - E_{l'n'}) \qquad (19.48)$$

Here $|l,n\rangle$ are the adiabatic eigenstates of $\mathcal{H}_{ad}$. The total transition rate is obtained by summing over all final phonon states and weighting thermally the initial states

$$W_{1',1}(k_BT) = \prod_i (1 - \exp(-h\nu_i/k_BT)) \times \sum_{n',n} W_{1'n',1n} \exp(-(\sum n_j h\nu_j/k_BT)) \qquad (19.49)$$

The author has calculated the transition rate from the CB to the ground state of the F center. Acoordingly, s-type wave functions (19.3) were used to construct the final state, while the CB electron wave function

$$\psi_{CB}(r) = V^{-\frac{1}{2}} \exp(i\underline{k}\cdot\underline{r}) \qquad (19.50)$$

was taken simply as $V^{-\frac{1}{2}}$, since only transitions from the bottom of the band at $k = 0$ were assumed to matter. The dressed-phonon wave functions were introduced in the perfect lattice approximation. Without entering into the details, we will note that the calculated rates for a microblock volume of $V = 10^5\ d^3$ increase as T is raised from 50 to 300 K. Undoubtedly, these rates correspond to the eq.(4.30) case and describe the transition into a bound state. The particular numerical values obtained seem to be of little significance, since they depend markedly on the resonant conditions of the calculation. However, the displayed temperature trend is noteworthy as it lends a qualitative support to the general arguments in Section 4.1. concerning the low-temperature behavior of the electron-trapping coefficients (eq.(4.29)).

The electron-capture cross-section of an anion vacancy in NaCl has also been calculated for room temperature by Ruiz-Mejia (1970). He defines the cross-section by means of the thermal-ionization rate of a ground-state F electron using the detailed-balance principle (see Section 4.3.). To calculate that rate, a simple point-ion vibronic model-potential is considered, while the electron-transfer matrix element is computed using Simpson's hydrogen-like wave functions of the (1s) ground state and the (2p) first excited state of the F center, assuming that the system will automatically autoionize, once lifted to its excited state. The F center electron couples to a continuum of lattice phonons. Nevertheless, despite a number of oversimplifying presumptions, the calculation seems to yield the correct order of magnitude for the cross section.

19.3. The F' band

It has very early been appreciated that from a theoretical point of view the F' center should be regarded as an analogue of the H^- ion, embedded in a crystalline medium (Seitz (1946)). H^- is the predominant source of opacity in most stellar photospheres. The F' band resembles the continuum H^- absorption, arising from electronic transitions between a bound level and a continuum of ionization states (see Fowler (1968)). Accordingly, calculations on H^- are expected to be helpful in clarifying the theoretical problem. The calculated H^- spectrum has proved sensitive to the accuracy of both the initial- and final- state wave functions. There is no bound excited singlet state but the existence of a bound triplet state along with the ground singlet state is not ruled out. At the same time, however, there are important differences arising from the presence of a strongly polarizable environment. For instance, the Hartree-Fock approximation, which excludes the electron-electron correlation, leaves no room for an H^- bound state because of the strong e-e repulsion (Bethe and Salpeter (1957)). Nevertheless, the F' center has a more extended wave function leading to a decrease of the e-e repulsion. As a result, a bound F' state can occur, even without the aid of the e-e correlation (Pekar (1951)). Further, the presence of an extra electron (the effective negative charge) on the F' center should produce a static polarization which will generate an attractive force thus strengthening the binding (Fowler (1968)). Consequently, the possible existence of a bound singlet excited F' state in absorption cannot again be ruled out as easily as in the H^- case. However, if populated by light absorption, this state might autoionize after lattice relaxation.

The theoretical works on the F' center can be classified into several categories depending on the model used to describe the environment (Stoneham (1975)). Some authors use a continuum polarizable medium and take the central electrostatic potential of the anion vacancy either in the form of a purely Coulomb potential (continuum approach) or preserve the Coulomb type only partially assuming a spherical-well central potential inside the vacancy (semicontinuum approach). Other workers apply basically the point-ion potential with or without taking into account lattice-polarization effects. All works assume a 1s-type F' singlet ground state and many use a symmetric-coordinate wave function

$$\Psi(\underline{r}_1,\underline{r}_2) = \Psi(\underline{r}_1)\Psi(\underline{r}_2) \qquad (19.51)$$

This form, however, does not account for the e-e correlation, which has to be included in the calculations by other means. Instead, others prefer to use trial wave functions containing the e-e correlation explicitly. In nearly all the aspects except for the e-e interaction, the calculations resemble those of the F center. Among the quantities that are being calculated are: the threshold F' optical and thermal ionization energies, the optical absorption cross-section of the F' band, and the F' thermal ionization rate.

19.3.1. Continuum approach

The earliest theoretical work on the F' center has been done by Pekar and co-workers (Pekar (1951)) within the frameworks of Pekar's polaron approach. The F' center is considered to be the result of adding an extra electron to an F center embedded in a polarizable continuum. Accordingly, the F' Hamiltonian is taken in the form

$$\mathcal{H} = -\frac{h^2}{2m^*}(\Delta_1 + \Delta_2) + V(\underline{r}_1) + V(\underline{r}_2) - \frac{Ze^2}{\varepsilon_o}(\frac{1}{r_1} + \frac{1}{r_2}) + \frac{e^2}{\varepsilon_\infty|\underline{r}_1 - \underline{r}_2|} + \mathcal{H}_{LO} \tag{19.52}$$

where

$$V(\underline{r}) = e\int\frac{(\underline{r} - \underline{r}')\cdot\underline{P}(\underline{r}')}{|\underline{r} - \underline{r}'|^3}\,d\underline{r}' \tag{19.53}$$

is the potential energy of an F' electron at $\underline{r}$ in the polarization field $\underline{P}(\underline{r})$. $\mathcal{H}_{LO}$ is the Hamiltonian of the LO phonons, coupled to the electronic subsystem:

$$\mathcal{H}_{LO} = (2\pi/c)\int(\underline{P}^2(\underline{r}) + \frac{1}{\omega^2}\,\dot{\underline{P}}^2(\underline{r}))d\underline{r} \tag{19.54}$$

where

$$c = \frac{1}{\varepsilon_\infty} - \frac{1}{\varepsilon_o} \tag{19.55}$$

$\underline{P}(\underline{r})$ does not include the static polarization induced by the vacancy. The fourth term in (19.52) is the Coulomb energy of the two F' electrons in the vacancy field, the fifth is the e-e repulsion energy. The Coulomb approximation to the vacancy field will only be justified if the effective radius of the electron cloud exceeds largely the ionic radii. In this sense, Pekar's macroscopic theory would have a rather limited success when applied to small-size color centers, such as F', which must be considered as quantum-mechanical systems. Nevertheless, a great effort has been spent to examine the applicability of this theory to atomic-size centers.

The problem is further dealt with within the frameworks of the Born-Oppenheimer approximation: The F' electrons are assumed to follow adiabatically the ionic motion, the ions responding to the average electronic distribution only. Consequently, for any instantaneous ionic configuration there will be an electronic steady state $\Psi(\underline{r}_1,\underline{r}_2)$. The adiabatic wave function Ψ , as well as the induced self-consistent polarization $\underline{P}(\underline{r})$, are found independently by minimizing in Ψ and in $\underline{P}$ the functional

$$F\,[\Psi,\underline{P}\,] = \int \Psi^* \mathcal{H} \Psi \, d\underline{r}_1 d\underline{r}_2 \tag{19.56}$$

under the normalization condition

$$\int \Psi^* \Psi \, d\underline{r}_1 d\underline{r}_2 = 1 \tag{19.57}$$

The minimization has been carried out at 0 K, assuming $\dot{\underline{P}} = 0$ in the classical limit. Inserting (19.52) into (19.56) and introducing

$$\underline{D}\,[\Psi;\underline{r}) = -e \int |\Psi(\underline{r})|^2 \, (\underline{r} - \underline{r}')|\underline{r} - \underline{r}'|^{-3} \, d\underline{r}' \tag{19.58}$$

the $V(\underline{r})$ containing terms in (19.56) can be represented as $\int \underline{P}.\underline{D} \; d\underline{r}$. Minimizing now with respect to $\underline{P}$, we obtain

$$\underline{P}(\underline{r}) = (c/2\pi)\; D\,[\Psi;\underline{r}) \tag{19.59}$$

From (19.59) and (19.58), the self-consistent polarization $\underline{P}$ is created by the average electronic distribution, as required by the adiabatic approximation, due to an electron cloud of density $2\,|\Psi(\underline{r})|^2$.

According to the variational principle, (19.56) gives the total ionic potential energy of the crystal in the presence of F' centers, provided Ψ is an extremal adiabatic wave function. This energy is minimal when the ions are at equilibrium for $\underline{P}$ given by eq.(19.59). A self-consistent relationship is now established between the electronic and ionic counterparts in the crystal. The last term in (19.56) is the polarization energy of the crystal $(2\pi/c)\int \underline{P}^2 \, d\underline{r}$.

Inserting into eq.(19.56) $\underline{P}$ from (19.59) and assuming (19.51), we obtain after simple manipulations (Pekar and Tomasevich (1951)):

$$F\,[\Psi] = (\hbar^2/m^*)\int |\nabla\Psi|^2 \, d\underline{r} - (2Ze^2/\varepsilon_0)\int |\Psi(\underline{r})|^2 r^{-1} d\underline{r} - (c'/4\pi)\int D^2\,[\Psi;\underline{r})\, d\underline{r} \tag{19.60}$$

where

$$c' = 2c - 1/\varepsilon_\infty^2 \tag{19.61}$$

and $\Psi(\underline{r})$ is any of the single-electron wave functions entering in (19.51). The F' functional (19.60) differs from the F center functional in Pekar's theory only in that c is now replaced by c' and that (19.60) is twice as large. Accordingly, the same type of trial wave function is now used for the F' ground state:

$$\Psi_{1s,F'}(\underline{r}) = (\alpha^3/7\pi)^{\frac{1}{2}}\,(1 + \alpha r)\exp(-\alpha r) \tag{19.62}$$

Inserting into (19.60) and minimizing in α gives

$$\alpha = \frac{m^* e^2}{2\hbar^2}\left(c' + \frac{3Z}{\varepsilon_o}\right) \tag{19.63}$$

Inasmuch as the ionic kinetic energy in the self-consistent (equilibrium) state vanishes, the total energy of the system in ground state is given by the ionic potential energy. The latter is obtained by inserting (19.63) into (19.60). The result is

$$W_{F'} = -\frac{m^* e^4}{\hbar^2}\left(\frac{Z}{\varepsilon_o} + \frac{c'}{3}\right)^2 \tag{19.64}$$

Now, assuming that the F' center decomposes thermally into an F center and a free polaron, the F' thermal ionization energy is given by $E_{F'} = W_F + W_P - W_{F'}$. This yields (Pekar (1951)):

$$E_{F'} = \frac{m^* e^4}{9\hbar^2 \varepsilon_o^2}\left(\frac{\varepsilon_o}{\varepsilon_\infty} - \frac{3}{2}\right) - \frac{3}{2}\hbar\omega \tag{19.65}$$

The F' bound state is, therefore, energetically possible ($E_{F'} > 0$) for $\varepsilon_o/\varepsilon_\infty > 1.5$, which holds good for all the alkali halides (see Knox and Teegarden (1968)). Inserting the numerical values of the universal constants in (19.65), we obtain

$$E_{F'} = 3.046(m^*/m)\varepsilon_o^{-2}(\varepsilon_o/\varepsilon_\infty - 1.5) - 1.5\hbar\omega_{LO}$$

For KCl, $\varepsilon_o = 4.84$, $\varepsilon_\infty = 2.19$, we get $E_{F'} = 0.05$ if $m^*/m = 1$; however, it compares unfavorably with the experimental data.

By choosing a multiplicative form (19.51) for the trial wave function, Pekar and Tomasevich (1951) have completely ignored the mutual correlation in the motion of the two F' electrons. A more accurate trial wave function should also depend on $r_{12} = |\underline{r}_1 - \underline{r}_2|$. Tomasevich (1951) has done this by choosing

$$\Psi(\underline{r}_1,\underline{r}_2) = A(1 + \alpha r_1)(1 + \alpha r_2)(1 + \beta r_{12})\exp(-\alpha(r_1+r_2)) \tag{19.66}$$

Inserting into (19.56) and carrying out the minimization results in a negligible lowering, the order of 1 %, of $E_{F'}$. Generally, the inclusion of the e-e correlation is expected to lower the energy of the system. Now, the novel equation for $E_{F'}$ differs from (19.65) in that the numerical factor 3.046 is replaced by 3.0.

Generally, Pekar's theory disagrees with the experiment as far as the F' thermal ionization energy is concerned. The situation improves slightly if one abandons the polaron concept, setting formally $W_P = 0$ which would add some 0.1 eV to $E_{F'}$. This implies that the final state in the dissociation process should involve a CB electron rather than a polaron (Cheban (1963)). The unfavorable experimental checkup of Pekar's formulae does not solely confine to the F' center. As a matter of fact, this applies to the F center thermal ionization energy as well, $E_F = W_P - W_F$. Here the F ground state energy W_F is given by half the value from (19.64), in which c' has also been replaced by c, while the polaron energy W_P is obtained from W_F at $Z = 0$ minus $\frac{3}{2}\hbar\omega$ (Pekar (1951)). For KCl, $\omega_{LO} = 4.02 \times 10^{12}\ s^{-1}$ (Fowler (1968)) and the numerical calculation yields: $W_F = -1.1436$ eV, $W_P = -0.1322$ eV, $W_{F'} = -1.3297$ eV for $m^*/m = 1$. One obtains $E_F = 1.0114$ eV, while the experimental value is 2.05 eV (Markham (1966)). The discrepancy becomes even deeper if one takes $m^*/m = 0.6$ (see Lynch and Robinson (1968)).

In a further application of the polaron approach, Tomasevich (1951) has calculated the peak energy of the F' band using essentially the Hamiltonian (19.52) short of its last term $\mathcal{H}_{LO}$. $\underline{P}(\underline{r})$ in (19.53) is now the polarization corresponding to the F' ground state which does not change during the electronic transition, in accordance with the Franck-Condon principle. The final state in absorption is assumed to be an F' excited state, in which one F' electron is 1s, while the other is 2p. The variational calculation has yielded the following F' peak energies: 1.1 (NaCl), 0.8 (KCl), and 0.65 eV (KBr). All these estimates are less than half the corresponding experimental values. A self-consistent F' excited state has also been sought. In this state, the polarization is given by (19.59). However, the latter state has proved to lie higher in energy than the F' ionized state (ground F free polaron).

An attempt to calculate the F' band shape has been made by Ostroukhov and Tomasevich (1958) based on the Hamiltonian (19.52) again, in which the last term has been represented in its canonical form

$$\mathcal{H}_{LO} = \frac{1}{2} \sum_s \hbar\omega_s \left(q_s^2 - \frac{\partial^2}{\partial q_s^2}\right) \tag{19.67}$$

Here the canonical coordinates are related to $\underline{P}$ by way of

$$q_s = (4\pi/\hbar\,\omega_s c_s)^{\frac{1}{2}}\, P_s \tag{19.68}$$

where P_s are the Fourier expansion coefficients of $\underline{P}$:

$$\underline{P}(\underline{r}) = (2/L)^{\frac{1}{2}} \left(\sum_{s\leqslant o} \underline{P}_s \cos(sr) + \sum_{s>o} \underline{P}_s \sin(sr)\right) \tag{19.69}$$

The F' ground state wave function

$$|\,i\,\rangle = \Psi_{1s,F'}(r_1)\,\Psi_{1s,F'}(r_2)\,\Phi_{F'}(q) \tag{19.70}$$

is taken to describe the initial state in absorption, while the final state is assumed to be composed of a ground state F center and a CB electron:

$$|f> = (2\pi)^{-\frac{3}{2}}(\Psi_{1s,F}(r_1)e^{i\underline{k}\cdot\underline{r}_2} + \Psi_{1s,F}(r_2)e^{i\underline{k}\cdot\underline{r}_1})\Phi_F(q) \qquad (19.71)$$

Now, the calculated peak ($h\nu_{max}$) and threshold ($h\nu_{min}$) optical energies compare much more favorably with the experimental data, as taken from Mott and Gurney (1948).

Table 9

Comparison of calculated and experimental F' band energies*

	$h\nu_{max}$(th) (10 K)	$h\nu_{max}$(th) (140 K)	$h\nu_{max}$(ex)	$h\nu_{min}$(th) (10 K)	$h\nu_{min}$(th) (140 K)	$h\nu_{min}$(ex)
KCl	1.89	1.78	1.7	1.04	0.93	1.1
NaCl	3.11	2.88	2.4	1.50	1.27	
KBr	1.64	1.59	1.4	0.83	0.78	

*After Ostroukhov and Tomasevich (1958)

The F' band halfwidths have also been computed; however, the result is not so encouraging.

Further steps in examining the applicability of the polaron concept to the F center have been taken by Cheban. In a preliminary investigation of the F' photoionization rate (Cheban (1960)), the F' ground state wave function is taken in the form

$$|i> = \Psi_{1s,F'}(r_1)\Psi_{1s,F'}(r_2)\varphi_s(\xi)\prod_u{}' \Phi_{n_u}(q'_u - r_u) \qquad (19.72)$$

where $r_u = q_{u_{F'}} - q_{u_0}$, the q' are the normal lattice coordinates for a free lattice, a lattice with F', and a lattice with a polaron, respectively; Φ_{n_u} are the wave functions of the lattice harmonic oscillators, while φ_s is an electron-bound-to-polaron contribution, ξ is a polaron variable. The F' ground state energy is

$$E_{1s} = I_{F'} + \hbar\omega\sum_u(n_u + \tfrac{1}{2}) + \tfrac{3}{2}\hbar\omega , \qquad (19.73)$$

the second term giving the LO-phonon energy. In a subsequent paper (Cheban (1963)), the following approximations are used:

$$\Psi_{1s,F'}(r) = ((\alpha_0\gamma_1)^3/7\pi)^{\frac{1}{2}}(1 + \alpha_0\gamma_1 r)\exp(-\alpha_0\gamma_1 r) \qquad (19.74)$$

where

$$\gamma_1 = (c'/c)(1 + 3/\varepsilon_o c'), \quad \alpha_o = m^* e^2 c/2\hbar^2 \quad .$$

$$\varphi_s(\xi) = ((\alpha_o\lambda)^3/\pi)^{\frac{1}{2}} \exp(-\alpha_o\lambda\xi) \tag{19.75}$$

The final state in absorption is taken to be

$$|f> = \Psi_o(|\underline{r}_1-\underline{\xi}|)\Psi_F(r_2)\varphi_k(\xi)\prod_u{}' \Phi_{n_u'}(q_u'-q_{u_F}) \tag{19.76}$$

Ψ_o and Ψ_F are the wave functions of the electron bound to the polaron and the F center, respectively:

$$\Psi_{1s,F}(r) = ((\alpha_o\gamma)^3/7\pi)^{\frac{1}{2}} (1 + \alpha_o\gamma r)\exp(-\alpha_o\gamma r) \tag{19.77}$$

$$\Psi_o(r) = (\alpha_o^3/7\pi)^{\frac{1}{2}} (1 + \alpha_o r)\exp(-\alpha_o r) \tag{19.78}$$

where $\gamma_2 = 1 + 3/\varepsilon_o c$. $\varphi_k(\xi)$ is a solution to the Schrodinger equation

$$(-\frac{\hbar^2}{2M}\Delta_\xi + K(\xi))\varphi_k(\xi) = E_k\varphi_k(\xi) \tag{19.79}$$

where

$$K(\xi) = \frac{e^2}{\varepsilon_o}\left(\int\frac{\Psi_o^2(|\underline{r}_1-\underline{\xi}|)\Psi_F^2(\underline{r}_2)}{r_{12}}\, d\underline{r}_1 d\underline{r}_2 - \int\frac{\Psi_o^2(|\underline{r}-\underline{\xi}|)}{r}\, d\underline{r}\right) \tag{19.80}$$

is the effective field acting on the polaron, M is the polaron effective mass. φ_k is chosen p-type to secure an allowed optical transition to the ionized F' state:

$$\varphi_k = (\frac{3}{4\pi})^{\frac{1}{2}} \frac{\chi_k(\xi)}{\xi}\cos\vartheta \tag{19.81}$$

The final-state energy is

$$E_{f,k} = I_o + I_F + E_k + \hbar\omega \sum_u{}' (n_u' + \frac{1}{2}) \tag{19.82}$$

Here and above $I_{F'}$ is the minimized adiabatic F' potential, I_o and I_F are the ground state energies of the polaron and the F center, respectively.

The matrix element of the optical transition is

$$< f|U|i > = M_{12}\prod_u S_{n_u'n_u} \tag{19.83}$$

$$M_{12} = \int\varphi_k(\xi)\, z_{12}(\xi)\,\varphi_s(\xi)\, d\underline{\xi}$$

$$z_{12} = \int\Psi_o(|\underline{r}_1-\underline{\xi}|)\Psi_F(r_2)(z_1 + z_2)\Psi_{1s,F'}(r_1)\,\Psi_{1s,F'}(r_1)\, d\underline{r}_1 d\underline{r}_2$$

$$S_{n_u'n_u} = \int \Phi_{n_u'}(q_u'-q_{u_F})\Phi_{n_u}(q_u'-r_u)\, dq_u'$$

The dipole - field interaction operator is taken to be

$$U = -eE_z(z_1 + z_2) \tag{19.84}$$

where E_z is the z-component of the electric-field strength. Further, the differential multiphonon photoionization rate is

$$P_{sk} = (2\pi/\hbar\omega)e^2E^2|M_{12}(k)|^2 \exp(-\frac{a_{F'}}{2}\coth\frac{\beta}{2} - \frac{l\beta}{2})I_1(\frac{a_{F'}}{2\sinh\frac{\beta}{2}}) \sum_n \delta(l-n)$$

$$a_{F'} = \sum_u (r_u - q_{u_F})^2 \quad , \quad \beta = \hbar\omega/k_BT \tag{19.85}$$

$$l = \frac{\hbar k^2}{2M\omega} + \frac{E_{F'} - h\nu}{\hbar\omega}$$

Here $h\nu$ is the absorbed photon energy,

$$E_{F'} = I_o + I_F - I_{F'} - \frac{3}{2}\hbar\omega$$

is the F' thermal ionization energy, I( ) is Bessel's function. The total optical rate is

$$q_{F'} = \frac{L}{\pi}\int_0^\infty P_{sk}\, dk \tag{19.86}$$

The calculated spectral distribution of $q_{F'}$ resembles the F' band but is shifted to enormously long wavelengths. For instance, $q_{F'}$ exhibits a maximum at about 0.5 eV for NaCl which is highly unrealistic.

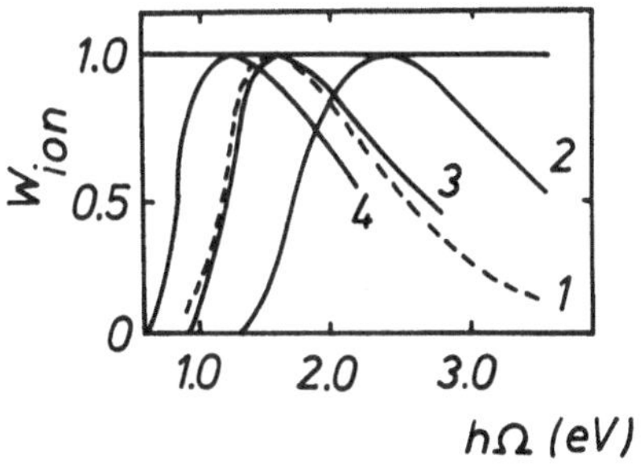

Figure 58: Calculated F' bandshape (normalized) for several alkali halides using the continuum approach: (2) NaCl, (3) KCl, and (4) KBr. Curve (1) (dashed) is experimental for KCl, as reported by Pick (1958). F' has been assumed to decompose into an F center and a conduction-band electron following the photon absorption. (After Cheban (1963)).

To improve the model, Cheban abandoned the polaron functions φ from both the initial (19.72) and final (19.76) state and replaced φ_o in the latter by a free electron wave $L^{-3/2}\exp(i\underline{k}\cdot\underline{r}_1)$; also, q_{u_o} in (19.72) was deleted, while $E_{F'} = I_F - I_{F'}$. The result was now more satisfying, as manifested in Fig.58 through a comparison with the KCl experimental F' band at 80 K (Kingsley (1961)). Table 10 presents some of Cheban's conduction-electron based results.

Table 10

Calculated F' thermal and peak optical energies*

Quantity	KCl	NaCl	KBr
$E_{F'}$ (eV)	0.30	0.53	0.36
$h\nu_{max}$ (eV)	1.7	2.44	1.5

*After Cheban (1963)

However, Cheban's theory of the F' band predicts too strong temperature effects, due to phonon broadening, as seen from eq's (19.85) and (19.86), which have not been observed experimentally.

19.3.2. Semicontinuum models

A semicontinuum model has first been used by Pincherle (1951) based on the following Hamiltonian:

$$\mathcal{H}_{F'} = -\frac{\hbar^2}{2m}\left(\frac{d^2}{dr_1^{\,2}} + \frac{d^2}{dr_2^{\,2}}\right) + \frac{e^2}{\varepsilon^* r_{12}} + \sum_{i=1,2}\begin{cases} eV_o & (r_i < R) \\ \dfrac{-Ze^2}{\varepsilon_\infty r_i} & (r_i > R)\end{cases} \qquad (19.87)$$

where $r_{12} = |\underline{r}_2 - \underline{r}_1|$, ε^* is an effective dielectric constant defined such that it is equal to ε_∞ for $r_2 < r_1$ and to ε_o for $r_2 > r_1$ when r_1 is held fixed. V_o is the square-well potential

$$V_o = -\frac{\alpha_M e^2 Z}{a} + \frac{Ze^2}{R}\left(1 - \frac{1}{\varepsilon_\infty}\right) \qquad (19.88)$$

R is the radius of the spherical cavity, assumed to represent the anion vacancy. A first-order Mott-Littleton calculation has yielded $R = 0.94a$ where a is the lattice half-spacing. The first term in (19.88) is the

Madelung potential, while the second one is the polarization energy induced by the vacancy, in Jost's approximation. To find the F' binding energy, Pincherle has used the variational method. The following trial wave function was chosen for the initial state:

$$\Psi_{F'}(\underline{r}_1,\underline{r}_2) = 4\gamma_{F'}^3 r_1 \exp(-\gamma_{F'} r_1)\, r_2 \exp(-\gamma_{F'} r_2) \qquad (19.89)$$

The variational parameter $\gamma_{F'}$ was determined from

$$\frac{d}{d\gamma_{F'}} \int \Psi_{F'}^* \mathcal{H}_{F'} \Psi_{F'}\, d\underline{r}_1 d\underline{r}_2 = 0$$

under the normalization condition

$$\int \Psi_{F'}^* \Psi_{F'}\, d\underline{r}_1 d\underline{r}_2 = 1$$

This gave the initial-state energy

$$I_{F'} = \int \Psi_{F'}^* \mathcal{H}_{F'} \Psi_{F'}\, d\underline{r}_1 d\underline{r}_2 \qquad (19.90)$$

To find the final-state energy, that is, the F center energy if the ionized CB electron was assumed to be at rest, the author has used the Hamiltonian

$$\mathcal{H}_F = -\frac{\hbar^2}{2m}\frac{d^2}{dr^2} + \begin{cases} eV_o & (r < R) \\ -Ze^2/\varepsilon_\infty r & (r > R) \end{cases} \qquad (19.91)$$

Using a trial wave function analogical to (19.89) and performing the variational calculation to find γ_F, he obtained

$$I_F = \int \Psi_F^* \mathcal{H}_F \Psi_F\, d\underline{r} \qquad (19.92)$$

Now, the F' binding energy is found as the difference between I_F and $I_{F'}$. The following numerical values have been obtained: 1.28 eV (NaCl), 0.85 eV (KCl). Interpreted as the threshold F' optical ionization energies, these compare favorably to within 30% with the corresponding experimental data. As seen from the foregoing, the e-e correlation has not been taken into account.

A refined semicontinuum calculation of the F' optical absorption cross-section has been made by Lynch and Robinson (1968). The assumed square-well potential is similar to (19.88) but the entering parameters are calculated self-consistently. V_o is now given by (see Fig.59):

$$V_o = -\frac{\alpha_M e^2}{a} + W + \chi \qquad (19.93)$$

where the second term is the polarization energy, while the third one is the electron affinity of the crystal, the zero-energy level lying in the vacuum at rest. A continuum approach proposed by Krumhansl and

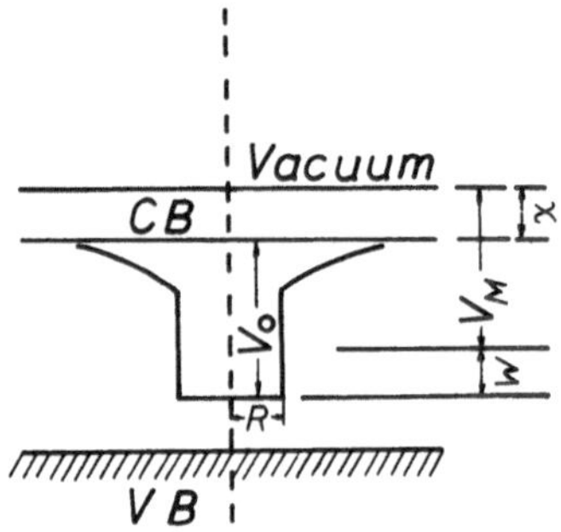

Figure 59: Potential type used in semicontinuum theories. (After Lynch and Robinson (1968)).

Schwartz (1953), more refined than the crude Jost method, has been used to tackle the polarization problem. W is split into ionic and optical parts:

$$W = W_{ion} + W_{opt} \tag{19.94}$$

W_{opt} is calculated as the work done in the polarization field $\varphi(q) = -(q/R)(1 - 1/\varepsilon)$ at the center of a spherical cavity of radius R on recharging the cavity from q_i to q_f :

$$W_{opt} = \int_{q_i}^{q_f} \varphi(q)dq = \tfrac{1}{2}(q_f\varphi(q_f) - q_i\varphi(q_i))$$
$$= -(1 - 1/\varepsilon_\infty)(q_f^2 - q_i^2)/2R \tag{19.95}$$

It is implied that the polarization, produced by the effective charge q of the cavity, follows the motion of the trapped electron ($\varepsilon = \varepsilon_\infty$). W_{ion} is the work done, due to ionic polarization, in removing the charges from the well:

$$W_{ion} = ce^2 \int_R^\infty \frac{Z - NP(r)}{r^2} d\underline{r} \tag{19.96}$$

where c is the polaron constant (19.55), N is the number of electrons in the well of effective charge Ze , P(r) is the fraction of the one-electron distribution inside a sphere of radius r :

$$P(r) = \int_0^r |\Psi(r)|^2 d\underline{r} \tag{19.97}$$

W_{ion}/e is the potential the electron sees at any instant due to the ionic polarization induced by its average distribution. The effective dielectric constant ε_{eff} outside the well is found by a continuum approach: The potential energy of a trap of effective charge Ze that has captured N electrons is

$$-\frac{e^2}{\varepsilon_{eff} r} = -\frac{Ze^2}{\varepsilon_o r} - ce^2 \int_r^\infty \frac{NP(r)}{r^2} dr$$

Consequently,

$$\frac{1}{\varepsilon_{eff}} = \frac{1}{\varepsilon_o} + (\frac{1}{\varepsilon_\infty} - \frac{1}{\varepsilon_o}) r \int_r^\infty \frac{NP(r)}{r^2} dr \qquad (19.98)$$

for $Z = 1$.

The F' Hamiltonian is now taken in the form

$$\mathcal{H}_{F'} = -\frac{\hbar^2}{2m}(\Delta_1 + \Delta_2) + 2V_o + \frac{e^2}{\varepsilon^* r_{12}} \quad (r_1, r_2 < R) \qquad (19.99)$$
$$= -\frac{\hbar^2}{2m}(\Delta_1 + \Delta_2) + V_o - \frac{e^2}{\varepsilon_{eff} r_2} + \frac{e^2}{\varepsilon^* r_{12}} \quad (r_1 < R,\ r_2 > R)$$
$$= -\frac{\hbar^2}{2m^*}(\Delta_1 + \Delta_2) - \frac{e^2}{\varepsilon_{eff} r_1} - \frac{e^2}{\varepsilon_{eff} r_2} + \frac{e^2}{\varepsilon^* r_{12}} \quad (r_1, r_2 > R)$$

Here ε^* is the dielectric constant of the Coulomb correlation potential. The trial F' center wave function used was a correlated Hyleraas function

$$\Psi(\underline{r}_1, \underline{r}_2) = (1 + \beta r_{12} - \gamma (r_1 - r_2)^2) \exp(-\frac{\alpha}{2}(r_1 + r_2)) \qquad (19.100)$$

This function has yielded good results for the H^- ion. $\Psi(\underline{r}_1, \underline{r}_2)$ being symmetric with respect to interchanging $\underline{r}_1$ and $\underline{r}_2$, it should describe a singlet state. At any instant of time the electronic motion is correlated, while each electron will have an s-symmetry about the origin over an extended period of time. Correlation was included because of the otherwise weak binding and the diffuse nature of the F' electron. Further, the ground-state parameters are found by minimizing

$$E(\alpha, \beta, \gamma) = \int \Psi^* \mathcal{H} \Psi \, d\underline{r}_1 d\underline{r}_2$$

under the normalization condition

$$\int \Psi^* \Psi \, d\underline{r}_1 d\underline{r}_2 = 1$$

This results in

$$E(\alpha, \beta, \gamma) = N^{-1}(V_{sq.well} + V_{coulomb} + V_{corr.} + T) \qquad (19.101)$$

where N is a normalization constant. Here

$$V_{sq.well} = 2V_o \langle \Psi | \Psi \rangle_{in,in} + 2(V_o + W_{opt}) \langle \Psi | \Psi \rangle_{in,out} \qquad (19.102)$$

where

$$V_o = -\alpha_M e^2/a + W_{ion} + \chi , \quad W_{opt} = e^2(1 - 1/\varepsilon_\infty)/2R \qquad (19.103)$$

'in' stands for integration over r inside and 'out' for outside the well, the first subscript referring to r_1 and the second one to r_2.

$$V_{coulomb} = -\frac{2e^2}{\varepsilon_{eff}} <\psi|\frac{1}{r_1}|\psi>_{out,all\ space}$$

$$V_{corr.} = <\psi|\frac{e^2}{r_{12}}|\psi>_{in,in} + \frac{2}{\varepsilon_\infty}<\psi|\frac{e^2}{r_{12}}|\psi>_{in,out} + \frac{1}{\varepsilon_\infty}<\psi|\frac{e^2}{r_{12}}|\psi>_{out,out}$$

$$T = -<\psi|\frac{\hbar^2}{2m}(\Delta_1 + \Delta_2)|\psi>_{all,all} \qquad (r_i < R)$$

$$= -<\psi|\frac{\hbar^2}{2m^*}(\Delta_1 + \Delta_2)|\psi>_{all,all} \qquad (r_i > R) \qquad (19.104)$$

Calculations on the F center have also been done using the Hamiltonian

$$\mathcal{H}_F = \begin{cases} -\frac{\hbar^2}{2m}\Delta + V_o & (r < R) \\ -\frac{\hbar^2}{2m^*}\Delta - \frac{e^2}{\varepsilon_{eff} r} & (r > R) \end{cases} \qquad (19.105)$$

and the trial functions

$$\psi_{1s} = (\alpha^3/7\pi)^{\frac{1}{2}} (1 + \alpha r) \exp(-\alpha r)$$
$$\psi_{2p} = (\beta^5/\pi)^{\frac{1}{2}} r \cos\vartheta \exp(-\beta r) \qquad (19.106)$$

following Fowler (1964).

In calculating the F' bandshape the initial state is described by (19.100), while the final state is composed of one bound s-like electron (19.106) and one free plane-wave electron: (19.107)

$$\psi_{final} = (2L^3)^{-\frac{1}{2}}(\psi_{1s}(r_1)\exp(i\underline{k}_e\cdot\underline{r}_2) + \psi_{1s}(r_2)\exp(i\underline{k}_e\cdot r_1))$$

Now, the absorption cross-section is calculated from

$$c_{F'} = \frac{(n^2 + 2)^2}{9n}\frac{2e^2L^3}{3mc^2}\frac{k_e}{k_p}| < i|\sum_j \nabla_j|f>|^2 \qquad (19.108)$$

where k_e and k_p are the emitted electron and absorbed photon wave vectors, respectively, related by $\hbar c k_p = A + \hbar^2 k_e^2/2m^*$ where A is the electron affinity of the F center left behind.

Calculations have been made using $m^* = 0.6$ m , R = 0.9 , and χ from literature data. The F' optical binding energy has been computed from $E_{F'optical} = E_{F'}(\alpha,\beta,\gamma) - E_F(\alpha)$. Results for several crystals are reproduced in Fig.1. The F' band for KCl was calculated by adjusting the potential well-depth V_0 to fit $E_{F'optical}$ to the experimental data. The open circles are experimental points. The weak second shoulder on the high-energy side of the experimental absorption curve is not explained by the above calculations. It may correspond to optical transitions to a final state composed of a 2p (or even 2s) electron and a plane wave, following reaction (5.a) (Lynch and Robinson (1968)). The authors have calculated the corresponding edge energies, marked by arrows in Fig.1. It is pointed out, however, that since the calculation has neglected the phonon broadening, the obtained ionization energies should be higher than the respective experimental values. Another feature of the experimental F' band in the potassium salts, the asymmetry of the main peak (a shoulder appears on its high-energy side) has tentatively been attributed to a low-lying d-like minimum in the conduction band. This could produce the asymmetry by increasing the density of the final states in absorption. The alternative would be to assume the existence of some additional absorption due to the transition to a localized state, split from the CB by the F' potential.

Arora and Wang (1969) have applied Wang's polaron approach tc the F' center in KCl, KBr, and NaCl. In this approach, the electronic and ionic polarizations are accounted for as exchange effects of excitons and phonons, respectively, between the two trapped electrons. The Hamiltonian is

$$\mathcal{H} = \sum_{i=1,2} \mathcal{H}_i + e^2/r_{12} + \mathcal{H}_\varepsilon + V_{q-\varepsilon} + \mathcal{H}_p + V_{q-p} \qquad (19.109)$$

where

$$\mathcal{H}_i = p_i^2/2m + \sum_u V(\underline{r}_i - \underline{R}_{u_0}) - V(\underline{r}_i) + V_{e_i-\varepsilon} \qquad (19.110)$$

for $|\underline{r}_i| < R$, and

$$\mathcal{H}_i = p_i^2/2m + \sum_u V(\underline{r}_i - \underline{R}_{u_0}) - V(\underline{r}_i) + V_{e_i-\varepsilon} + V_{e_i-p} \qquad (19.111)$$

for $|\underline{r}_i| > R$. Here R is the radius of a spherical region about the center of the vacancy, such that the i-th trapped electron does not polarize the lattice for $r_i < R$. R is numerically set equal to the Mott-Littleton radius of an anion vacancy. The second terms in (19.110) and (19.111) are the electron interaction energy with the perfect lattice, when all the nuclei are at their equilibrium positions and all core and valence electrons are in ground state. The third term is the interaction with the trapping center of effective charge q . The second term in (19.109) is the e-e Coulomb interaction energy. The ε-labelled terms account for the electronic polarization. $\mathcal{H}_\varepsilon$ is the Hamiltonian of the virtual exciton field describing the system of valence electrons. $V_{q-\varepsilon}$ and $V_{e_i-\varepsilon}$ stand for the interaction energy of the exciton field with the trapping center q and the i-th electron e_i, respectively. $\mathcal{H}_p$ is the Hamiltonian of the virtual phonon field describing the ionic (nuclear) subsystem. V_{q-p} and V_{e_i-p} stand for the trapping center - phonon and the i-th electron - phonon interaction energies, respectively. Note that the e_i-p term enters the F' Hamiltonian in (19.111) only, in accordance with the definition of R . The ε- and p- terms can be expressed by means of appropriate creation and annihilation operators for the two fields. Equation (19.109) then transforms into

$$\mathcal{H} = \sum_{i=1,2} \mathcal{H}_i^{\cdot} + e^2/r_{12} + \mathcal{H}_\varepsilon^{\cdot} \qquad (19.112)$$

where

$$\mathcal{H}_\varepsilon^{\cdot} = \sum_{w} B_w^+ B_w E_w + \sum_{j=1}^{2} \sum_{w} (V_w B_w e^{i\underline{w}\cdot\underline{r}_j} + \text{c.c.}) \qquad (19.113)$$

is the excitonic part expressed in terms of the operators B_w, related to the annihilation operators b_w by means of $B_w = b_w - V_w^+/E_w$, $\underline{w}$ being the exciton wave vector. Further,

$$\mathcal{H}_i^{\cdot} = p_i^2/2m - \alpha_M e^2/a + (1 - 1/\varepsilon_\infty)e^2/r_i\,, \quad r_i < R \qquad (19.114)$$

$$= p_i^2/2m^* - e^2/\varepsilon_o r_i + \mathcal{H}_{pi}^{\cdot}\,, \quad r_i > R \qquad (19.115)$$

and also

$$\mathcal{H}_{pi}^{\cdot} = \sum_k A_k^+ A_k \hbar\omega_k + \sum_k (V_k A_k e^{i\underline{k}\cdot\underline{r}_i} + \text{c.c.}) \qquad (19.116)$$

which is the p-term analogous to (19.113). In (19.112) the difference $\sum_u V(\underline{r}_i - \underline{R}_{uo}) - V(\underline{r}_i)$ has been replaced by the Madelung energy for $r_i < R$, while $p_i^2/2m + \sum_u V(\underline{r}_i - \underline{R}_{uo})$ is replaced by $p_i^2/2m^*$ for $r_i > R$ on grounds of the effective mass approximation, and also $V(r_i) = e^2/\varepsilon_o r_i$ for $r_i > R$, that is, the q-e_i interaction is assumed Coulombic outside the cavity.

Physically, the phonon part of the Hamiltonian is the self-energy of the trapped electrons due to their own phonon clouds and a term arising from the exchange of phonons between the electrons. The exchange effect reduces the repulsive e-e interaction. Following a method due to Wang, the effective F' Hamiltonian finally is

$$\mathcal{H}_{F',\text{eff}} = \sum_{i=1}^{2} \mathcal{H}_{i,\text{eff}} + e^2/\varepsilon_1 r_{12} + \langle F|\mathcal{H}_\varepsilon^{\cdot}|F\rangle \qquad (19.117)$$

where $\mathcal{H}_{i,\text{eff}}$ is given by eq.(19.114) inside the cavity, while in the outer part (19.115) $\mathcal{H}_{pi}^{\cdot}$ is now replaced by the matrix element $<\Phi_i|\mathcal{H}_{pi}^{\cdot}|\Phi_i>$. Here ε_1 accounts for the e-e phonon-exchange effect and is equal to 1 inside and to 1/c outside the cavity. F and Φ_i are the all-exciton and all-phonon wave functions assumed to be multiplicatives of the respective one-particle wave functions: $F = \prod_w f(w)$, $\Phi = \prod_k \varphi(k)$. The matrix elements have been found to be

$$<F|\mathcal{H}_{\varepsilon}^{\cdot}|F> = 4\sum_w \frac{|V_w|^2}{E_w}(\rho_w^*\rho_w - 2e^{i\underline{w}\cdot\underline{r}_j}\rho_w^*)$$
$$<\Phi_i|\mathcal{H}_{pi}^{\cdot}|\Phi_i> = \sum_k \frac{|V_k|^2}{\hbar\omega_k}(\rho_k^*\rho_k - 2e^{i\underline{k}\cdot\underline{r}_i}\rho_k^*) \qquad (19.118)$$

where

$$\rho_w = \int|\Psi_{F'}(r_1,r_2)|^2\, e^{i\underline{w}\cdot\underline{r}_j}\, d\underline{r}_1 d\underline{r}_2 \quad \text{, etc.}$$

Finally, the F'energy is found by minimizing the functional

$$I_{F'} = <\Psi_{F'}(\underline{r}_1,\underline{r}_2)|\mathcal{H}_{F',eff}|\Psi_{F'}(\underline{r}_1,\underline{r}_2)> \qquad (19.119)$$

Calculations on $I_{F'}$ have been made for the $^1S(1s,1s)$ ground singlet and for the $^1P(1s,2p)$ excited singlet. The following 1s trial wave functions were checked:

$$\Psi_{1s}(\underline{r}) = (\alpha^3/\pi)^{\frac{1}{2}}\exp(-\alpha r) \qquad (19.120)$$

$$\Psi_{1s}(\underline{r}) = (\gamma^3/7\pi)^{\frac{1}{2}}(1+\gamma r)\exp(-\gamma r) \qquad (19.121)$$

Based on any of these, the ground state wave function was built:

$$\Psi_g(\underline{r}_1,\underline{r}_2) = N(\Psi_{1s}^{\alpha_1}(r_1)\Psi_{1s}^{\alpha_2}(r_2) + \Psi_{1s}^{\alpha_1}(r_2)\Psi_{1s}^{\alpha_2}(r_1)) \qquad (19.122)$$

Here the two values of the variational parameter were expected to allow for two different 1s orbitals, one for each of the F' electrons. However, the variational procedure has in fact led to $\alpha_1 = \alpha_2$. The 2p-type function checked was

$$\Psi_{2p}(r) = (\beta^3/\pi)^{\frac{1}{2}}\, r\cos\vartheta\exp(-\beta r) \qquad (19.123)$$

and the excited-state wave function was constructed by means of

$$\Psi_e(\underline{r}_1,\underline{r}_2) = 2^{-\frac{1}{2}}(\Psi_{1s}(r_1)\Psi_{2p}(r_2) + \Psi_{1s}(r_2)\Psi_{2p}(r_1)) \qquad (19.124)$$

The excited-state energy has been computed from the functional

$$I_{F',e} = <\Psi_e(\underline{r}_1,\underline{r}_2)|\mathcal{H}_{eff}|\Psi_e(\underline{r}_1,\underline{r}_2)> + e^2(1-\frac{1}{\varepsilon_\infty})/a \qquad (19.125)$$

to secure that $I_{F',g}$ and $I_{F',e}$ are both measured from the same reference level.

An alternative Hamiltonian has also been examined which differs from

(19.117) in the exciton matrix element. Each F' electron is now assumed to have its own exciton cloud. Correspondingly, the matrix element is the self-energy of the electron due to its own cloud, while the term $-(1 - 1/\varepsilon_\infty)/r_{12}$ arises as an attractive Coulomb potential due to the exchange of excitons between the two clouds. The whole exciton cloud is now treated as the superposition of two independent clouds, each surrounding a particular electron. The resulting Hamiltonian is

$$\mathcal{H}_{eff} = \sum_{i=1,2} \mathcal{H}_{i,eff} + e^2/\varepsilon_2 r_{12} \qquad (19.126)$$

with (19.127)

$$\mathcal{H}_{i,eff} = p_i^2/2m - \alpha_M e^2/a + (1 - 1/\varepsilon_\infty)e^2/r_i + \langle F_i | \mathcal{H}'_{\varepsilon i} | F_i \rangle \quad (r_i < R)$$

$$= p_i^2/2m^* - e^2/\varepsilon_0 r_i + \langle F_i | \mathcal{H}'_{\varepsilon i} | F_i \rangle + \langle \Phi_i | \mathcal{H}'_{pi} | \Phi_i \rangle \quad (r_i > R)$$

Results of these calculations are presented in Table 11, where $\Delta E = I_{F',e} - I_{F',g}$, computed using two different trial functions (19.120) and (19.121) for the initial ^1S state (19.122) and the two Hamiltonians (19.117) and (19.126), are listed in eV.

Table 11

Results of F' calculation by Wang's polaron approach*

Crystal	Hamiltonian (19.117)		Hamiltonian (19.126)		Ionization
	(19.120)	(19.121)	(19.120)	(19.121)	energy
KCl	1.68	2.26	1.66	1.70	1.96
KBr	1.60	2.35	1.58	1.81	1.77
NaCl	2.04	2.95	1.99	1.70	2.33

*After Arora and Wang (1969)

It may be seen that both approaches to the exciton cloud, that is, to the electronic polarization, yield the same binding energy if the hydrogenic 1s- trial function (19.120) is used. At the same time, using the modified 1s- orbital (19.121) yields $\beta \sim 0$ which suggests a direct F' ionization. However, columns 3 and 5 are both in poor agreement with the experimental F' optical binding energies, in contrast to columns

2 and 4 where the agreement is better. The modified 1s- orbital is more compact and leads to larger discrepancies for Hamiltonian (19.120). Hamiltonian (19.121) is, therefore, considered more appropriate for treating the electronic polarization of deeper (more compact) centers.

The F' ionized state has also been studied by setting $\beta = 0$ and minimizing in α alone. The energy of this state was found giving 1.96 (KCl), 1.77 (KBr), and 2.33 eV (NaCl) for the optical ionization energy. If one now uses the data from the second column in Table 11, the optical depth of the F' excited singlet is obtained to be 0.28 eV (KCl), 0.17 eV (KBr), and 0.29 eV (NaCl) below the conduction band. It is concluded that the F' center could have a bound excited state (1s,2p) below the CB bottom which state is presumably final in the optical absorption within the F' band. The estimated optical depths of (1s,2p) compare favorably with Crandall's data for the F' thermal depth in KBr.

19.3.3. Point-ion models

Perhaps the earliest point-ion calculation of the F' optical binding energy is due to Raveche (1965). His approach is very crude: Although the point-ion potential is used, the ions are assumed to be unpolarizable. The Hamiltonian is of the Gourary & Adrian type

$$\mathcal{H}(1,2) = -\tfrac{1}{2}(\Delta_1 + \Delta_2) + \sum_{\underline{r}_i \neq \underline{o}}{}' (-1)^{x_i+y_i+z_i} \left((\underline{r}_1 - d\underline{r}_i)^2 + (\underline{r}_2 - d\underline{r}_i)^2 \right)^{-\frac{1}{2}} \quad (19.128)$$

in Hartree's atomic units, d is the lattice constant. This Hamiltonian can also be put in the form

$$\mathcal{H}(1,2) = \mathcal{H}(1) + \mathcal{H}(2) + \frac{1}{r_{12}} \quad (19.129)$$

The third term, representing the e-e interaction energy, is further taken as a perturbation to $\mathcal{H}(1) + \mathcal{H}(2)$. The wave function is chosen in the form (19.51) with

$$\begin{aligned} \Psi(r) &= A\, J_0(\xi r/d) \exp(-\eta)/\eta & r < d \\ \Psi(r) &= A\, J_0(\xi) \exp(-\eta r/d)/(\eta r/d) & r > d \end{aligned} \quad (19.130)$$

$J_0(\;)$ is the spherical Bessel function of zeroth order, A is a normalization constant. ξ is a variational parameter, while η is determined by the requirement that Ψ and $d\Psi/dr$ be continuous at $r = d$.

The first-order energy is $E = 2E^o + E'$:

$$E = 2 < \Psi(r_1)| \mathcal{H}(1)| \Psi(r_1)> + < \Psi(\underline{r}_1, \underline{r}_2)|r_{12}^{-1}| \Psi(\underline{r}_1, \underline{r}_2)> \quad (19.131)$$

since

$$<\mathcal{H}(1)> = <\mathcal{H}(2)> \quad .$$

E^o is Gourary & Adrian's ground-state energy of the F center (Gourary and Adrian (1960)), E' is the perturbation energy. The F' binding energy is now given by $E^o - (2E^o + E') = -(E^o + E')$. Calculations have yielded 1.16 eV (KCl), 1.29 eV (NaCl), which are not too far from the experimental data, especially for KCl.

Another step forward in the effort to understand the nature of the F' binding energy has been undertaken by La and Bartram (1966). They took advantage of the fact that the success of the semicontinuum approach to the F center is largely due to the similarity between the square-well potential and the more realistic point-ion potential at short distances from the center of the vacancy, as seen in Fig.60. This suggests

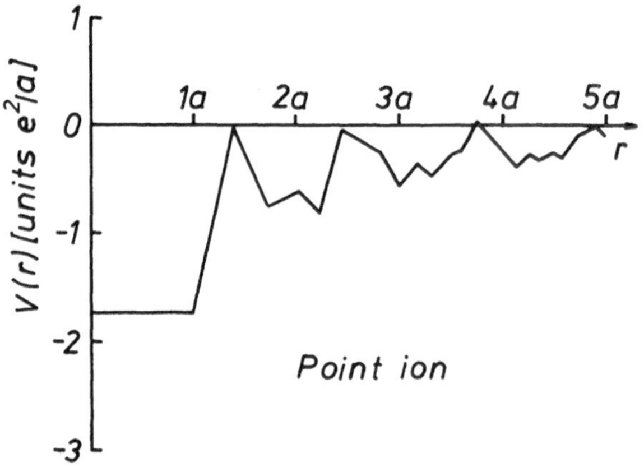

Figure 60: The point-ion potential in reduced units. (Reproduced from Fowler (1968)).

treating the F center problem by the square-well model as a first approximation and then introducing the remainder of the point-ion potential (the point-ion correction) as a perturbation. The point-ion correction is

$$V_i(r) = V_o - e \sum_k q_k(r^{-1} - r_k^{-1}) \quad , \quad r > a \quad (19.132)$$

where

$$V_o = \alpha_M e^2/a \qquad (19.133)$$

is the Madelung potential, $\alpha_M = 1.7476$ is Madelung's constant and a is the interionic distance. q_k is the charge of the k-th point ion at a distance r_k from the vacancy. The sum is over all ions within a sphere of radius r about the vacancy. Now, the wave functions and the energy levels for the square-well potential are well-known (see Schiff (1955)). If the ground-state wave function is written as

$$\psi_F = (4\pi)^{-\frac{1}{2}} \frac{1}{r} P_F(r) \qquad (19.134)$$

and E_F^o is the ground-state energy to zeroth order, then the corrected ground-state energy will be given by $E_F = E_F^o + E_F^i$, where

$$E_F^i = \int_a^\infty 4\pi r^2 |\psi_F| V_i(r)\, dr = \int_a^\infty P_F(r)^2 V_i(r)\, dr \qquad (19.135)$$

is the point-ion correction by first-order perturbation theory.

Further, the F' problem is treated by the Hartree-Fock approach and the electrostatic e-e interaction is only taken into account when both electrons are inside the well. Thus, the Hartree-Fock equation for the ground state of two electrons in a well is

$$\left(-\frac{\hbar^2}{2m}\frac{d^2}{dr^2} - V_o - E + e^2\int_0^a \frac{P_{F'}(r')^2}{r_1}\, dr'\right) P_{F'}(r) = 0 , \quad r < a$$
$$\left(-\frac{\hbar^2}{2m}\frac{d^2}{dr^2} - E\right) P_{F'}(r) = 0 , \quad r > a \qquad (19.136)$$

where again

$$\psi_{F'}(r) = (4\pi)^{-\frac{1}{2}} \frac{1}{r} P_{F'}(r) \qquad (19.137)$$

is the F' ground-state wave function, $r_1 = \max(r,r')$. The ground-state energy of the unperturbed problem is

$$E_{F'}^o = 2E - \langle g \rangle_1 \qquad (19.138)$$

where

$$\langle g \rangle_1 = e^2 \int_0^a \int_0^a dr\, dr'\, P(r)^2 P(r')^2 \frac{1}{r_1} \qquad (19.139)$$

is the electrostatic interaction of both electrons inside the well. Eq.(19.136) has been solved by expanding $P(r)$ into a power series for $r < a$ whose coefficients have been determined by means of an iterative procedure under the requirement that $P(r)$ is continuous in both slope and value at $r = a$. At $r > a$

$$P(r) = B\beta^{-1}\exp(-\beta r) , \quad \beta = (2m\hbar^{-2}|E|)^{\frac{1}{2}} \qquad (19.140)$$

B is a normalization constant. The e-e interaction, when one of the electrons is outside the well, is regarded as a perturbation given by

$$<g>_2 = 2\frac{e^2}{\varepsilon_\infty}\int_0^a P(r')^2\,dr'\int_a^\infty \frac{1}{r}P(r)^2\,dr \qquad (19.141)$$

Now, the dominant part of the ground-state energy is

$$E_{F'} = 2E - <g>_1 + E^i_{F'} + <g>_2 \qquad (19.142)$$

where

$$E^i_{F'} = 2\int_a^\infty P(r)^2\,V_i(r)\,dr \qquad (19.143)$$

is the point-ion correction. Finally, the F' binding energy is obtained from $\Delta E_{F'} = E_F - E_{F'}$.

The above procedure gives the dominant part of the F' binding energy, interpreted by the authors as the peak rather than threshold energy. To improve the agreement with the experiment, several corrections have been added accounting for the polarization effects, completely ignored by the main terms. These have been made following the Krumhansl & Schwartz method which treats the vacancy as a spherical cavity of radius R in a dielectric continuum, characterized by ε_0 or by ε_∞. Outside the cavity, the vacancy appears as a point positive charge. The corrections under consideration are:

$$\text{(i)}\quad E_{pF} = \int_0^\infty P_F(r)^2\,V_F(r)\,dr \qquad (19.144)$$

arising from the Franck-Condon principle which requires that the F center formed in the optical ionization process will be at the nuclear configuration of the F' ground state. Here

$$\begin{aligned} V_F(r) &= e^2\left(1-\frac{1}{\varepsilon_\infty}\right)r^{-1} - e\varphi(r)\,, \qquad r > a \\ &= e^2\left(1-\frac{1}{\varepsilon_\infty}\right)(2a)^{-1} - e\varphi(a)\,,\quad r < a \end{aligned} \qquad (19.145)$$

$$\text{(ii)}\quad -E_p = -\int_0^\infty P(r_1)^2\,P(r_2)^2\,V(r_1,r_2)\,dr_1dr_2\;, \qquad (19.146)$$

due to the interaction of each electron with the infrared polarization, the optical polarization induced by the vacancy, and with its own contribution to the optical polarization. Here

$$\begin{aligned} V(r_1,r_2) &= e^2\left(1-\frac{1}{\varepsilon_\infty}\right)(r_1^{-1}+r_2^{-1}) - e(\varphi(r_1)+\varphi(r_2))\;,\; r_1,r_2 > a \\ &= e^2\left(1-\frac{1}{\varepsilon_\infty}\right)\left(r_1^{-1}+\frac{1}{2a}\right) - e(\varphi(r_1)+\varphi(r_2))\;,\; r_1>a, r_2<a \\ &= -2e\varphi(a)\;,\; r_1 < a\;,\; r_2 < a \end{aligned} \qquad (19.147)$$

(iii) $$-E_e = -2\frac{e^2}{\varepsilon_\infty}\int_a^\infty P(r)^2 \frac{1}{r} p(r)\, dr \qquad (19.148)$$

is the residual e-e interaction when one or both electrons are outside the well diminished by optical polarization. Here

$$\begin{aligned} p(r) &= 1 - q(r) \\ q(r) &= \int_r^\infty P(r)^2\, dr \end{aligned} \qquad (19.149)$$

$q(r)$ is the fraction of charge outside a sphere of radius r. $\varphi(r)$ is the electrostatic potential due to the infrared polarization of the lattice which does not follow the electron motion

$$\begin{aligned} \varphi(r) &= -ec\int_r^\infty (2q(r') - 1)\, r'^{-2}\, dr' \quad , \ r > a \\ &= \varphi(a) \quad , \ r < a \end{aligned} \qquad (19.150)$$

where c is the polaron constant (19.55).

(iv) Another correction arises from the F' photoionization creating a conduction rather than free electron: $-\chi$. A final correction of comparable size but of opposite sign results from the orthogonalization of the color center wave functions to the ion-core orbitals (Gourary and Adrian (1960)). Both corrections are introduced by adding χ to the potential energy inside the well and measuring the energy from the bottom of the conduction band. $E_{F'}$ is then increased by $2p_{F'}\chi$, while E_F increases by $p_F\chi$.

Finally,

$$E_{F'corr} = \Delta E_{F'} + E_{pF} - E_p - E_e + (p_F - 2p_{F'})\chi .$$

The polarization corrections have been found to contribute as little as 0.5 eV to the F' binding energy. The computed F' band energies are: 2.40 eV (NaCl), 2.42 eV (KCl), and 1.91 eV (KBr).

Berezin (1967) has based his approach to the F' center on the point-ion potential model too, taking into account the polarization effects (a polarizable point-ion lattice). The ground state F' wave function is taken in the (19.51) multiplicative form with a Pekarian 1s- trial function $\Psi(r)$ from (19.62). The final-state wave function (photo-electron + F center) is composed of a ground-state F center wave function $\varphi(r)$ and a free-electron wave:

$$\Psi_{\underline{k}}(\underline{r}_1,\underline{r}_2) = \varphi(\underline{r}_1)e^{i\underline{k}\cdot\underline{r}_2} + \varphi(\underline{r}_2)e^{i\underline{k}\cdot\underline{r}_1} \qquad (19.151)$$

The F' ground-state energy is obtained by minimizing the functional

$$E = \langle \Psi_{F'}(\underline{r}_1,\underline{r}_2)|\mathcal{H}_{PI}|\Psi_{F'}(\underline{r}_1,\underline{r}_2)\rangle + W \qquad (19.152)$$

where $\mathcal{H}_{PI}$ is the point-ion lattice F' Hamiltonian, while W is the energy of the system of dipoles induced by the F' field upon the polarizable lattice ions. Inasmuch as W depends on α (the variational parameter) through its dependence on $\psi_{F'}$, the inclusion of W into (19.152) implies a self-consistent account of the optical polarization. The ionic polarization has been accounted for by considering the displacement of the six nn cations, the ionic interaction potential including Coulombic, Born-Mayer, and Van-der-Waals terms. In F' ground state, the nn cations have been displaced towards the vacancy at about 3 to 5 percent of the lattice parameter. This lowers the F' ground-state energy by about 1.5 eV.

The F' photoionization cross-section was calculated from

$$\sigma(\hbar\omega) \propto \hbar\omega(\hbar\omega - \hbar\omega_0)^{3/2}/(\hbar\omega - \hbar\omega_0 + \hbar^2\alpha^2/2m)^8 \qquad (19.153)$$

for $\omega \geqslant \omega_0$. $\hbar\omega_0$ is the red photoionization edge: this is the difference between the energies of the final state (F center + photoelectron at the bottom of the conduction band) and the F' ground state. The results of Berezin's calculations are presented in Table 12.

Table 12

Comparison with experiment of Berezin's calculation*
(All energies are in electron-volts)

Crystal	$E_{1s}(F')$	$\hbar\omega_0$(th)	$\hbar\omega_0$(exp)	$\hbar\omega_{max}$(th)	$\hbar\omega_{max}$(exp)
NaCl	-8.78	1.1	1.2	2.0	2.4
KCl	-8.34	1.6	1.1	2.1	1.7
KBr	-7.92	0.9	0.9	1.6	1.4

*After Berezin (1967)

The experimental values are from Pick (1938). The rigid (unpolarizable) point-ion lattice model yields too small values for $\hbar\omega_0$ (e.g. 0.4 eV for NaCl). Polarization is, therefore, essential for building up the F' binding.

The F' optical binding energy in KCl and NaCl has also been calculated by Bennett (1970) based on the point-ion lattice model ignoring completely the electronic polarization (rigid ions). The ionic polarization has been accounted for classically (lattice vibrations neglected), considering only the displacement of the nn ions in the breathing-mode approximation. The Hamiltonian used reads

$$\mathcal{H}(\underline{r}_1,\underline{r}_2;\sigma) = -\frac{\hbar^2}{2m*}(\Delta_1 + \Delta_2) + V_{PI}(\underline{r}_2,\sigma) + V_{PI}(\underline{r}_1,\sigma) + e^2/r_{12} + \Delta E_L \qquad (19.154)$$

Here, σ stands for the relative magnitude of the ionic displacement, V_{PI} is the spherically-symmetrical part of the point-ion potential, while the lattice polarization energy is $\Delta E_L = \Delta E_e + \Delta E_r$, where the first term is the change in electrostatic energy and the second one is the change in repulsive energy taking into account the Pauli exclusion principle between the i-th and j-th cores. The latter term is expressed by means of the Born-Mayer exponential. Schrödinger's equation

$$\mathcal{H}\Psi_\eta(\underline{r}_1,\underline{r}_2) = E\,\Psi_\eta(\underline{r}_1,\underline{r}_2) \qquad (19.155)$$

is solved approximately by the Hartree-Fock self-consistent method, using Slater's representation for the exchange energy term to simplify solving the variational equations. The e-e correlation energy, missing in the Hartree-Fock approach, has also been included following Mittler's formula. The importance of including the e-e correlation term for the F' singlet state is particularly stressed. The corresponding equation for the radial Hartree-Fock function

$$\left(-\frac{\hbar^2}{2m*}\frac{d^2}{dr^2} + \frac{\hbar^2}{2m*}\frac{l(l+1)}{r^2} + V_{PI}(r;\sigma) + V_c(r) + W(r)\right)P_{nl}(r) = E_{nl}P_{nl}(r) \quad , \qquad (19.156)$$

where W is the e-e term and V_c is the Coulomb energy, has been solved numerically for various values of σ at a given electronic configuration η and the steady-state energy E found by inserting the extremal σ, which has minimized the energy. The F' ground state is assumed to be of the singlet ${}^1S(1s,1s)$ type: This is the initial state in optical absorption. Now, the optical binding energy is obtained from

$$E^{F'}_{\text{opt. binding}} = E(F';{}^1S(1s,1s);\sigma=0) - E(F;1s;\sigma=0) \qquad (19.157)$$

Here, the F center ground-state energy is given by (19.158)

$$E(F;1s;\sigma=0) = E_L(\sigma) + \int\Psi_{HF,1s}(r)\,\mathcal{H}_0(r;\sigma=0)\,\Psi_{HF,1s}(r)\,dr$$

where

$$\mathcal{H}_0(r,\sigma) = -\frac{\hbar^2}{2m^*}\Delta + V_{PI}(r,\sigma) \qquad (19.159)$$

and Ψ_{HF} is the solution to the Hartree-Fock-Slater equation (19.156) at $V_c = W = 0$. The calculated F' optical binding energy was 0.0496 (KCl) and 0.0505 (NaCl), in atomic units. The possibility that a higher excited bound state (an F' singlet) of the $^1P(1s,2p)$ type may exist has also been checked but the outcome is uncertain. Nevertheless, computer calculations have indicated that while the 1s counterpart is compact, the 2p component is too diffuse and of a small eigenvalue. The latter is, therefore, more likely to be a CB state rather than a localized state. The presumed uniqueness of the ground singlet F' state in some (conventional) alkali halides seems to have been confirmed once again.

Apparently, the most refined calculation so far of the F' band based on the point-ion model has been made by Strozier and Dick (1969). The polarization effects have been accounted for by means of the Krumhansl & Schwartz semicontinuum method. The following additional interactions have been included: (i) the Coulomb repulsion between the F' electrons and the core electrons of the nn and nnn ions (i.e. the finite size of the ion cores); (ii) orthogonalization of the vacancy-centered F' wave function to the nn and nnn ion core orbitals; (iii) the correlation energy of the F' electrons.

The following Hamiltonian has been used:

$$\mathcal{H} = \mathcal{H}_0 + e^2/r_{12} \qquad (19.160)$$

where

$$\mathcal{H}_0 = \sum_{i=1,2} \left(-\frac{\hbar^2}{2m}\Delta_i + V_L(r_i) + V_C(r_i) - V_{LO}(r_i) - V_{CO}(r_i)\right) + V_{PI}(r_1,r_2) + V_{PE}(r_1,r_2) \qquad (19.161)$$

The F' system consists of: lattice, negative-ion vacancy, and two F' electrons. Here

$$V_L(r) = \sum_j \frac{Z_j}{|\underline{R}_j - \underline{r}|} \qquad (19.162)$$

is the point-ion potential of the perfect lattice,

$$V_C(r) = \sum_j \left(\frac{n_j}{|\underline{R}_j - \underline{r}|} - \sum_c^{n_j} \int \frac{|\Psi_{jc}(r')|^2}{|\underline{r} - \underline{r}'|}\, d\underline{r}' + V_j(\text{exch.})\right) \qquad (19.163)$$

The first term in (19.163) is due to the core electrons centered at ion sites, the second one results from a positive point charge at each ion site equal in magnitude to the number of core electrons at that site (Ψ_{cj} is the c-th atomic orbital on the j-th ion, n_j is the number of electrons on the j-th ion), V_j(exch.) is the core exchange operator. V_{L0} and V_{C0} are the point-ion and core potentials of the ion at the origin (center of the vacancy) in a perfect crystal. The latter terms are subtracted in the Hamiltonian, since that ion has been removed to form the vacancy. In other words, $V_L'(r) = V_L(r) - V_{L0}(r)$ and $V_C'(r) = V_C(r) - V_{C0}(r)$ are, respectively, the point-ion and the ion-core potentials at an anion vacancy in the imperfect lattice. V_{PI} and V_{PE} are the ionic and electronic polarization terms, respectively, describing the lattice relaxation around the F' center. These terms are computed using the semicontinuum model in which the lattice containing a vacancy is replaced by a dielectric continuum with a spherical cavity of diameter equal to the nn distance. Both the F' electrons and the effective positive charge of the vacancy polarize the medium and the resulting polarization potential of the F' electrons is described by V_{PI} and V_{PE}. These are found by an extension of the Krumhansl & Schwartz approach to two electrons in a vacancy.

Considering V_{PI} first, the authors assume that the ions respond to the time-average of the electronic distribution only. The ions "see" a potential arising from the F' electrons' static average charge distribution and from the effective charge of the cavity. This potential is found by solving Poisson's equation for a spherical well

$$\Delta \Phi(r) = -4\pi \rho(r)/\varepsilon \tag{19.164}$$

with

$$\rho(r) = -e\left(\int |\Psi_{F'}(\underline{r}_1,\underline{r}_2)|^2(\delta(\underline{r}-\underline{r}_1) + \delta(\underline{r}-\underline{r}_2))d\underline{r}_1 d\underline{r}_2 + \delta(\underline{r})\right) \tag{19.165}$$

Subtracting from Φ the potential due to ρ itself, one finds

$$V_{PI} = e\left(\Phi(r) - \frac{1}{\varepsilon}\int \frac{\rho(r')d\underline{r}'}{\underline{r} - \underline{r}'}\right) \tag{19.166}$$

The dielectric constant of the medium is set equal to the polaron value $1/\varepsilon = 1 - c$, where c is given by (19.55), when dealing with the ionic polarization.

To find V_{PE}, imagine first the electrons to be far removed from the center. The potential inside the cavity due to electronic polarization of the lattice will be

$$\varphi(Z) = -(1 - 1/\varepsilon_\infty)\frac{Z}{R_0}$$

where Z is the vacancy effective charge. One electron is next brought and placed in the center. The polarization potential becomes

$$\varphi(Z-1) = -(1 - 1/\varepsilon_\infty)\frac{Z-1}{R_0}$$

This shows that the electronic potential at the center should be a linear function of the charge inside; consequently, the potential energy of an electron inside the cavity will be

$$W(Z_f, Z_i) = -(1 - 1/\varepsilon_\infty)(Z_f^2 - Z_i^2)/2R_0 \quad , \tag{19.167}$$

the difference $Z_f - Z_i$ being the charge of the electron whose potential energy is sought. From W the authors construct two one-electron operators

$$\begin{aligned} V_1(r_1) &= P_2\, W(-1,0) + (1 - P_2)\, W(0,1) \\ V_2(r_2) &= P_1\, W(-1,0) + (1 - P_1)\, W(0,1) \end{aligned} \tag{19.168}$$

where

$$P_j = \int_0^{R_0} |\Psi_j(r)|^2\, d\underline{r} \tag{19.169}$$

is the probability that the j-th electron is inside. The total electronic polarization operator is then $V_{PE} = V_1 + V_2$ giving

$$V_{PE} = e^2(1 - 1/\varepsilon_\infty)((1 - 2P_2) + (1 - 2P_1))/2R_0 \;, \; r_1, r_2 < R_0 \tag{19.170}$$

the first term within the square brackets acting only on e_1, the second one - on e_2. Outside the vacancy, one has for e_1:

$$V_{PE}(r_1) = e^2(1 - 1/\varepsilon_\infty)\left(-\frac{1}{r_1} + \int \frac{|\Psi_2(r')|^2}{\underline{r}_1 - \underline{r}'}\, d\underline{r}'\right) - \chi, \; r_1 > R_0 \tag{19.171}$$

The first term comes from the effective charge of the vacancy, while the second one is the electronic polarization shielding of the other F' electron (e_2). A corresponding expression holds good for e_2. The electron affinity χ is accounted for by its experimental value 0.413 eV (KCl).

Further, the authors employ a two-electron Hartree-Fock procedure to find approximate eigenfunctions of (19.160). The energy functional to be minimized reads

$$E = \langle \Psi_v | \mathcal{H}_{HF} | \Psi_v \rangle \tag{19.172}$$

where

$$\begin{aligned} \mathcal{H}_{HF}^{(s)} &= \mathcal{H}_0^{\cdot} + A^2 \int \frac{|\Psi_v(r')|^2}{|\underline{r} - \underline{r}'|}\, d\underline{r}' \\ \mathcal{H}_{HF}^{(t)} &= \mathcal{H}_0^{\cdot} + A^2 \int \frac{|\Psi_v(r')|^2}{|\underline{r} - \underline{r}'|}\, d\underline{r}' - A'^2 \int \frac{\Psi_v(r')\Psi_v^{\cdot}(r')}{|\underline{r} - \underline{r}'|}\, d\underline{r}' \end{aligned} \tag{19.173}$$

are the two-electron Hartree-Fock Hamiltonians for the F' singlet and triplet states, respectively. In both cases, Ψ_v (the trial function) is subjected to the condition that it should be orthogonal to a set of core states Ψ_c . This is guaranteed by taking Ψ_v in the form

$$\Psi_v = \varphi_v - \sum_c \langle \Psi_c | \varphi_v \rangle \Psi_c \tag{19.174}$$

where φ_v is a normalized "smooth" trial function. The resulting Hartree-Fock equations are

$$\begin{aligned} \mathcal{H}_{HF}^{(s)} \Psi_v &= E_v \Psi_v \\ \mathcal{H}_{HF}^{(t)} \Psi_v' &= E_v' \Psi_v' \end{aligned} \tag{19.175}$$

In (19.173) $\mathcal{H}_0'$ is the one-electron part of the F' Hamiltonian

$$\mathcal{H}_0' = -\frac{\hbar^2}{2m}\Delta_1 + V_L'(r_1) + V_C'(r_1) + V_{PI}(r_1) + V_{PE}(r_1) \tag{19.176}$$

A and A' are normalization constants for Ψ_v and Ψ_v' . Substituting (19.173) into (19.172) and using (19.174) results in lengthy expressions for the ground singlet, the triplet, and the excited singlet states of the F' center. The last one of these was obtained by dropping the third (exchange) term in (19.173). The core states Ψ_c were assumed identical to the free-ion states for K^+ and Cl^-. Although the Hartree-Fock method neglects the e-e correlation, the authors consider the latter an important contribution to the binding energy, experimentally the order of the correlation energy for atoms and molecules (about 1 eV per electron pair of opposite spins). The correlation potential is taken in the form

$$V_{corr} = 0.44(7.8 + (3/8\pi)^{1/3} |\rho(r)|^{-1/3})^{-1} \tag{19.177}$$

The one-electron trial wave functions φ_v were taken in the form of the lowest eigenfunctions of a spherical well: 1s- type

$$\varphi_g = \begin{cases} A\sin(k_1 r/R_0)/(k_1 r/R_0) & r \leqslant R_0 \\ B(R_0/l_1 r)\exp(-l_1 r/R_0) & r > R_0 \end{cases} \tag{19.178}$$

for the ground singlet state, and 2p- type

$$\varphi_e = \begin{cases} C\left(\dfrac{\sin(k_2 r/R_0)}{(k_2 r/R_0)^2} - \dfrac{\cos(k_2 r/R_0)}{(k_2 r/R_0)^2}\right)\cos\vartheta\ , & r \leqslant R_0 \\ iD((l_2 r/R_0)^{-2} + (l_2 r/R_0)^{-2})\exp(-l_2 r/R_0)\cos\vartheta\ , & r > R_0 \end{cases} \tag{19.179}$$

as one of the wave functions for the triplet and the excited singlet state, the other being Ψ_g . The constants are determined by matching by slope and value at $r = R_0$. k_1 and k_2 are variational parameters.

After minimization, equation (19.172) gives the electronic energy of F', that is, the work necessary to remove the F' electrons to infinity without letting the lattice to relax. Relaxing the lattice would yield the total energy of the F' configuration. Assuming the ionic polarization to be proportional to the local field at any point, the total energy is

$$E_{total} = E_e + \frac{1}{2}\int V_{PI}\,\rho_f\,dv \qquad (19.180)$$

Here E_e is that part of the energy which is explicitly ionic-polarization independent; it contains the kinetic, perfect lattice, Coulomb interaction, and electronic-polarization terms. ρ_f is the total free-charge density. Eq.(19.180) can be rewritten as follows

$$E_{total} = E_e + \int V_{PI}\,\rho_e\,dv + \frac{1}{2}\int V_{PI}(\rho_+ - \rho_e)\,dv \qquad (19.181)$$

$\rho_f = \rho_e + \rho_+$, ρ_e, and ρ_+ are the electronic and vacancy charge densities, respectively. The last term is interpreted as a purely lattice energy

$$E_L = \frac{1}{2}\int V_{PI}(\rho_+ - \rho_e)\,dv\ , \qquad (19.182)$$

the former two constituting the electronic energy. Now, the F' thermal dissociation energy is given by

$$E_{Th}(F') = E_g(F') - E_g(F) + E_L(F') - E_L(F) \qquad (19.183)$$

$E_g(F)$, as well as $E_{exc}(F)$, have been computed by dropping all terms involving the second electron in (19.173) and using an ionic polarization appropriate to the F center. The F' calculation has been carried out by minimizing (19.172) with $\mathcal{H}_{HF} = \mathcal{H}_{HF}^{(s)}$ using (19.178) for the ground singlet (this yielded $E_g(F')$), and with $\mathcal{H}_{HF} = \mathcal{H}_{HF}^{(t)}$ using (19.179) for the ground triplet (which yielded $E_t(F')$). For the F' excited singlet minimization has been performed with $\mathcal{H}_{HF} = \mathcal{H}_{HF}^{(s)}$ using (19.179) which yielded E_{exc}. The low-energy edge of the optical ionization energy is

$$E_{Opt} = E_g(F') - E_g(F)^{*} + \chi \qquad (19.184)$$

where the asterisk marks ionic polarization appropriate to F'. The following binding energies have been computed for the F' ground singlet: $E_{Th}(F') = 0.491$ eV, $E_{Opt}(F') = 1.05$ eV. These compare very favorably with most experimental data for KCl. The calculations made have also shown that while a ground singlet F' state does exist in KCl, the triplet and the excited singlet do not, since their respective energies, even though negative, are higher than the sum of the energies of an F center and a CB electron. Consequently, such states, if formed, would decay spontaneously. Table 13 presents the contribution of the various energy terms to the F' binding energy.

Table 13

Values of matrix elements (eV)*

Energy	F' center	F center Polarization F'	F center appropriate to F
Kinetic	3.222	1.859	2.121
Point-ion	-14.316	-7.347	-7.470
Ionic polarization	-2.225	-1.112	0.005
Electronic polarization	0.533	1.293	1.265
Coulomb	5.967	-	-
Correlation	-0.920	-	-
Orthogonalization	3.907	1.956	1.811
Total energy	-5.367	-4.320	-2.900

*After Strozier and Dick (1969)

Strozier and Dick have also calculated the F' optical absorption cross-section. For any first-order optical transition from a discrete level to a continuum of states in the dipole approximation:

$$\sigma(\omega) = 4\pi^2 \alpha_f \omega \, |\langle f | \underline{r} \cdot \underline{e} | i \rangle|^2 \, \delta(\hbar\omega - E - I) \, \frac{dn(E_f)}{d\omega} \qquad (19.185)$$

where ω and $\underline{e}$ are the angular frequency and the polarization vector of the incident radiation, $dn/d\omega$ is the density of final states, E_f is the final-state energy. The initial state in absorption is the above-calculated F' ground-state wave function

$$|i\rangle = 2^{-\frac{1}{2}} \left(\Psi_{1s}(1)\,\alpha(1)\,\Psi_{1s}(2)\,\beta(2) - \text{interchange } (1,2) \right) \qquad (19.186)$$

For simplicity, no orthogonalization to the core states is included, since it is concluded to have little effect on the matrix element. The final state is constructed as the antisymmetrized product of an F center wave function Ψ_{1s} and a square-well eigenfunction φ_c :

$$|f> = 2^{-\frac{1}{2}} (\Psi_{1s}(1)\alpha(1)\varphi_c(2)\beta(2) \text{ - interchange } (1,2)) \qquad (19.187)$$

More specifically, φ_c are the continuum eigenstates of

$$\mathcal{H}_0 = -\frac{\hbar^2}{2m} + \begin{matrix} V_o & , r \leqslant R_0 \\ 0 & , r > R_0 \end{matrix} \qquad (19.188)$$

They are believed to approximate well the ionized state of the F' electron following the optical absorption, because of the shielding of the vacancy effective charge by the remaining electron. This is more correctly so outside the vacancy. α and β are the spin factors. For a

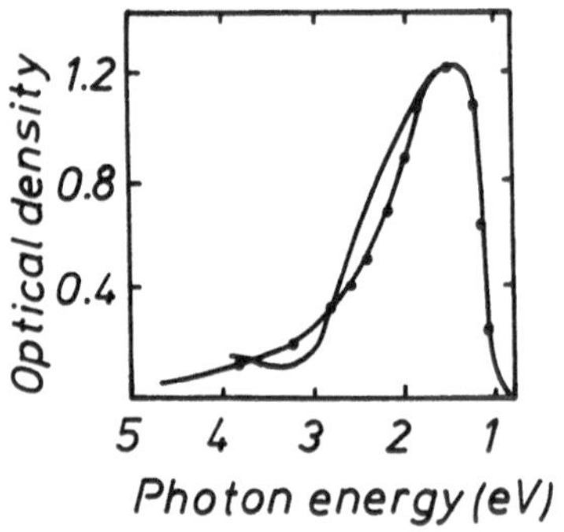

Figure 61: Calculated F' bandshape in KCl using a point-ion approach (continuous line), as compared with an experimental curve by Lüty (1961b) (triangles). (After Strozier and Dick (1969).

calculation of the density of states $dn/d\omega = (dn/dk)(dk/d\omega)$, it is noted that $E_f = \hbar^2 k^2/2m + I = E_k + I$ giving $dk/d\omega = m/\hbar k$. On the other hand, $dn/dk = R/\pi$ where R is the radius of the "normalization sphere". The authors finally got

$$\sigma(E_p) = 12 \frac{1}{2^{\frac{1}{2}}\hbar^3} R_0^5 m^{3/2} \alpha_f E_p (E_p - I)^{\frac{1}{2}} |F(R_0, V_0, E_p - I)|^2 \qquad (19.189)$$

where I is the F' ionization threshold, F( ) is a slowly varying function of the energy. The F' bandshape calculated by Strozier and Dick at $I = 1$ eV and $V_o = 2$ eV is shown in Fig.61, in good agreement with experimental data by Luty (1961b).

19.3.4. Semi-empirical models

Berezin and Kirii (1969) have used the delta-potential model of Bethe and Peierls which approximates the potential of short-range forces. They argue that the model should be applicable to negatively charged color centers, such as the F', where the potential field of the neutral core (F center) acting on the loosely bound electron falls down rapidly as the distance r from the center of the vacancy increases. This model considers a bound state with eigenvalue $E < 0$ in the field of a spherical well of radius R_0 and depth V_0 which turns delta-like by setting $R_0 \rightarrow 0$ and $V_0 \rightarrow \infty$ so that the product $V_0 R_0^2$ remains finite while E is unchanged. The δ- potential induces a single bound eigenstate

$$\Psi(r) = (\lambda/2\pi)^{\frac{1}{2}} \frac{1}{r} \exp(-\lambda r) \qquad (19.190)$$

of energy

$$E = -(\hbar^2/2m)\lambda^2 \qquad (19.191)$$

Next, the photoionization cross-section is computed according to

$$\sigma(\omega) = L \frac{2^{7/2}}{3} \pi \alpha_{hf} \lambda (\hbar\omega - \hbar\omega_o)^{3/2}/(\hbar\omega)^3 , \quad \omega \geqslant \omega_o \qquad (19.192)$$

Here $\omega_o = (\hbar/2m)\lambda^2$. $\sigma(\omega)$ exhibits a maximum at $\omega_{max} = 2\omega_o$ $(\hbar/m)\lambda^2$. Consequently,

$$\lambda = (m\omega_{max}/\hbar)^{\frac{1}{2}} = (2m\omega_o/\hbar)^{\frac{1}{2}} \qquad (19.193)$$

Results of a calculation using (19.192) and (19.193) in which $\hbar\omega_o$ is taken from experiment (Pick (1938)) are presented in Table 14.

Table 14

Delta-potential F' calculation*

Crystal	$\hbar\omega_o$ (eV)	$\hbar\omega_{max,th}$ (eV)	$\hbar\omega_{max,exp}$ (eV)	$\sigma_{max,th}$ ($\AA^2$)	$\sigma_{max,exp}$ ($\AA^2$)
NaCl	1.2	2.4	2.4	0.15	1.1
KCl	1.1	2.2	1.7	0.16	1.0
KBr	0.9	1.8	1.4	0.20	1.0

*After Berezin and Kirii (1969)

In another semiempirical model, Berezin (1971) again considers the F' center problem as that of a weakly bound electron in the field of a neutral F center. Following the analogy between F' and the H^- ion having similar optical binding energies (about 1 eV (F') and 0.75 eV (H^-)), he proposes the Hulthen potential

$$V(r) = -\frac{be^{-ar}}{1 - e^{-ar}} \qquad (19.194)$$

to approximate for the actual F center field. As a matter of fact, Hulthen's potential has proved to be a good approximation to the field created by the H atom core in the case of the H^- ion. The exact eigenfunction of Hulthen's Hamiltonian is given by

$$\Psi(r) = N\frac{1}{r}(1 - e^{-ar})e^{-\gamma r} \qquad (19.195)$$

with eigenvalue

$$E = -\gamma^2/2 \qquad (19.196)$$

Here the parameters are

$$\begin{aligned} b &= \gamma a + a^2/2 \\ N &= B(\gamma/2\pi)^{\frac{1}{2}} \qquad (19.197) \\ B &= (1 + 3(\gamma/a) + 2(\gamma/a)^2)^{\frac{1}{2}} \end{aligned}$$

in atomic units. It is clear that the F' ground-state energy E in this model should give the F' optical binding energy $\hbar\omega_0$; thus,

$$\gamma = (2\hbar\omega_0)^{\frac{1}{2}} \qquad (19.198)$$

Similar arguments applied to the F center yield

$$a = (8E_{F,peak}/3)^{\frac{1}{2}} \qquad (19.199)$$

where $E_{F,peak}$ is the energy at the F band peak. Further, Berezin derives the F' photoionization cross-section in the form

$$\sigma(\omega) = L\frac{2^{7/2}}{3}\pi\alpha_{hf}\gamma B^2\frac{(\hbar\omega - \hbar\omega_0)^{3/2}}{(\hbar\omega)^3}\left(1 - \left(1 + \frac{b}{\hbar\omega}\right)^{-2}\right)^2 \qquad (19.200)$$

where L is the reduced Lorentz factor

$$L = (n^2 + 2)^2/9n \qquad (19.201)$$

The results of Berezin's calculation are presented in Table 15, where the experimental data are again by Pick (1938)). Comparing with the delta-potential calculations in Table 14 reveals an improved agreement of Hulthen's potential model with the experiment. This implies that Hulthen's wave function (19.195) has a better asymptotic behavior than

(19.190) because the main contribution to the cross-section comes from large r regions.

Table 15

Hulthen-potential F' calculations*

Crystal	$\hbar\omega_{o,exp}$ (eV)	$E_{F,peak}$ (eV)	$\hbar\omega_{max,th}$ (eV)	$\hbar\omega_{max,exp}$ (eV)	$\sigma_{max,th}$ (A^2)	$\sigma_{max,exp}$ (A^2)
NaCl	1.2	2.77	2.3	2.4	0.46	1.1
KCl	1.1	2.31	2.1	1.7	0.49	1.0
KBr	0.9	2.06	1.7	1.4	0.63	1.0

*After Berezin (1971)

Good asymptotics is perhaps the main factor explaining the remarkable agreement of semi-empirically calculated F center lifetimes with experimental data for NaCl, KCl, KBr, and RbCl (Berezin (1969b)).

Generally, the theoretical approaches to calculating the electronic structure of color centers employed by the Leningrad school are reviewed by Petrashen et al. (1970a,b). While pointing out that F' and H^- are similar to each other, the authors also stress some essential differences: Inasmuch as the F' radius (5 atomic units) exceeds the H^- radius (2 atomic units), the e-e electrostatic repulsion is smaller for F', and one obtains an F' bound state without accounting for the e-e correlation, otherwise mandatory for the H^- ion.

19.4. F' thermal ionization rate

Cheban and co-authors have so far been the only workers to intentionally seek a theoretical expression for the F' thermal-dissociation rate $\tau_{F'}^{-1}$. Cheban (1961) based his calculation on the polaron (continuum) approach to the color centers. The F' thermal ionization was assumed to result from the interaction with the acoustical phonons of the lattice. The rate $\tau_{F'}^{-1}$ is given by

$$\tau_{F'}^{-1} = \int P_{sk}\, \rho(k)\, dk \qquad (19.202)$$

with

$$P_{sk} = (2\pi/\hbar)\ \mathrm{Av}\Big(\sum_{N_q'} \sum_{n_u'} |< k\,|\,W\,|\,s >|^2\ \delta(E_k - E_s)\Big) \qquad (19.203)$$

Here

$$W = -(\underline{u}_1 \cdot \mathrm{grad}\, w(\underline{r}_1) + \underline{u}_2 \cdot \mathrm{grad}\, w(\underline{r}_2)) \qquad (19.204)$$

is the interaction operator, assumed to be of the Bethe-Sommerfeld type,

$$\underline{u} = (d/L)^{3/2} \sum_{q,j} \underline{I}_{qj}(a_{qj}\, e^{i\underline{q}\cdot\underline{r}} + a_{qj}^{*}\, e^{-i\underline{q}\cdot\underline{r}}) \qquad (19.205)$$

is the acoustic phonon displacement, a_{qj} and a_{qj}^{*} are the acoustic phonons' creation and annihilation operators, respectively, of wave vector $\underline{q}$ and polarization vector $\underline{I}$, $w(\underline{r})$ is the periodic potential of the frozen-in lattice. The eigenstates are

$$|i> = |s> = \Psi_{F'}(r_1)\,\Psi_{F'}(r_2)\,\varphi_s(\xi) \prod_u \Phi_{n_u}(q_u' - r_u)\,\eta(N_{qj}) \qquad (19.206)$$

$$|f> = |k> = \Psi_p(|\underline{r}_1 - \underline{\xi}|)\,\Psi_F(\underline{r}_2)\,\varphi_k(\underline{\xi}) \prod_u \Phi_{n_u'}(q_u' - q_{u_F})\,\eta(N_{qj}') \qquad (19.207)$$

with eigenvalues, correspondingly,

$$E_s = I_{F'} + \hbar\omega \sum_u (n_u + \tfrac{1}{2}) + \sum_{qj} \hbar\omega_{qj}(N_{qj} + \tfrac{1}{2}) + \tfrac{3}{2}\hbar\omega \qquad (19.208)$$

$$E_k = I_F + I_P + \hbar\omega \sum_u (n_u' + \tfrac{1}{2}) + \sum_{qj} \hbar\omega_{qj}\,(N_{qj}' + \tfrac{1}{2}) + E_k' \qquad (19.209)$$

Here Ψ_F and $\Psi_{F'}$ are the one-electron wave functions of F and F', φ_k is the free-polaron wave function, φ_s and φ_p are the polaron-electron contributions to F' and F, respectively, the I's are the minimized values of the adiabatic functionals in either case, η is the acoustic phonon amplitude, q_u' are the equilibrium lattice coordinates (for a free lattice, an F' center, a polaron), $r_u = q_{uF} - q_{uo}$. The following trial functions have been used:

$$\Psi_F = (\alpha_F^3/\pi)^{\frac{1}{2}} \exp(-\alpha_F r) \qquad \text{(F electron)}$$

$$\Psi_{F'} = (\alpha_{F'}^3/\pi)^{\frac{1}{2}} \exp(-\alpha_{F'} r) \qquad \text{(F' electron)}$$

$$\varphi_p = (\alpha_p^3/\pi)^{\frac{1}{2}} \exp(-\alpha_p r) \qquad \text{(polaron electron)}$$

$$\varphi_s = (\alpha_s^3/\pi)^{\frac{1}{2}} \exp(-\alpha_s r) \qquad \text{(polaron electron)}$$

and also

$$\varphi_k = L^{-3/2} \exp(i\underline{k}\cdot\underline{\xi}) \qquad \text{(free polaron)}$$

E_k' is the free-polaron kinetic energy. Cheban arrives at

$$P_{sk} = \omega Q \beta^{-1} I_e(z) \exp(-\frac{a_{F'}}{2}\coth\frac{\beta}{2} - \frac{l\beta}{2}) , \quad \beta = \hbar\omega/k_BT \qquad (19.210)$$

Here

$$Q = \frac{4MG^2L^3k}{9\pi^2\hbar^2 c_{11}}(A_o^2|A_o'|^2 + A_oA(A_o'^*A_1' + A_o'A_1'^*) + A_1^2|A_1'|^2) \qquad (19.211)$$

in which the following notations have been used:

$$A_o = \int\Psi_F(r)\Psi_{F'}(r)d\underline{r} , \quad A = \int\Psi_F(r)\Psi_{F'}^2(r)d\underline{r} ,$$
$$A_1^2 = \int\Psi_F^2(r)\Psi_{F'}^2(r)d\underline{r} , \quad A_1' = \int\varphi_k(\xi)I_o(\xi)d\underline{\xi} ,$$
$$A_o' = \int\varphi_k(\xi)\varphi_s(\xi)\Psi(\xi)d\underline{\xi} \qquad (19.212)$$

G is the Bethe-Sommerfeld constant, c_{11} is the elastic modulus. is the LO-phonon frequency, ω_{qj} and N_{qj} are the frequencies and the occupation numbers of acoustic phonons at given (q,j), n_u is the number of LO phonons at given u . $I_e(z)$ is Bessel's function of an imaginary argument ($e \sim m_o$)

$$z = a_{F'}/2\sinh(\beta/2) \qquad (19.213)$$

where

$$a_{F'} = \sum_u (q_{u_{F'}} - (q_{u_F} + q_{u_o}))^2 ,$$

while $m_o\hbar\omega = I_F + I_P - I_{F'} - \frac{3}{2}\hbar\omega$. In fact, the right-hand side of the last equation should give the F' thermal-dissociation energy: $m_oh = E_{F'}$. The quantity l in (19.210) is defined by

$$\hbar\omega l = E_{F'} + \hbar^2k^2/2M \qquad (19.214)$$

where M is the polaron effective mass. $\hbar\omega(l - m_o)$ is, therefore, the excess energy over $E_{F'}$ due to the polaron kinetic energy.

Next, the effective density of states in k-space is determined by

$$\rho(k)dk = 4\pi(L/\pi)^3 k^2 dk \qquad (19.215)$$

Using (19.214) this turns into

$$\rho(k)dk = 4(L^3/\pi^2)(M\omega/h)k\, dl \qquad (19.216)$$

We finally insert (19.210), (19.216), and (19.214) in (19.202) to obtain $\tau_{F'}^{-1}$.

Cheban presents the results of his numerical calculations for RT: $\tau_{F'}^{-1} = 7 \times 10^{12}$ s^{-1} (NaCl) and 5×10^{10} s^{-1} (KCl) . Both misfit by orders of magnitude the experimental values (e.g. $\sim 10^3$ s^{-1} (KCl)). Apparently, the enormously high rates calculated result from the very low thermal ionization energies predicted by the polaron theory. This is supported by calculations of the trapping cross-section of the F center based on eq.(4.59) using the above estimates. They give σ_F = 3×10^{-14} cm^2 (NaCl) , 6.2×10^{-15} cm^2 (KCl) . In the latter calculation, the F' thermal ionization energy cancels.

The reverse process of direct multiphonon polaron capture by an F center with the formation of an F' in ground state has also been dealt with by Perlin, Cheban, and Tsukerblat (1961). However, Cheban (1961a) had insisted that the F' center should have at least one bound excited state, corresponding to polaron motion in some effective short-range field. Accordingly, Tsukerblat and Cheban (1964) have considered a two-step capturing mechanism with the intermediate formation of an excited F' center, followed by nonradiative transition to the F' ground state, in addition to the direct capture process. Polaron scattering by F centers has also been dealt with. The corresponding capture cross-sections are given by

$$\begin{aligned} \sigma_{F,\text{capture}} &= \sigma_F^* (1 - \eta_{F',\text{ion}}) \\ \sigma_{F,\text{scatter}} &= \sigma_F^* \eta_{F',\text{ion}} \end{aligned} \qquad (19.217)$$

Here $\eta_{F',\text{ion}}$ is the probability for thermal ionization of the F' excited state. The authors first derive the presumed F'* thermal ionization rate, assuming one-phonon transitions to free-polaron states due to LA-phonon coupling, and then use the detailed balance (4.59) to calculate σ_F^* .

Kitada, Kakui, and Tomura (1968) and also Tomura and Kitada (1967) have applied Kubo & Toyozawa's theory to explain their data on the temperature dependence of the thermal lifetime of F' centers in KBr and F_A' centers in KCl:Na crystals, respectively. A two-slope temperature dependence in Arrhenius coordinates was observed in both cases and explained by means of eq's (19.28) and (19.29), correspondingly, for the higher and the lower temperature ranges. The agreement between theory and experiment is shown in Figures 8 and 62, respectively. Kubo and Toyozawa's theory refers to a configurational-coordinate situation of the weak-coupling type depicted in Fig.51. The values of the parameters in equations (19.28) and (19.29) obtained from the two fits are summarized in Table 16.

Although these values can be considered to be generally satisfactory, the small E_0 suggests thermal excitation to a higher excited rather than ionized state. There seems to be no other cases reported in the literature so far of an observed two-slope temperature dependence for the F' thermal lifetime.

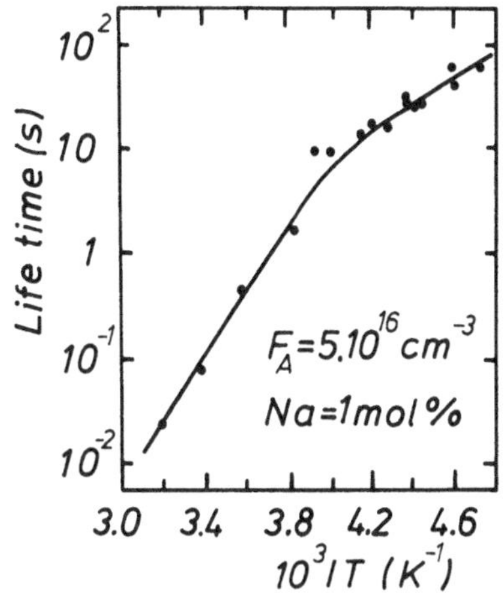

Figure 62: Temperature dependence of the F'_A(Na) thermal lifetime in KCl. The solid line is the best fit of Kubo&Toyozawa's equations (19.28) and (19.29). (After Tomura and Kitada (1967)).

Table 16

F' thermal lifetime parameters*

Crystal	E_o (eV)	E* (eV)	ω (s^{-1})	γ	γ'	γ''
KBr	0.21	0.52	3.4×10^{14}	0.3	0.3	0.3
KCl:Na	0.28	0.60	2.25×10^{13}	0.31	0.16	0.23

*After Tomura and Kitada (1967)

19.5. Bound-polaron theory of the F-F' conversion

A series of very recent papers have been published aimed at working out a novel insight into the nature of the F-F' conversion in alkali halides. The physical background is based on two puzzling occurrences at the F center, already mentioned:

(i) the apparent ionizability of the relaxed excited state (F*) at low temperature (Brown (1966), Bosi, Podini, and Spinolo (1968));

(ii) the optical appearance of "ghost" polaron states related to the F center (Park and Faust (1966), Borms and Jacobs (1971), Kondo and Kanzaki (1975), Carlier and Jacobs (1978a,b), Jacobs (1985), Brandt (1972)

The point is that you cannot at all explain (i) in terms of the usual transition to a conduction band (CB) state, for this would require a positive reaction heat rendering the (endothermic) process physically impossible at low temperature. Then, what if you regard ionization as merely the transition to some appropriate polaron state assuming that (i) and (ii) are interrelated? Clearly, (i) would follow immediately, provided the latter transition requires negative or no reaction heat at all.

The above viewpoint gained considerable support on constructing a configurational-coordinate diagram of the KCl F center using empirical data on the optical absorption, emission, and ionization energies of F and F*, as mentioned in 19.1.1. and shown in Fig.53 (Georgiev, Gochev, Christov, and Kyuldjiev (1982)). Definitely, an exothermic "ionization" reaction emerges for F* which involves the transition to a virtual polaron state. A similar configurational-coordinate diagram resulted from a theoretical calculation based on a simple vibronic extension of the semicontinuum potential of the F center in NaI, as seen in Fig.63 (Georgiev (1984a,1985b)).

19.5.1. Nature of the polaron state

Semicontinuum electronic potentials for the F center have been discussed in 19.3.2. The vibronic extension assumes coupling to a A_{1g}-mode through modulating both the depth V_0 and width r_0 of the spherical well by the phonon coordinate q, while leaving the Coulomb tail unchanged:

$$V_e(r,q) = -V_0(r_0,q)(1 - N(r-r_0-q)) - (e^2/\varepsilon r)N(r-r_0-q) \qquad (19.218)$$

where $N(\ )$ is Heaviside's step function, and r is the electron radial coordinate. V_0 contains both Madelung's energy, and the vacancy self-energies relative to electronic polarization and to ionic polarization, induced by the electron cloud outside the cavity. The effective dielectric constant ε was calculated by means of Krumhansl & Schwartz's prescription, even though using a value averaged over the outer portions of the static electronic wave functions at $q = 0$ to linearize the

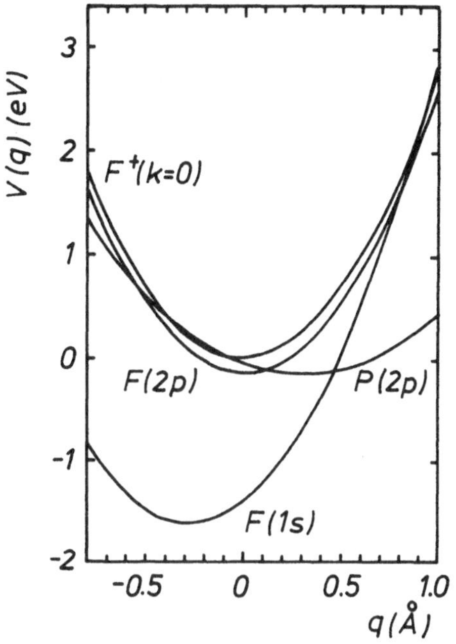

Figure 63: Calculated configurational-coordinate diagram of the F center in NaI depicting the diabatic potentials (parabolas) of the 1s- and 2p- like F center states, and of the 2p-like bound polaron state versus the coordinate q of an assumed promoting A_{1g}- type vibration. The diabatic potential of the lowest conduction band state pertaining to the F center is also shown. (After Georgiev (1984a)).

Schrodinger equation. Nevertheless, the static eigenvalue problem remained self-consistent because of the wave function dependence of V_o and ε . The static polaron problem was treated likewise but more simply, since now $\varepsilon = \varepsilon_p = (1/\varepsilon_\infty - 1/\varepsilon_o)^{-1}$, the reciprocal high-frequency and static dielectric constants, respectively, and the bottom potential is simply $V_o(r_o,q) = e^2/\varepsilon_p(r_o + q)$. Once all this has been done and the static eigenstates $|t>$ and eigenvalues E_t^o found, the diabatic vibronic potentials were calculated from

$$V_t(q) = \frac{1}{2} M_t\omega_t^2 q^2 + b_t q + E_t^o \tag{19.219}$$

where M_t and ω_t stand for the mass and angular frequency of the electron-coupled oscillator, while

$$b_t = < t|\partial V_e(r,q)/\partial q_{(q=0)} | t > \tag{19.220}$$

is the diagonal matrix element of the linear electron-lattice coupling operator. A breathing-mode oscillator (BMO) comprising the synchronous vibrations of the six cations nearest-neighboring the vacancy was assumed to couple to the bound F center states, while the polaron state coupled to the A_{1g} mode of the whole lattice. The resulting diabatic parabolae for the 1s- like ground and 2p- like excited F center states, as well as the one of the 2p- like bound polaron state, are shown in Fig.63. The degeneracy of the p- like states was not taken into account. Also shown therein is the lowest (k=0) CB state parabola in BMO coupling, for the sake of comparison.

However, the reliability of these diabatic energy calculations should not be overestimated, mainly due to uncertainties in the electronic nature and vibronic coupling of the excited F state (Georgiev (1984b)). Too simple as it may be, the semicontinuum vibronic model described above, while allowing for an unified description of the excited F center, pertaining to both the Franck-Condon state and the relaxed-excited state, has sacrificed the diffuse nature of the latter. Nevertheless, the model displays several key features, some of which are in a more-than-qualitative agreement with experiments:

(i) There are two conceivable emission bands at $E_E^I = 1.27$ eV and $E_E^{II} = 0.55$ eV due to vertical F(2p)-F(1s) and P(2p)-F(1s) transitions, respectively. F(2p) populates through optical absorption at $E_A = 1.69$ eV from the ground F(1s) state, while P(2p) does so through subsequent horizontal nonradiative transition from F(2p).

(ii) There is a crossover barrier of 1.93 eV between $V_{F(1s)}(q)$ and $V_{F(2p)}(q)$. Horizontal transitions through this barrier may contribute to quenching E^I . Although the present calculation shows the barrier to be too high to render such transitions competitive at all, there are indications that its actual height may be lower.

(iii) A crossover barrier of 0.02 eV forms between $V_{F(2p)}(q)$ and $V_{P(2p)}(q)$. At the same time, the reaction heat $-E^o_{F(2p)}$ associated with a F(2p)-CB transition is considerably higher (0.14 eV). This implies that the F* ionization may mainly occur through nonradiative transitions to the polaron state. These transitions, effective even at low temperature via tunneling through the barrier, will tend to populate P(2p) at the expense of F(2p) thereby giving rise to E^{II} at the expense of E^I . From experimental data (Baldacchini, Pan, and Luty (1981)) the zero-point F*-polaron transition time can be estimated to be about 2×10^{-7} s , based on an extrapolated low-temperature residual ionizability. Whatever the numerical value, experiments on the delay of E^{II} relative to E^I can be crucial in checking the virtues of the model.

(iv) There is a low barrier (0.03 eV) between $V_{P(2p)}(q)$ and $V_{F(1s)}(q)$. Transitions across it can bring about a depopulation of the polaron state through deexcitation to the ground state.

Giving credit to the model is the observation of two emission bands at 0.75 eV and 0.55 eV, pumped by optical absorption at 2.08 eV in the NaI F center (Baldacchini, Pan, and Lüty (1981)). One is tempted to ascribe them to E^{I} and E^{II}, respectively. While the numerical comparison is quite favorable for E_E^{II}, it is not that good for E_E^{I} and E_A, over- and under- estimated by the model, respectively. The two misfits may be interrelated, both arising from an underestimation of the lattice-displacement energy $Sh\nu$ (S- the Huang-Rhys factor) by the model calculation of $V_{F(2p)}(q)$. Introducing changes in both the force constant $M_{F(2p)}\omega^2_{F(2p)}$ and the electron-phonon coupling coefficient $b_{F(2p)}$ one could displace the excited-state parabola so as to fit the empirical data (Georgiev (1984b)). This displacement also results in lowering the $V_{F(2p)}$-$V_{F(1s)}$ crossover barrier down to 0.32 eV. Inasmuch as the point reached on absorption from F(1s) now is at 0.73 eV above the $V_{F(2p)}(q)$ minimum, deexcitation to the ground state through nonequilibrium horizontal transitions (Dexter-Klick-Russell mechanism) is also conceivable in addition to equilibrium ones (see Section 5.1).

The Dexter-Klick-Russell (DKR) mechanism has inspired renewed interest as a possible source of nonradiative luminescence quenching at isolated F centers (Bartram and Stoneham (1975)). It would lead to the virtual nonattainability of the relaxed excited state, for the system would deexcite to the ground electronic state while relaxing. Recently, arguments have been presented against the presumed efficiency of the process, considering it to be a low-chance competitor of the vibrational relaxation in the excited electronic state (Georgiev (1984b,c,1985a). That relaxation ultimately leads to an equilibrium thermal population of the vibronic levels in the latter state. Once equilibrium has been attained, one can calculate the rate τ^{-1}_{nrad} of nonradiative transitions across the $V_{F(2p)}(q)$-$V_{F(1s)}(q)$ barrier. Clearly, the DKR process, if effective enough, would lead to disagreement between calculated equilibrium rate and experiment. On its hand, miscalculating the nonequilibrium and equilibrium F* deexcitation rates would result in the inaccurate determination of the F*-polaron transfer rate. Nevertheless, the latter rate has been dealt with on the premise of it being largely superior to the former two at the F center in NaI (Georgiev (1984a)).

With the prediction of a bound-polaron state coupled inherently to the F* ionization process and resulting straightforwardly from the vibronic model, the foregoing considerations of the merits of that model can also be expected to throw some light on the nature of the polaron state involved. The presumed vibronic potential of the polaron

$$V_p(r,q) = -(e^2/\varepsilon_p(r_o + q))(1 - N(r-r_o-q)) - (e^2/\varepsilon_p r)N(r-r_o-q) \tag{19.221}$$

is merely the A_{1g}- modulated extension (perhaps not unique) of the truncated-Coulomb polaron potential considered elsewhere by Mott and Davis (see Georgiev (1984a)). What is also important is that the 'polaron radius' r_o has been chosen to match the spherical-cavity radius of the semicontinuum F center potential. In the terms of Mott's empirical approach, this implies a bound-to-defect polaron (to an F center in the particular case) rather than a free-polaron entity whose radius is determined through variational considerations. It seems that as though the 'bound-to-the-F-center polaron' identification for the entity described by $V_p(r,q)$ is physically appropriate. The word is of a neutral F center, for the Coulomb field of the empty vacancy would add up otherwise.

The calculations leading to Fig.63 have shown the bound-polaron state to couple strongly with the excited F state to some appropriate 'ionization-promoting mode'. The latter can well be the effective vibration resulting from the superposition of a local mode at the F* center and the LO lattice mode at the polaron site (Georgiev and Mladenov (1986)). Just how far that site is from the anion vacancy can be estimated from the value of the electron-exchange matrix element V_{12} of the F*-polaron transfer obtained by comparing the computed zero-point rate with the experimental residual ionizability, as explained in 19.1.1. The order of 4 to 5 lattice spacings has been obtained (Georgiev (1984a)). Earlier quasiclassical estimates had yielded about 12 lattice spacings for KCl (Georgiev, Gochev, Christov, and Kyuldjiev (1982)). Recent more accurate quantum-mechanical calculations have produced the order of the average F center separation for a statistically-distributed dilute F centered system (Georgiev and Mladenov (1986)). This last estimate lends strong support to the polaron-bound-to-an-F-center interpretation, as opposed to an alternative free-polaron option (cf. Cheban (1963)).

19.5.2. F-F' conversion in dilute systems

The foregoing suggestions have shaped the following scenario for the optical conversion of F centers into F' centers in dilute F centered systems: An excited F center electron, formed by photobleaching in the

F band, passes by virtue of the coupling to some promoting vibration into a polaron state, bound to another F center, before forming an F' with the latter center. The F*-polaron transfer is effective over large distances because of the large spatial extension of the electron clouds at F* and the polaron. Unlike the F-F' conversion scheme involving a thermal ionization transition to CB states and a subsequent electron capture by another F center at random (see Section 4.1.), the detachment of the electron cloud from the original anion vacancy now involves configurational tunneling in an exothermic process which occurs spontaneously and does not need any thermal agitation. This leads automatically to finite conversion rates even at low temperature.

Whether through thermal agitation or through configurational tunneling, the F-F' conversion efficiency will obtain as

$$\eta_{F\text{-}F'} = 2k_{dis}/(k_{dis} + \tau_F^{-1}) \tag{19.222}$$

where k_{dis} is the dissociation rate of the original F center, and τ_F is the F center lifetime. Georgiev and Mladenov (1986) express k_{dis} by means of the reaction-rate method, in the spirit of the polaron concept, and compare eq.(19.222) to experimental F-F' efficiencies in five alkali halides. They use (see eq.(19.31)):

$$k_{dis}(T) = (1 - \exp(-h\nu/k_BT))(k_{dis}^0 + \sum_{n=1}^{m} W_{el,sub}(E_n)W_{conf}(E_n) \times \exp(-nh\nu/k_BT) + \sum_{n=m+1} W_{el,over}(E_n)\exp(-nh\quad/k_BT) \tag{19.223}$$

where $E_n = (n + \frac{1}{2})h\nu$, $m = INT(E_b/h\nu - 1/2)$, E_b is the barrier height at the $V_{F*}(q)$-$V_P(q)$ crossover, ν is the promoting-mode frequency. $W_{el}(E_n)$ is the electron-transfer probability at sub- ($E_n < E_b$) and over- ($E_n > E_b$) barrier levels E_n, respectively. $W_{conf}(E_n)$ is the configurational-tunneling probability at E_n. Introducing Landau-Zener's parameter

$$\gamma_n = (V_{FP}^2/2h\nu)\, E_r^{-\frac{1}{2}}\, |E_n - E_c|^{-\frac{1}{2}} \tag{19.224}$$

$W_{el}(E_n)$ reduces to

$$W_{el,sub}(E_n) = 2\pi\gamma_n \tag{19.225}$$

$$W_{el,over}(E_n) = 4\pi\gamma_n$$

for a nonadiabatic electron transfer ($\gamma_n \ll 1$)(Christov (1982)). Here E_r is the configurational-displacement energy, related to the barrier height E_b and the crossover energy E_c by means of

$$E_c = (E_r + Q)^2/4E_r$$
$$E_c = E_b + V_{FP} \tag{19.226}$$

where $2V_{FP}$ is the resonance crossover splitting of the F* and polaron adiabatic potentials, Q is also the reaction heat at 0 K (see also Georgiev (1985a,b)).

For an exothermic process ($Q < 0$):

$$W_{conf}(E_n) = (F_{op}^2/2^p p!)\exp(-p^2 h\nu/E_r - E_r/h\nu) \tag{19.227}$$

with

$$F_{op} = \xi^o H_p(\xi^c - \xi^o) + 2pH_{p-1}(\xi^c - \xi^o)$$
$$p = -Q/h\nu \; , \; \xi^o = (2E_r/h\nu)^{\frac{1}{2}} \; , \; \xi^c = (2E_c/h\nu)^{\frac{1}{2}} \tag{19.228}$$

Accordingly, the zero-point rate has been calculated from

$$k_{dis}^o = \nu \, W_{el,sub}(E_o) \, W_{conf}(E_o) \tag{19.229}$$

At higher E_n $(n \geq 1)$, $W_{conf}(E_n)$ has been calculated quasiclassically

$$W_{conf}(E_n) = (1 + \exp(K_1(E_n) + K_2(E_n))^{-1} \tag{19.230}$$

where $K_j(E_n)$ are given by

$$K_j(E_n) = (2E_c^j/h\nu)(a_j^{\frac{1}{2}} - (1 - a_j)\ln((1 - a_j)^{\frac{1}{2}}/(1 - a_j^{\frac{1}{2}}))) \tag{19.231}$$

with $E_c^j = E_c$ $(j = 1)$, $= E_c - Q$ $(j = 2)$; $a_j = (E_c - E_n)/E_c^j$. Equations (19.224) through (19.231) hold good for harmonic potential-energy profiles, which can only be close to the actual adiabatic energy sheets at small $V_{FP} \ll E_b$. Such is the presumption leading to equations (19.225) for nonadiabatic electron transfers, one of the premises of the authors' particular calculation.

The comparison between theory and experiment is shown in Fig.64.for dilute F centered systems in five crystals. The solid lines therein have been obtained by fitting eq.(19.222) to the experimental temperature dependences of $\eta_{F-F'}$. This yields values for the four fitting parameters ν, V_{FP}, E_b, and Q. A constant τ_F was used, equal to the radiative lifetime in each particular case, which implies that: (i) any eventual nonradiative deexcitation to the ground F center state has been discarded, and (ii) the whole F-F' formation efficiency has been attributed to the F*-polaron transfer rate, as given by eq.(19.223) thus neglecting any possible contribution of the CB channel at higher temperatures. The obtained fitting parameters are listed in Table 17 (the last four columns).

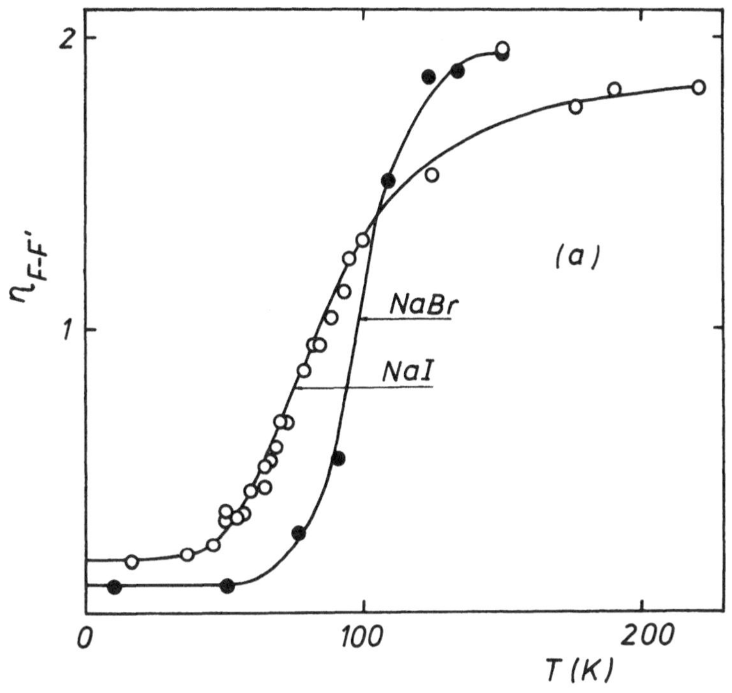
2
1
0
100
200
T (K)
$\eta_{F-F'}$
(a)
NaBr
NaI

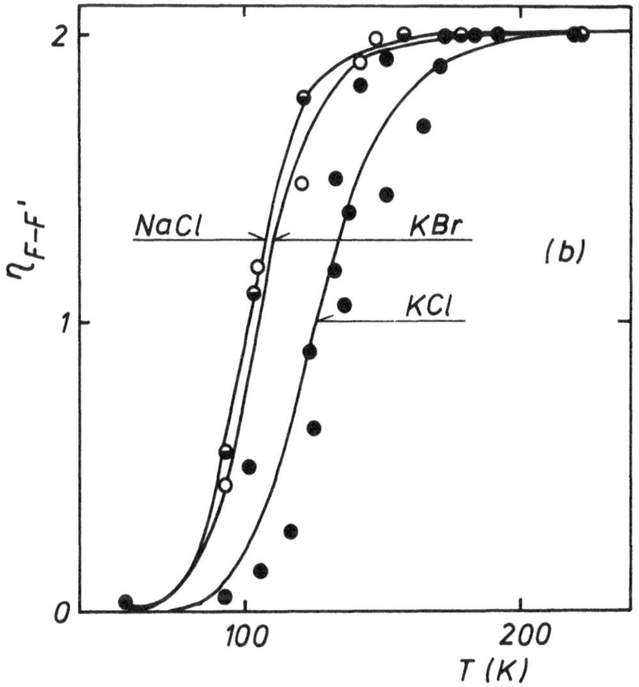
2
1
0
100
200
T (K)
$\eta_{F-F'}$
(b)
NaCl
KBr
KCl

Figure 64: Comparison between experiment and bound-polaron theory of the F-F' conversion efficiency in several alkali halides. Experimental points are depicted by circles. Solid lines are best fits of eq.(19.222) with k_{dis} from (19.223). The fitting parameters are those listed in Table 17. A non-adiabatic electron transfer from excited F centers to bound polarons in a dilute F centered system is presumed. (After Georgiev and Mladenov (1986)).

Table 17

Fitting parameters of the F-F' conversion rate*

Crystal	τ_F (s)	E_b (eV)	V_{FP} (eV)	$h\nu$ (eV)	$Q/h\nu$
NaCl	1.0×10^{-6}	0.1550	0.01134	0.0240	-2
NaBr	1.6×10^{-7}	0.1580	0.02409	0.0248	-4
NaI	2.0×10^{-8}	0.0423	0.00024	0.0175	-2
KCl	5.7×10^{-7}	0.1800	0.00704	0.0260	-6
KBr	1.1×10^{-6}	0.1533	0.00800	0.0213	-6

*After Georgiev and Mladenov (1986)

19.5.3. Electron-exchange matrix element

From the viewpoint of identifying the nature of the presumed F*-polaron transfer, the fitted values of the electron-exchange matrix element V_{FP} are perhaps most informative. They all are in a nearly perfect accord with the assumed nonadiabaticity leading to (19.225). Small electron-coupling terms imply a nearly perfect orthogonality of the electronic states involved. With the predicted equal parity of the electronic wave functions pertaining to the process in NaI (Georgiev (1984a)), the nonadiabaticity results from the large F center - polaron separations leading to small wavefunction overlap. Inasmuch as such are the average F center separations in lightly colored crystals, the non-adiabatic occurrence thus gives credit to the bound-polaron, rather than to the free-polaron preposition, the latter not necessarily requiring large separations.

Recently, the electron resonance energy of a distant donor-acceptor pair has been derived as a function of the pair separation R by means of a quasiclassical approach assuming Coulomb potentials (Georgiev, Gochev, Christov, and Kyuldjiev (1982)). It reads

$$V_{FP}(R) = (h\nu_e/\pi)\exp(-\beta(R)) \qquad (19.232)$$

with

$$\begin{aligned} \beta(R) &= 4R_o(-2m_eE_e/\hbar^2)^{\frac{1}{2}}\left(\frac{E(k)}{1-k^2} - K(k)\right) \\ R_o &= -e^2/\varepsilon E_e \qquad (19.233) \\ k &= (1 - 4R_o/R)^{\frac{1}{2}} \end{aligned}$$

$K(k)$ and $E(k)$ are the complete elliptic integral of first and second kind, respectively, E_e is the electron energy. The electron beating frequency is

$$\nu_e = (-E_e^3)^{\frac{1}{2}}/2\pi e^2(2m_e)^{\frac{1}{2}} \qquad (19.234)$$

For a more rigorous static-basis quantum-mechanical approach, Schrodinger's equation has first been solved

$$\begin{aligned} H_e^t|t\rangle &= E_t^o|t\rangle \\ H_e^t &= \underline{p}_e^2/2m_e + V_e^t(r,0) \end{aligned} \qquad (19.235)$$

where $t = F$ or $t = P$ for the F center and the polaron, respectively, $V_e^F(r,0)$ and $V_e^P(r,0)$ being correspondingly given by (19.218) and (19.221) at $q = 0$. The radial bound-state wave functions are:

$$\varphi(r) = \begin{cases} A_{Kl}\, j_l(Kr) & \text{, in} \\ B_{\alpha l}\, R_{nl}(\alpha r) & \text{, out} \end{cases} \qquad (19.236)$$

where $j_l(\;)$ are the spherical Bessel functions, while R_{nl} are the radial hydrogenic functions. The incorporated parameters A_{Kl}, $B_{\;l}$, K, and α are determined from

$$\begin{aligned} \langle t|t\rangle_{\text{all space}} &= 1 \\ |t\rangle_{in} &= |t\rangle_{out} \qquad \text{at } r = r_o \qquad (19.237) \\ \frac{d}{dt}|t\rangle_{in} &= \frac{d}{dt}|t\rangle_{out} \qquad \text{at } r = r_o \\ \alpha &= m_e e^2/\varepsilon n\hbar^2 \quad . \end{aligned}$$

We next define an adiabatic Hamiltonian, one for each case,

$$H_{ad}^t = H_e^t + H_{eL}^t + \frac{1}{2}\sum M_l\,\omega_l^2 q_l^2 \qquad (19.238)$$

in which the second term is the electron-lattice interaction Hamiltonian, while the third one is the lattice potential energy in the harmonic approximation. Using (19.238) and (19.236) has led to calculating the diabatic vibronic potentials (19.219) as the diagonal matrix elements of H_{ad}^{t} , for the F center and polaron states separately. To find the adiabatic energy sheets, however, a common adiabatic Hamiltonian of the F center - polaron system must be constructed:

$$H_{ad} = H_{ad}^{F} \, | F \rangle \langle F | + H_{ad}^{P} \, | P \rangle \langle P | \qquad (19.239)$$

where $| F \rangle$ and $| P \rangle$ will be assumed orthonormal.

The matrix elements of H_{ad} are:

$$\begin{aligned} V_{tt}(q) &= \langle t | H_{ad} | t \rangle = \langle t | H_{ad}^{t} | t \rangle \\ V'_{st}(q) &= \langle s | H_{ad} | t \rangle = \langle s | H_{ad}^{t} | t \rangle = \langle s | H_{eL}^{t} | t \rangle \end{aligned} \qquad (19.240)$$

We further solve Schrodinger's equation

$$H_{ad} \, | FP \rangle = E \, | FP \rangle \qquad (19.241)$$

by means of the linear combination

$$| FP \rangle = C_F(q) | F \rangle + C_P(q) | P \rangle \qquad (19.242)$$

to obtain the adiabatic eigenvalues

$$E_{U/L}(q) = \tfrac{1}{2}(V_{FF}(q) + V_{PP}(q) \pm ((V_{FF}(q) - V_{PP}(q))^2 + 4G^2 V'_{FP}(q) V'_{PF}(q))^{\frac{1}{2}}) \qquad (19.243)$$

These represent two potential energy sheets for the vibronic problem, lower $E_L(q)$ and upper $E_U(q)$, q standing for the manifold of lattice (configurational) coordinates. The energy splitting between $E_L(q)$ and $E_U(q)$ at crossover ($V_{FF}(q) = V_{PP}(q)$) amounts to twice $| V_{FP} |$, the electron-exchange term, now given by

$$V_{FP}(q) = G(V'_{FP}(q) V'_{PF}(q))^{\frac{1}{2}} \qquad (19.244)$$

Assuming linear electron-lattice coupling to a single mode q ,

$$H_{eL}^{t} = b^{t}(q) q$$

The b-operators in A_{1g}-coupling are (from (19.220)):

$$\begin{aligned} b^F(r_F) &= (V_o + \chi) r_o^{-1} (1 - N(r_F - r_o)) - (V_o - e^2/\varepsilon r_o) \delta(r_F - r_o) \\ b^P(r_P) &= (e^2/\varepsilon_p r_o^2)(1 - N(r_P - r_o)) \end{aligned} \qquad (19.245)$$

$V_{FP}(q)$ now obtains from (19.244) and from

$$V'_{PF}(q) = qr_o^{-1}(V_o + \chi) < P|F >_{out,in} - q(V_o - e^2/\varepsilon r_o)r_o^{-1} < P|F >_{r_F = r_o}$$

$$V'_{FP}(q) = qr_o^{-1}(e^2/\varepsilon_p r_o) < F|P >_{out,in} \qquad (19.246)$$

Here $V_o = V_o(r_o, 0)$, χ is the electron affinity (Georgiev (1984a)). Using (19.246), eq.(19.244) gives V_{FP} as a function of the F center - bound polaron separation R .

The G factor in (19.244) arises for the following reason: According to Huang, the total potential energy of the lattice in the presence of an external electric field $\underline{E}(\underline{r})$ is

$$\frac{M}{2}\int(\dot{\underline{u}}_l^2 + \dot{\underline{u}}_t^2 + \omega_l^2\underline{u}_l^2 + \omega_t^2\underline{u}_t^2)\frac{dv}{v_a} - \omega_l(\frac{Mv_a}{4\pi\varepsilon_p})^{\frac{1}{2}}\int \underline{u}_l\cdot\underline{E}(\underline{r})\,\frac{dv}{v_a}$$

where the subscripts l and t label the longitudinal and transversal modes, respectively, while $M = M_+M_-/(M_+ + M_-)$ is the reduced mass of the ions in an unit cell of volume v_a , $\omega_l = (\varepsilon_o/\varepsilon_\infty)^{\frac{1}{2}}\omega_t$ (Huang and Rhys (1950)). Comparing with (19.220) and using $E(r) = -(d/dq)V_p(r,q)/(-e)$ at $q = 0$, we get

$$G = \omega_{LO}(M_{LO}v_a/4\pi\varepsilon_p e^2)^{\frac{1}{2}}\,\frac{10^{-8}}{1.6 \times 10^{-12}} , \qquad (19.247)$$

turning back to the 'LO' subscripts to mark the longitudinal-optic mode. The numerical ratio to the right is due to the convention to measure the electron-phonon coupling coefficients in eV/A. Otherwise, G arises because on using $M_{LO}\omega_{LO}^2$ as a force constant in the case of A_{1g}-lattice mode coupling account must be taken of the fact that this coupling is the interaction with many lattice oscillators, while M_{LO} is the mass of but one of them.

Further, using 2p- like states for both the F center and the polaron

$$j_1(Kr) = (Kr)^{-1}(\frac{\sin(Kr)}{Kr} - \cos(Kr)) , \text{ in}$$

$$R_{21}(\alpha r) = \alpha r\exp(-\alpha r) , \text{ out} \qquad (19.248)$$

with $r = r_F$ for the F center and $r = r_P$ for the polaron, where $\underline{r}_P = \underline{r}_F - \underline{R}$, to compute the matrix element in (19.246), we obtain

$$< P\ F >_{out,in} = A_F B_P \int_0^{r_o} drr^2(Kr)^{-1}(\frac{\sin(Kr)}{Kr} - \cos(Kr)) \times$$

$$\alpha_P r_P \exp(-\alpha_P r_P) , (r = r_F) \qquad (19.249)$$

$$< F\ P >_{out,in} = A_P B_F \int_0^{r_o} drr^2(Kr)^{-1}(\frac{\sin(Kr)}{Kr} - \cos(Kr)) \times$$

$$\alpha_F r_F \exp(-\alpha_F r_F) , (r = r_P) \qquad (19.250)$$

and also

$$\langle P|F\rangle_{r_F=r_o} = \int_0^{\infty} dr_F r_F^2 \varphi_F(r_F)\varphi_P(r_P)\delta(r_F-r_o) \qquad (19.251)$$

These expressions hold good for $2r_o < R$, the intercenter separation. For distant pairs $2r_o \ll R$, and one can substitute R for r_P in the first and third matrix elements, and for r_F in the second one. This results in

$$V'_{PF} = (q/r_o)A_F B_P r_o^3 u_P v_F^{-1}((V_o + \chi)v_F^{-2}(2(1 - \cos v_F) - v_F \sin v_F) - (V_o - e^2/\varepsilon r_o)(v_F^{-1}\sin v_F - \cos v_F))(R/r_o)\exp(-u_P(R/r_o))$$

$$V'_{FP} = (q/r_o)A_P B_F r_o^3 u_F v_P^{-3}(e^2/\varepsilon_p r_o)(2(1 - \cos v_P) - v_P \sin v_P) \times (R/r_o)\exp(-u_F(R/r_o)) \qquad (19.252)$$

where $u_{F,P} = \alpha_{F,P}\, r_o$, $v_{F,P} = K_{F,P}\, r_o$. Calculated values of the underlying parameters for NaI (r_o = 3.237 A) are listed in Table 18.

Table 18

Parameters for calculating V_{FP} in NaI*

State	ε	V_o (eV)	u	v	A ($\AA^{-3/2}$)	B ($\AA^{-3/2}$)
F(1s)	3.10	5.00290	0.98529	1.5615	0.20092	0.34464
F(2p)	3.51	4.55741	0.43510	1.4298	0.04151	0.05690
P(2p)	4.90	0.90607	0.31145	1.4	0.02061	0.03446

*After Georgiev (1984a)

Using (19.252) and (19.244), $V_{FP}(q)$ obtains in the form

$$V_{FP}(q) = C_{FP}(q/r_o)(R/r_o)\exp(-(u_F + u_P)(R/r_o)/2) \qquad (19.253)$$

where C_{FP} is given by

$$C_{FP} = G(A_F A_P B_F B_P r_o^6 u_F u_P v_F^{-1} v_P^{-1}((V_o + \chi)v_F^{-2}(2(1 - \cos v_F) - v_F \sin v_F) - (V_o - e^2/\varepsilon r_o)(v_F^{-1}\sin v_F - \cos v_F))(e^2/\varepsilon_p r_o)v_P^{-2}(2(1 - \cos v_P) - v_P \sin v_P))^{\frac{1}{2}} \qquad (19.254)$$

Now, from $\chi = 1$ eV and the tabulated data we get $C_{FP} = 24.17167$ eV (NaI). V_{FP} should be calculated at $q = q_c$, the crossover configuration, where eq.(19.253) is only meaningful.

It is to be noted that the product $V'_{FP}(q)V'_{PF}(q)$ in eq.(19.243) should be positive, for otherwise the adiabatic energy based on eq. (19.239) would turn imaginary at the crossover point. However, the above calculation of $V'_{PF}(q)$ based on the tabulated NaI data yields a negative result, while $V'_{FP}(q)$ obtains positive. To preserve consistency, we assume that the LO-coupling factor G in (19.244) has different signs for V'_{PF} and V'_{FP}, thus compensating for the minus in the particular case (NaI). As a matter of fact, the sign of any given electron-phonon coupling matrix element may be expected to depend on the choice of origin along the configurational coordinate. The NaI calculation presumes $q = 0$ for the configuration around the empty vacancy (Georgiev (1984a). There is an outward equilibrium displacement of the nearest-neighbor cations relative to the perfect lattice, due to the net positive charge of the vacancy. On adding an electron in ground state of the F center, the nearest neighbors will undoubtedly displace inwardly which means $q < 0$. As the electronic state gets less compact on going from 1s to 2p, the equilibrium q will get closer to 0, that is, less negative, and may even surpass 0 to become positive, if the electronic state is largely extended, even though still localized. With these suggestions in mind, one may expect a change of sign of the diagonal coupling coefficient, as one goes from F(2p) to P(2p), the equilibrium cavity radius evolving from sub- to over- vacancy values, as this seems to occur in NaI (Georgiev (1984a)).

One way or the other, the electron-exchange matrix element V_{FP} will next be averaged over the pair separations R. Using the pair-distribution function (6.5), one obtains

$$V_{FP}(q) = \int_{R_o}^{\infty} 4\pi R^2 F \exp(-\tfrac{4}{3}\pi R^3 F)\, V_{FP}(q,R)\, dR \tag{19.255}$$

$$= (q/r_o)C_{FP}(4\pi r_o^3 F) \int_{R_o/r_o}^{\infty} (R/r_o)^3 \exp(-(\tfrac{4}{3}\pi r_o^3 F)(R/r_o)^3 - \tfrac{1}{2}(u_F + u_P)(R/r_o))\, d(R/r_o) \tag{19.256}$$

where $V_{FP}(q,R)$ is given by (19.253), while R_o is the smallest separation at which the distant-pair approximation is still meaningful. For an estimate we take $R_o = 10r_o$ to be a good basis, for it corresponds to a color center density of $7 \times 10^{18}\ \mathrm{cm}^{-3}$, which is close to the experimentally attainable limit, and provides a reasonable approximation for calculating the integrals in (19.251). On its hand, r_0 in NaI corresponding to 7×10^{21} cm^{-3} of color center density, the F containing exponential in (19.256) is close to unity, and the integral therein can be approximated by means of

$$\int_{R_0/r_0}^{\infty}(R/r_0)^3\exp(-u(R/r_0))d(R/r_0) = \frac{1}{u}\exp(-u(R_0/r_0))((R_0/r_0)^3 + (3/u)(R/r_0)^2 + (6/u^2)(R_0/r_0) + 6/u^3) .$$

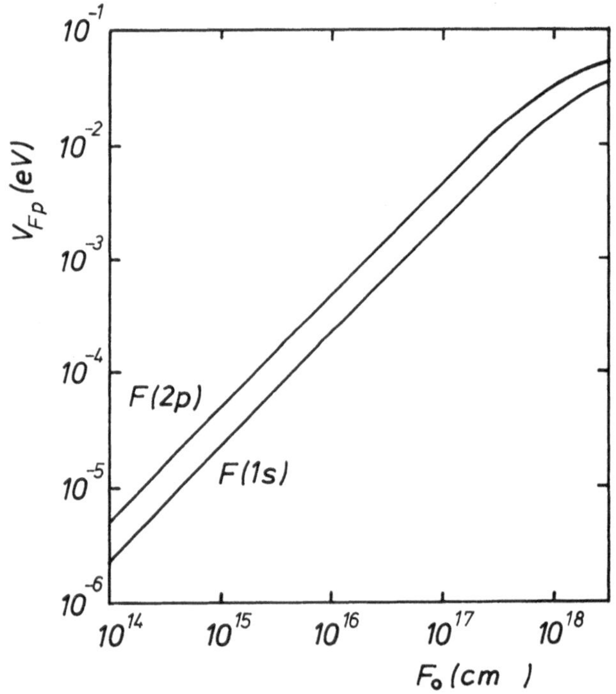

Figure 65: Electron-exchange matrix element V_{FP} of the F center to bound-to-another-F-center polaron transfer versus the F center density. The calculation is exemplified for NaI. The upper curve is obtained by means of eq.(19.256) assuming a 2p-like F center, while the lower one results from eq.(19.264) for a 1s-like F center. A distant-pair approximation is used.

Here $u = \frac{1}{2}(u_F + u_P)$. Setting $F = 5 \times 10^{15}$ cm^{-3}, well within the dilute F center range, we get $V_{FP}(q) = 2.4 \times 10^{-4}$ eV at $q = q_c$ = 0.1 A (Georgiev (1984a)) which compares quite favorably with experimentally based data (Georgiev and Mladenov (1986)). By numerical integration one finds that the linear-in-F approximation in (19.256) holds good up to concentrations over 10^{17} cm^{-3} in NaI (Fig.65).

The above static-basis estimate for V_{FP} is very sensitive to the choice of a value for the crossover configuration q_c. Inasmuch as the quasiclassical expression (19.232) is independent of q_c, any straightforward comparison would be informative, insofar as eq.(19.232) may be expected to work well for distant pairs. Using $\varepsilon = \varepsilon_\infty = 2.93$ for NaI and $m_e = 0.5m_e^o$, the rest electron mass, we obtain at $E_e = -0.11$ eV : $\nu_e = 4.9615 \times 10^{12}$ s^{-1} and $R_o = 44.6178$ A (Georgiev and Mladenov (1986)). Under these conditions, eq.(19.232) gives $R/r_o = 66$ corresponding to $(4\pi R^3/3)^{-1} = 2 \times 10^{16}$ F/cc . Thus, both the quasi-classical and the static-basis quantum-mechanical approaches lead to convergent results for the F center - polaron separation, assuming $q_c = 0.1$ A . This estimate for the crossover coordinate had earlier been arrived at based on semicontinuum diabatic-potential calculations of 2p- like F center and bound polaron (Georgiev (1984a)). However, in view of some inherent uncertainties involved (Georgiev (1984b)), the above should be regarded as an order-of-magnitude agreement only.

Anticipating further needs, we shall also calculate the P(2p)-F(1s) off-diagonal matrix element, the importance of which is inferred from Fig.63. Using the 1s- like wave functions

$$\begin{aligned} j_0(Kr) &= (Kr)^{-1} \sin(Kr) \quad &&\text{, in} \\ R_{10}(\alpha r) &= \exp(-\alpha r) \quad &&\text{, out} \end{aligned} \tag{19.257}$$

and (19.248), we obtain for distant pairs

$$< P|F >_{out,in} = B_P A_F \int_0^{r_o} dr r^2 \ (Kr)^{-1}\sin(Kr)\, \alpha_P r_P \exp(-\alpha_P r_P)$$

$$= A_F B_P r_o^3 u_P v_F^{-3}(\sin v_F - v_F \cos v_F)(R/r_o)\exp(-u_P(R/r_o)) \tag{19.258}$$

$$< P|F >_{r=r_o} = A_F B_P u_P v_F^{-1} \sin v_F \ r_o^2 (R/r_o)\exp(-u_P(R/r_o)) \tag{19.259}$$

$$< F|P >_{out,in} = B_F A_P \int_0^{r_o} dr r^2 \ (Kr)^{-1}\left(\frac{\sin(Kr)}{Kr} - \cos(Kr)\right)\exp(-\alpha_F r_F)$$

$$= A_P B_F r_o^3 v_P^{-3}(2(1 - \cos v_P) - v_P \sin v_P)\exp(-u_F(R/r_o)) \tag{19.260}$$

Now, equations (19.246) yield

$$\begin{aligned} V'_{PF}(q) = {} & (q/r_o)A_F B_P r_o^3 u_P v_F^{-1}((V_o + \chi)v_F^{-2}(\sin v_F - v_F\cos v_F) - \\ & (V_o - e^2/\varepsilon r_o)\sin v_F)(R/r_o)\exp(-u_P(R/r_o)) \end{aligned} \tag{19.261}$$

$$V'_{FP}(q) = (q/r_o)A_P B_F r_o^3 v_P^{-3}(2(1 - \cos v_P) - v_P \sin v_P) \times$$

$$\times\ (e^2/\varepsilon_p r_o)\exp(-u_F(R/r_o))\ .$$

Inserting into (19.244) we arrive at

$$V_{FP}(q) = C_{FP}(q/r_o)(R/r_o)^{\frac{1}{2}}\ \exp(-(u_F + u_P)(R/r_o)/2) \qquad (19.262)$$

with (19.263)

$$C_{FP} = G(A_F A_P B_F B_P r_o^6 u_P v_F^{-1} v_P^{-3}((V_o + \chi)v_F^{-2}(\sin v_F - v_F\cos v_F) -$$
$$(V_o - e^2/\varepsilon r_o)\sin v_F)(2(1 - \cos v_P) - v_P\sin v_P)(e^2/\varepsilon_p r_o))^{\frac{1}{2}}$$

The values of the underlying parameters are listed in Table 18. We obtain C_{FP} = 224.89091 eV with the same considerations of the signs under the square root holding good as the ones in (19.254). Equation (19.262) is to be next averaged over the pair separations, namely,

$$V_{FP}(q) = (q/r_o)(4\pi r_o^3 F)C_{FP}\int_{R_o/r_o}^{\infty} (R/r_o)^{5/2}\ \exp(-\tfrac{4}{3}\pi r_o^3 F(R/r_o)^3 - \tfrac{1}{2}(u_F + u_P)(R/r_o))d(R/r_o) \qquad (19.264)$$

Results of a numerical calculation under the conditions outlined before is also presented in Fig.65. q_c is taken to be 0.6 A .

19.5.4. F-F' conversion in densely colored crystals

We consider the following set of quasichemical reactions to account for the F-F' conversion via bound polarons:

$h\nu_F + F — F^*$	, $q_F F$	(19.a)
$F^* — F$	, F^*/τ_F	(19.b)
$F^* + F — P_F + \alpha$	, F^*/τ_1	(19.c)
$P_F + \alpha — F^* + F$	, P/τ_3	(19.d)
$P_F + \alpha — F + F$	, P/τ_5	(19.e)
$P_F — F'$	, P/τ_4	(19.f)
$F' + \alpha — F + F$	, F'/τ_2	(19.g)

These are described by corresponding rate equations, which are linear due to the assumed tunneling character of reactions (19.c) through (19.g):

$$\dot{F}^* = q_F F + P/\tau_3 - (\tau_F^{-1} + \tau_1^{-1})F^*$$
$$\dot{P} = F^*/\tau_1 - (\tau_3^{-1} + \tau_4^{-1} + \tau_5^{-1})P$$
$$\dot{F}' = P/\tau_4 - F'/\tau_2 \qquad (19.265)$$
$$\dot{F} = -q_F F + (\tau_F^{-1} - \tau_1^{-1})F^* + (\tau_3^{-1} + 2\tau_5^{-1})P + 2F'/\tau_2$$

Further, assuming

$$\dot{F}^* = 0 \ , \quad F^* = (q_F F + P/\tau_3)/(\tau_F^{-1} + \tau_1^{-1}) \qquad (19.266)$$

and introducing

$$\eta_1 = (1 + \tau_1/\tau_F)^{-1} \qquad (19.267)$$

we obtain

$$\dot{P} = \eta_1 q_F F - ((1 - \eta_1)\tau_3^{-1} + \tau_4^{-1} + \tau_5^{-1})P \qquad (19.268)$$
$$\dot{F} = -2\eta_1 q_F F + (\tau_3^{-1} + 2\tau_5^{-1})P + 2F'/\tau_2$$

Clearly, τ_1^{-1} is the same as k_{dis} of eq.(19.222). In densely colored systems, the second term of the second equation (19.268) could also play a significant role in determining the F-F' quantum yield. To evaluate it we set

$$\dot{P} = 0 \ , \quad P = \eta_1 q_F F/((1 - \eta_1)\tau_3^{-1} + \tau_4^{-1} + \tau_5^{-1}) \qquad (19.269)$$

Inserting into (19.268) we arrive at

$$\dot{F} = -2\eta_1(1 - (\tau_3^{-1} + 2\tau_5^{-1})/2((1 - \eta_1)\tau_3^{-1} + \tau_4^{-1} + \tau_5^{-1}))q_F F + 2F'/\tau_2 \qquad (19.270)$$

From the latter equation we derive the quantum efficiency

$$\eta_{F-F'} = -(\dot{F}/q_F F)_{t=t_0} \qquad (19.271)$$
$$= 2\eta_1((\tfrac{1}{2} - \eta_1)\tau_3^{-1} + \tau_4^{-1})/((1 - \eta_1)\tau_3^{-1} + \tau_4^{-1} + \tau_5^{-1})$$

neglecting the F' containing term.

It should be noted that τ_3^{-1}, the rate of the reverse reaction (19.d), is related to τ_1^{-1} by way of

$$\tau_3^{-1} = \tau_1^{-1} \exp(Q/k_B T) \quad \text{(endothermic rate)} \qquad (19.272)$$

where Q is the reaction heat (negative) of the exothermic reaction (19.c). τ_3^{-1} is therefore vanishing at low temperature, and (19.271) reduces to

$$\eta_{F-F'} = 2\eta_1 \tau_4^{-1}/(\tau_4^{-1} + \tau_5^{-1}) \qquad (19.273)$$

If τ_4, the relaxation time to an F' centered state, should be F-concentration independent, at least for distant pairs, so is τ_F. Unlike these, τ_1 and τ_5 are very definitely concentration-dependent. At sufficiently large Q, the low-temperature approximation (19.273) may well extend to higher temperatures, eventually covering the entire range of practical interest. This is illustrated in Fig.66

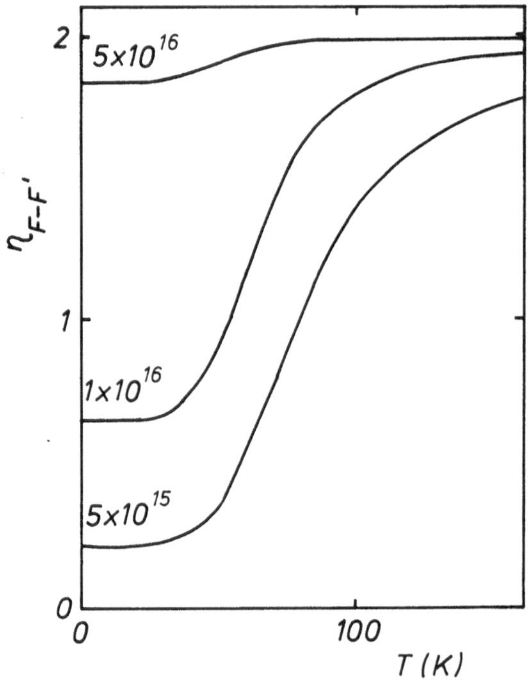

Figure 66: Calculated F-F' conversion efficiency vs. temperature dependence from eq.(19.267) at three different F center densities, indicated in cubic centimeters. The calculation is again exemplified for NaI using Fig.65 data on V_{FP}, the only concentration-dependent parameter of the theory. The curves are obtained neglecting the reverse polaron - F center reactions.

and Fig.67 using the V_{FP} data from Fig.65. Figure 66 shows the calculated temperature dependences of η_1 from (19.267) at three different F center densities. Nonadiabaticity is assumed, so that τ_1^{-1} increases as F^2 up to about 5×10^{17} F centers/cc . Correspondingly, the zero-point level rises with F , as the three curves all converge to the ultimate value of 2 at the highest temperatures. Figure 67 is cal-

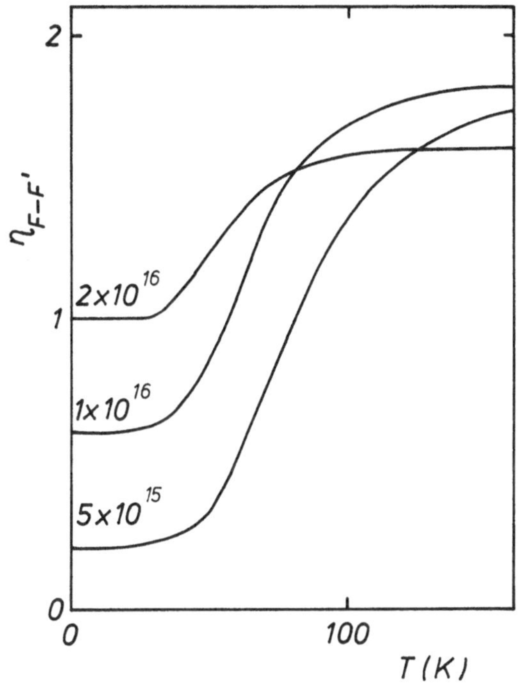

Figure 67: Same as Figure 66 but using equation (19.273) which takes into account the reverse reactions. The polaron-to-F' center transfer rate is somewhat arbitrarily set equal to 10^8 s^{-1} .

culated by means of eq.(19.273), using a tentative value of $\tau_4^{-1} = 10^8$ s^{-1} and $\tau_5^{-1} = \frac{1}{4}\tau_1^{-1}(T=0)$. The latter is a rough estimate based on Figures 63 and 65 indicating that the P(2p)-F(1s) barrier is approximately the same as the F(2p)-P(2p) barrier, and that V_{FP} in F(1s) is about half the value it has in F(2p), respectively. For the illustrative purpose aimed at, incorporating any temperature dependence of τ_5^{-1} would have been an unnecessary complication. Nevertheless, the qualitative resemblance of Luty's experimental data in Fig.24 is remarkable. Unlike the Fig.66 curves, the high-temperature level now drops down as a result of the increasing role of the reverse reaction (19.e) as F is increased.

It should be stressed that the concentration dependence of the F-F' efficiency in the present theory arises from the dependence of the electron-transfer matrix element V_{FP} (between states belonging to two different centers) on the pair separation, and from the subsequent averaging over the pair separations. The procedure employed herein involves taking first the average of V_{FP}, and then calculating $\eta_{F-F'}$ by means of that average, as done in Figures 66 and 67. An alternative way would be to calculate $\eta_{F-F'}$ for any given pair separation R, using the $V_{FP}(R)$ dependence, with the subsequent averaging of $\eta_{F-F'}(R)$ over R. The method adopted herein is technically simpler, even though its accuracy may be questioned. Nevertheless, with its qualitative capacity displayed in Figure 67, it can be used for checking up the theory, as further experimental data become available.

Based on the above premises, calculations have recently been carried out of the intercenter quenching of the F center luminescence in alkali halides (Georgiev and Staikova (1987)). Electron transfer between 2s-like F center and bound-polaron states was considered. The computed $\eta_L(F) = 1 - \eta_1(F)$ concentration-quenching curve for KCl was found to agree fairly well to within a factor of 2 with available experimental data by Miehlich (1963).

19.5.5. Bound-polaron versus conduction-band F' center formation

A conceivable configurational-coordinate diagram that accounts for reactions (19.a) through (19.g) of the bound-polaron approach can be sketched. The channels labelled 'F' and '4' can, in principle, incorporate both nonradiative and radiative deexcitation. Channel '2' (reaction (19.g) is believed to be exothermic in most of the halides, with two notable exceptions (NaI and NaBr), where the observed F' stability may result from that channel being endothermic (Baldacchini, Pan, and Lüty (1981)). Also, channel '5' may couple either weakly or strongly to the promoting vibration, as calculated for NaI and KCl, respectively (see Figures 63 and 53 for a comparison). A radiative polaron - F center transition, expected to occur in the former case, has been identified with an observed emission band in NaI, as explained in 19.5.1.

It should be stressed that the bound-polaron approach does not by any means preclude the conduction-band F'-formation mechanism, as described in Section 4. As a matter of fact, reconciliation can easily be achieved by regarding the bound-polaron concept as merely describing the dynamics of the tunneled F-F' conversion (cf. Section 6.2.1.). However, as far as no alternative theory has yet been developed, and the experimental data are rather scarce, any further speculations on the matter would be premature. Nevertheless, assuming both mechanisms to work, these can be expected to account for two complementary amounts of the overall F-F' efficiency. It seems surprising, therefore, that the bound-polaron approach can do that alone, and in reasonable physical terms, as shown in Fig.64.

In a sense, what is displayed in Fig.64 is the maximum contribution of the bound-polaron mechanism. Now, while the conduction-band F-F' yield (4.14) is virtually independent of the F center density, the bound-polaron contribution will drop steadily as F is decreased, due to the proportional decrease of the average electron-exchange matrix element (see Fig.65). As a result, the polaron-based F-F' efficiency vs. temperature dependence may transfigure considerably on going from one F center density to another, even though the ultimate efficiency of 2 may sooner or later be attained in both cases. However, there are two decisive options to distinguish experimentally between the conduction-band (CB) and bound-polaron (BP) mechanisms: (i) the low-temperature F-F' efficiency, predicted to be finite (BP) and vanishing (CB), and (ii) the low-temperature trapping cross-section of the anion vacancy which is expected to be finite (CB) or vanishing (BP), as far as electron capture in the excited state of the F center is concerned. BP predicts the latter quantity to be proportional to τ_3^{-1}, which is endothermic, and is, therefore, vanishing at low temperature. However, care must be taken in interpreting the data, since at the same time the trapping rate in ground state (proportional to τ_5^{-1}) may be finite in both cases. One way or the other, any experimental checkup should involve highly sensitive and time-resolved techniques, available presently. The use of free-electron lasers is strongly recommended (Baldacchini, Gallerano, and Grassano (1985)).

20. F' related nonlinearities

20.1. Pekar's trapping coefficients

In what follows we shall be speculating on the nature of F' related nonlinearities occuring as light-intensity dependent trapping coefficients and electron transition rates at the anion vacancy. Some of the experimental background has already been discussed in Section 4.

We begin by reproducing Pekar's arguments leading to his general equation (4.29) for the electron trapping coefficient. Imagine that each of the N_t traps per unit volume is encircled by a tiny sphere of radius r, assumed small compared with the average trap separation but large with respect to the mean free path of the electrons. Under these conditions the trapping rate z, the number of electrons to cross inward the manifold of the above spheres per unit volume per unit time will be

$$z = S n \mu_e \frac{dV}{dr} + S D_e \frac{dn}{dr} \tag{20.1}$$

where $S = 4\pi r^2 N_t$ is the total surface of the spheres in an unit volume, while $V(r)$ is the central field acting on the electron around each given trap. In steady state, z is assumed independent of r and the solution of eq.(20.1) is

$$n(r) = \left(n(\infty) - \frac{z}{4\pi N_t D_e} \int_0^{1/r} \exp(eV(r)/k_BT)\, d\tfrac{1}{r}\right)\exp(-eV(r)/k_BT) \tag{20.2}$$

When the electron approaches the trapping center as closely as r_o, the bound state radius, there will be a finite probability that the electron will get trapped. The number of transitions into the bound state per unit volume per unit time will be $\gamma_b N_t n(r_o) = z$. Combining with eq.(20.2) yields

$$z = \frac{\gamma_b\, e^{-eV(r_o)/k_BT}}{1 + \frac{\gamma_b}{4\pi D_e} e^{-eV(r_o)/k_BT} \int_0^{1/r_o} e^{+eV(r)/k_BT}\, d\frac{1}{r}} N_t n(\infty) \tag{20.3}$$

At the same time, however, $z = \gamma N_t n(\infty)$. Now, comparing with (20.3) eq.(4.29) results.

For $-eV(r_o)/k_BT >> 1$, the upper integration limit in (4.29) can be set ∞ ($r_o = 0$). Two notorious examples of trapping potentials (4.22) and (4.26) leading to equations (4.23) and (4.27), respectively, have been considered in Section 4.1.

Another example is the screened Coulomb potential

$$V(r) = -(e/\varepsilon r)\exp(-\chi r) \tag{20.4}$$

where $1/\chi = r_D$ is Debye's screening radius. For a one-component Coulomb gas of charged particles, e.g. conduction electrons in a crystal,

$$\chi = (4\pi n e^2/\varepsilon k_BT)^{\frac{1}{2}} \tag{20.5}$$

where n is the free-electron concentration. Physically, eq.(20.4) reflects the fact that at high free-electron densities the force exerted on a CB electron by an anion vacancy will be reduced significantly due to the polarization field induced by that vacancy in the free electron gas (Debye polarization). Inserting (20.4) into (4.29), the integral in the denominator is easily computed by expanding into a power series around $\chi = 0$:

$$\int_0^{\infty} \exp\left(-\frac{e^2}{\varepsilon k_B T r} e^{-\chi r}\right) d\frac{1}{r} = \sum_{m=0}^{\infty} \frac{1}{a} \frac{a^m \chi^m}{m!} = \frac{1}{a} e^{a\chi} , \quad a = \frac{e^2}{\varepsilon k_B T}$$

Now, setting the $r_o = 0$ limit in the remaining terms in eq.(4.29) we obtain

$$\gamma^*_{\alpha\chi} = \frac{4\pi e \mu_e}{\varepsilon} \exp(-e^2\chi/\varepsilon k_B T) = \gamma^*_\alpha \exp(-e^2\chi/\varepsilon k_B T) \tag{20.6}$$

where γ^*_α , given by eq.(4.23), is the trapping coefficient for a dilute electron gas. Equation (20.6) may be useful when dealing with the trapping behavior of anion vacancies at high excitation levels.

While the reduced electron trapping coefficient $\gamma^*_\alpha(\chi)$ in (20.6) is due to a collective effect, that of the interaction within the Coulomb gas of anion vacancies and free electrons, it is conceivable to seek a change in the trapping coefficient γ_F from (4.27) based on a pair interaction. Suppose there is a disturbing ionized partner at a distance r from the F center, placed in the center of a sphere. Let this partner be an F' center. It will induce a dipole moment $\alpha_F e/\varepsilon r^2$ on the F center. Averaging over the pair separations by means of (6.5) we obtain

$$p_{F'} = \alpha_F \int_0^{\infty} \frac{e}{\varepsilon r^2} 4\pi r^2 F' \exp(-\tfrac{4}{3}\pi r^3 F')\, dr = \alpha_F \frac{e}{\varepsilon} (\tfrac{4}{3}\pi F')^{2/3} \Gamma(\tfrac{1}{3})$$

$$= \alpha_F \Gamma(\tfrac{1}{3})(e/\varepsilon r^2_{F'}) , \quad r_{F'} = (\tfrac{4}{3}\pi F')^{-1/3} \tag{20.7}$$

where $r_{F'}$ is the average separation between the F' centers. The total dipole moment on the F center will then be

$$p(r) = p_{F'} + \alpha_F \frac{e}{\varepsilon r^2} = \alpha_F\left(\frac{e}{\varepsilon r^2_{F'}} + \frac{e}{\varepsilon r^2}\right) , \tag{20.8}$$

the second term being the moment induced by an approaching electron at r . According to electrostatics, the field produced by that dipole at r will be

$$E(r) = -2p(r)/\varepsilon r^3 = -(2e\,\alpha_F/\varepsilon^2)(\Gamma(\tfrac{1}{3})/r^2_F . r^3 + 1/r^5) , \tag{20.9}$$

while the corresponding potential is

$$V(r) = -g\left(\frac{2f}{r^2} + \frac{1}{r^4}\right) \tag{20.10}$$

with

$$g = \alpha_F e/2\varepsilon^2 , \quad f = \Gamma(\tfrac{1}{3})/r_{F'}^2.$$

Now, the integral in (4.21) is

$$I = \int_0^{1/r_o} \exp(-\frac{eg}{k_BT}(\frac{2f}{r^2} + \frac{1}{r^4}))\, d\frac{1}{r} = \frac{1}{4}(2f)^{\frac{1}{4}} \exp(\frac{egf^2}{2k_BT})\, K_{\frac{1}{4}}(\frac{egf^2}{2k_BT}) \quad (r_o = 0)$$

Introducing $z = egf^2/2k_BT$, this integral turns into

$$I = \frac{1}{4}\,(k_BT/eg)^{\frac{1}{4}}\, 8^{\frac{1}{4}}\, e^z\, z^{\frac{1}{4}}\, K_{\frac{1}{4}}(z) \tag{20.11}$$

where

$$z = (\alpha_F e^2/4\varepsilon^2 k_BT)(\Gamma(\tfrac{1}{3})/r_{F'}^2)^2 \tag{20.12}$$

At small $z << 1$

$$8^{\frac{1}{4}}\, z^{\frac{1}{4}}\, K_{\frac{1}{4}}(z) = \pi(2^{\frac{1}{2}}/\Gamma(\tfrac{3}{4}) - z^{\frac{1}{2}}/\Gamma(\tfrac{5}{4})) .$$

From (4.21) we now have

$$\gamma_F(F') = \gamma_F(0)\, \frac{2^{\frac{1}{2}}}{e^z 8^{\frac{1}{4}} z^{\frac{1}{4}} K_{\frac{1}{4}}(z)\Gamma(\frac{3}{4})} \tag{20.13}$$

Here

$$\gamma_F(0) = \frac{4}{\Gamma(\frac{1}{4})}\, 4\pi\mu_e\, (k_BT)^{\frac{3}{4}}\, (\alpha_F/2e^2\varepsilon^2)^{\frac{1}{4}} \tag{20.14}$$

coincides with (4.27) because $\Gamma(\frac{1}{4})\Gamma(\frac{3}{4}) = \pi 2^{\frac{1}{2}}$.

If the disturbing partner is an α center, the sign on the induced dipole moment in (20.7) should be reversed. We shall now present a general expression for the integral in (4.21) to cover both cases. Expanding $\exp(\mp 2egf/k_BTr^2)$ into a power series in r^{-2} and integrating, we obtain

$$\int_0^{\infty} \exp(-\frac{eg}{k_BT}(\frac{\pm 2f}{r^2} + \frac{1}{r^4}))\, d\frac{1}{r} = \frac{1}{4}(\frac{k_BT}{eg})^{\frac{1}{4}} \sum_{m=0}^{\infty} \frac{(\mp 1)^m}{m!}\, (2f(eg/k_BT)^{\frac{1}{2}})^m\, \Gamma(\frac{m}{2} + \frac{1}{4})$$

Comparing with (20.12) yields

$$I = \frac{1}{4}(2\varepsilon^2 k_BT/\alpha_F e^2)^{\frac{1}{4}} \sum_{m=0} \frac{(\mp 1)^m}{m!}\, 8^{\frac{m}{2}}\, \Gamma(\frac{m}{2} + \frac{1}{4})\, z^{\frac{m}{2}} \tag{20.15}$$

and therefrom

$$\gamma_F = 16\pi\mu_e(k_BT)^{\frac{3}{4}}(\alpha_F/2e^2\varepsilon^2)^{\frac{1}{4}} \Big/ \sum_{m=0} \frac{(\mp 1)^m}{m!}\, 8^{\frac{m}{2}}\, \Gamma(\frac{m}{2} + \frac{1}{4})\, z^{\frac{m}{2}} \tag{20.16}$$

where the $-(+)$ refers to perturbation by $F'(\alpha)$ centers. For small z , then

$$\gamma_F\binom{F'}{\alpha} = \gamma_F(0) \Big/ \sum_{m=0} \frac{(\mp 1)^m}{m!} 8^{\frac{m}{2}} \frac{\Gamma(\frac{m}{2}+\frac{1}{4})}{\Gamma(\frac{1}{4})} z^{\frac{m}{2}} = \gamma_F(0)/S^{\mp}(z) \quad (20.17)$$

can be approximated by

$$\gamma_F\binom{F'}{\alpha} = \gamma_F(0)\left(1 \pm 8^{\frac{1}{2}} \frac{\Gamma(\frac{3}{4})}{\Gamma(\frac{1}{4})} z^{\frac{1}{2}}\right) \quad (20.18)$$

Here and above

$$z = \frac{\alpha_F e^2}{4\varepsilon^2 k_B T} \Gamma(\tfrac{1}{3})^2 \left(\tfrac{4}{3}\pi \tbinom{F'}{\alpha}\right)^{\frac{4}{3}} , \quad (20.19)$$

the upper (lower) symbol in (20.19) corresponding to the upper (lower) sign in (20.15) through (20.18). It may be seen that the presence of a nearby F' (α) center exerts a stimulating (depressing) effect on the electron capture rate by F centers. This is a direct consequence from the presumed trapping mechanism, since in the former case the induced dipole adds to the one produced by the approaching electron, thereby enhancing the attractive force, and vice versa.

Equation (20.17) for γ_F can be complemented to account for the possibility that the electron may get trapped by an F* center. An expression for γ_{F*} can be derived similar to γ_F using a different polarizability α_{F*} of the F center in an excited state.

20.2. F' yields

Imagine that a square pulse of F band light is shed onto a crystal containing initially F centers at random, of an average density F_0. The F' band growth will be described by eq.(4.13) which transforms into

$$\dot{F}' = a_0(F' - F_1')(F' - F_2')/(\gamma_F F_0 + (\gamma_\alpha - 2\gamma_F)F') \quad (20.20)$$

with

$$a_0 = -\gamma_\alpha \beta_{F'}(1 - 4R^2)$$

$$F_{1,2}' = F_0 R/(2R \mp 1) \quad (20.21)$$

$$R = (\eta_i q_F \gamma_F / \gamma_\alpha \beta_{F'})^{\frac{1}{2}}$$

where q_F and $q_{F'}$ are finite and constant while the light is on. In arriving at eq.(20.20), equations (4.9) and (4.10) have been used neglecting n and $F*$. Comparing with eq.(4.16), one sees that $F_2' =$

F'_{max} . The solution of eq.(20.20) which vanishes at $t = 0$ is

$$a_o t = A \ln(1 - F'/F'_1) + B \ln(1 - F'/F'_2) \qquad (20.22)$$

with

$$A = -(\gamma_F F_0 + F'_1(\gamma_\alpha - 2\gamma_F))/(F'_2 - F'_1)$$

$$B = (\gamma_F F_0 + F'_2(\gamma_\alpha - 2\gamma_F))/(F'_2 - F'_1)$$

At sufficiently small R near room temperature,

$$a_o = -\gamma_\alpha \beta_{F'}$$

$$F'_{1,2} = \mp F_0 R \; , \quad F'_2 - F'_1 = 2F_0 R$$

$$A = -\gamma_F/2R + (\gamma_\alpha - 2\gamma_F)/2 \sim -\gamma_F/2R$$

$$B = \gamma_F/2R + (\gamma_\alpha - 2\gamma_F)/2 \sim \gamma_F/2R \quad ,$$

equation (20.22) is easily solved in F' to give

$$F'(t) = F'_{max} \tanh(t/\vartheta) \qquad (20.23)$$

with

$$\vartheta = (\gamma_F/\eta_i q_F \gamma_\alpha \beta_{F'})^{\frac{1}{2}} \qquad (20.24)$$

(cf. eq.(4.48)).

From the viewpoint of the conventional F-F' phenomenology, which assumes that the trapping coefficients and the lifetimes are all independent of the excitation-light intensity, eq.(20.23) displays intensity-time reciprocity if $\beta_{F'} = q_{F'}$. As a matter of fact, under this condition $F'_{max} = F_0 R$ is independent of, while ϑ^{-1} is linear in the intensity. However, experiments carried out in the F' thermal stability range (Costicas and Grossweiner (1962)) or at higher temperatures for $t < \tau_{F'}$ (Georgiev and Todorov (1976b)) have disclosed some remarkable deviations from reciprocity at high excitation levels. In the latter case, two light pulses of vastly different durations but with the same total energy $H = \int I(t)dt$ were found to produce very different F' yields near RT, as shown in Fig.68. The effective durations of these pulses were $t_{SF} = 0.01$ ms and $t_{LF} = 1$ ms , respectively.

Both F' yields increased as $I^{\frac{1}{2}}$ with the excitation-light intensity and decreased with the temperature as $\exp(E/k_B T)$, where the ratio $E_{SF}:E_{LF} = 2:1$. Saturation occurred at very high I resulting in F' peak absorption coefficients exceeding 20 at RT (Fig.10). On the other hand, the obtained E_{SF} agreed within experimental errors with the thermal energy derived from γ_F / γ_α vs. temperature data, extracted from the decay kinetics of the F' band in the dark, which data also

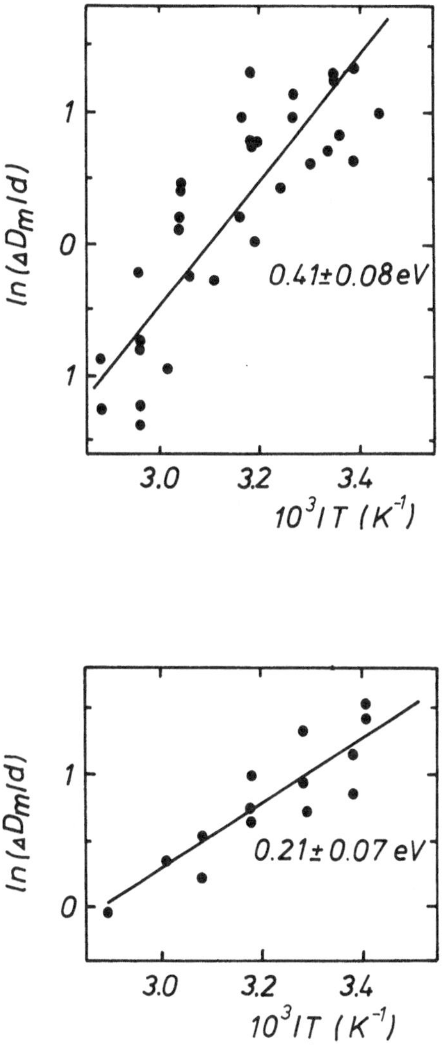

Figure 68: Temperature dependences of the F' yield in KCl using two different light flashes, SF (0.01 ms)(upper) and LF (1 ms) (lower), with nearly the same total energy. Note the spectacular difference of the activation energies involved. (After Georgiev and Todorov (1976b)).

followed a $\exp(E/k_BT)$ dependence. All these general features were found to reproduce for the F_A' band. This suggested strongly the following interpretation:

Assuming $t_{SF} < \vartheta < t_{LF}$, the two F' yields will be given by

$$F'_{mSF} = F'_{max}\, t_{SF}/\vartheta = \eta_i q_F F_0 t_{SF}$$
$$F'_{mLF} = F'_{max} = (\eta_i \gamma_F q_F/\gamma_\alpha \beta_{F'})^{\frac{1}{2}} F_0 \qquad (20.25)$$

This implies a relative stability of ϑ to light-intensity variations, in contrast to the conventional phenomenology. Inasmuch as $t_{LF} < \tau_{F'}$ near RT, $\beta_{F'} = q_{F'}$ and the observed $F'_m \propto I^{\frac{1}{2}}$ dependences are interpreted to give

$$\eta_i \propto I^{-\frac{1}{2}}$$
$$(\eta_i/\gamma_\alpha)\gamma_F \propto I \qquad (20.26)$$

and consequently

$$\gamma_F/\gamma_\alpha \propto I^{3/2} \qquad (20.27)$$

Under these conditions the time-constant ϑ is really independent of the excitation-light intensity I.

To unravel the physical nature of the nonlinearities under (20.26), equations (4.7), (4.8), and (4.60) have been used. It has been assumed that under the conditions of a high excitation during a light pulse the F* lifetime is in effect given by the F* deexcitation time (cf.(4.3)):

$$\tau_F = \tau_r \qquad (\tau_r << \tau_i) \qquad (20.28)$$

Near RT the total F* deexcitation rate is mainly due to transitions across the 'crossover barrier' (see Figure 51):

$$\tau_r^{-1} = \tau_{ro}^{-1} \exp(-E_b/k_BT), \qquad E_b = E^* - E_o \qquad (20.29)$$

Inserting (20.29) and (4.60) into (20.28) we obtain

$$\gamma_\alpha^* << (N_c \tau_{ro})^{-1} \exp(-(E_b - E_i)/k_BT) \qquad (20.30)$$

At low excitation levels, however, $\tau_r >> \tau_i$, $\tau_F = \tau_i$, and the inequality in (20.30) is reversed. (20.30) imposes too severe a limitation on the electron-trapping coefficient of the anion vacancy which could materialize through a corresponding drop of γ_α^* due to the presumed nonlinear effect. Thus, the decrease of γ_α^* as I is increased could well be one possible appearance of the nonlinearity at high I. In the quest for some quantitative indications we insert (20.28), (20.29) and (4.60) into (4.7) to obtain

$$\eta_i = \gamma_\alpha^* N_c \tau_{ro} \exp((E_b - E_i)/k_BT) \qquad (20.31)$$

Comparing with (20.26) we find

$$\gamma_\alpha^* \propto I^{-\frac{1}{2}} \tag{20.32}$$

at high I . From (4.8) and (20.28) we have $\gamma_\alpha = \gamma_\alpha^*$; now (20.26) gives

$$\gamma_F \propto I \tag{20.33}$$

The complete temperature and intensity dependences of the F' yields can be obtained by inserting into (20.25). This gives

$$\begin{aligned} F'_{mSF} &= F_0 N_c \tau_{ro} t_{SF} q_F \gamma_\alpha^* \exp((E_b - E_i)/k_B T) \\ F'_{mLF} &= F_0 (N_c \tau_{ro})^{\frac{1}{2}} (\gamma_F q_F/\beta_{F'})^{\frac{1}{2}} \exp((E_b - E_i)/2k_B T) \end{aligned} \tag{20.34}$$

It is now clear that two nonlinearities (20.32) and (20.33), as well as the resulting inequalities (20.30) and (20.28), can in principle bring about the observed intensity and temperature dependences of the F' yields.

20.3. Speculative nature of nonlinearity

When a flash of light is shone on an F centered crystal, the illuminated body is initially flooded with free electrons (Tomura, Murase, Takebayashi, and Kitada (1964)). This, as well as the observed intensity dependence proportional to $I^{\frac{1}{2}}$ in lieu of I , suggest strongly that the main contribution to the F' yields near RT may have come from CB- rather than tunnel-produced F' centers (cf. eq.(6.13)). At sufficiently high excitation-light intensities the free-electron concentration during the flash will be high which may considerably affect the electron-trapping coefficients.

Consider the anion vacancy first. The effective positive charge on the vacancy induces a Debye polarization on the free electron gas. The resulting electrostatic potential centered at the vacancy can be found in a self-consistent way by solving the Poisson-Boltzmann equation

$$V(r) = -\frac{4\pi en}{\varepsilon}(\exp(eV(r)/k_B T) - 1) \tag{20.35}$$

Assuming $eV(r) \ll k_B T$ (20.35) reduces to

$$V(r) = \frac{d^2 V(r)}{dr^2} + \frac{2}{r}\frac{dV(r)}{dr} = \chi^2 V(r) \tag{20.36}$$

At small r the solution is required to exhibit the Coulomb character. Then it reads (20.4). Here χ is defined by eq.(20.5), $n = \int \eta_i q_F F dt$

being the common free-electron - anion vacancy concentration produced by the flash. The corresponding electron-trapping coefficient of a screened vacancy has been derived and presented by eq.(20.6). Using some averaged values over the light pulse we have

$$n = \eta_i q_F F t_{pulse} \tag{20.37}$$

Now, inserting into (20.6) and using (20.5) and (20.31), we obtain a self-consistent equation for $\gamma^*_\alpha(\chi)$ of the form

$$\gamma = \gamma_o \exp(-p(\gamma q_F)^{\frac{1}{2}}) \tag{20.38}$$

where γ_o and p are independent of the excitation-light intensity. It may be seen that if eq.(20.38) is to be satisfied by $\gamma \propto q_F^s$, it will do so for negative s such that $|s| < 1$.

In addition to depressing the trapping ability of the anion vacancy, flooding with free electrons during a light pulse can exert a stimulating effect on the trapping rate by F center via the mechanism leading to eq.(20.17). Now, arguments similar to those in Section 20.1. suggest that the presence of a nearby electron may enhance the capture probability at an F center. At small z , eq.(20.18) applies with z given by (20.19), where F' should be replaced by n from eq.(20.37). Assuming $n \propto q_F^{\frac{1}{2}}$, z is proportional to $q_F^{2/3}$, $z^{\frac{1}{2}} \propto q_F^{1/3}$. However, the pair interaction in (20.17) does not seem to produce any sizeable effect on the electron trapping coefficient. As a matter of fact, setting $_F = 2 \times 10^{-23}$ cm^3 (Costicas and Grossweiner (1962)), $n = 10^{17}$ cm^{-3} we obtain $z = 2.67 \times 10^{-6}$ (KCl). This is to be compared with the magnitude of the collective effect in (20.6), where under the same conditions $a\chi = 1.38$. Both are RT estimates. More promising for accounting for the presumed linear-in-I dependence (20.33) of the electron-trapping coefficients of F centers seems to be eq.(4.44), for, many of these centers will be in the F* state under the conditions of a high optical excitation. Assuming a larger polarizability of F* relative to F, the F* induced term in (4.44) may contribute significantly to the effective trapping coefficient if the excitation-light intensity exceeds some critical level.

Alternatively, one can seek the origin of the F' related nonlinearities in an intensity dependence of the nonradiative deexcitation rate τ_r^{-1} at the F center. It is conceivable that configurational-coordinate diagrams of the Figure 51 type can undergo changes in the presence of high local fields due to ionized species. For a quantitative analysis, a model based on an electric-field dependence of the cross-over half-splitting may eventually be worked out.

21. Prospects

We believe to have manifested, as far as we could, the complexity of the F' problem. Yet, in spite of the large amount of experimental and theoretical work that has been done since 1938, and especially during the post-war years, there still seem to be quite a few "open ends". In what follows, we shall try to summarize some of these in the hope that they may stimulate further research:

(i) The electronic energy structure of the F' center is not understood completely. Almost nothing is known of the vibronic properties of the center, let alone its "configurational-coordinate diagram". More specifically, we do not know what modes couple to the F' electron in ground state. A related problem is that of the F' bandshape. It has commonly been assumed that the F' band is broad and insensitive to temperature variations. Based on experimental evidence collected for "orthodox salts" (KCl, NaCl, KBr), both assertions are now challenged by recent observations on "odd crystals" (NaBr, NaI, LiCl). It has also been believed that the F' binding is weak leading to a ground state level which is shallow relative to the F center ground state energy. This is again questioned by recent NaBr and NaI data placing the observed F' band peak to the high-energy side of the F band. It has been suggested that such "odd occurrences" may arise in crystals of a small cation to anion radii ratio $(r_+/r_- < 0.5)$. Anyway, the nature of the F' binding is not yet comprehended exhaustively.

(ii) No satisfactory theory has been created so far of the F' thermal lifetime. While the underlying problem relates to (i), there are certain specific difficulties. Such is the mixing between F' bound and delocalized states.

(iii) Another related puzzle still unresolved by the theory is whether or not there are any F' bound excited states. While most theories provide a negative net result, there also seems to be a sound experimental basis for ruling out the existence of such states active emission, at least for F centers in "orthodox crystals". Recent experiments on "odd crystals", however, reveal a bell-shaped F' band which suggests a bound excited state active in absorption. Such a Franck-Condon state may automatically ionize during lattice relaxation to account for the observed high photoionization efficiency of the F' center at low temperature. If so, it would be very interesting to study the autoionization mechanism, "odd crystals" apparently providing an experimental opportunity.

(iv) The optical F-F' conversion is not yet understood well from a theoretical point of view. One puzzling occurrence is the apparent residual ionizability of the relaxed excited F center at low temperature, even in lightly colored crystals. An attempt is being made for some time to reconcile experiment and theory through reviving a twenty year old concept of an F-F' conversion via (bound) polarons. 2p-like polarons have so far been invoked in a semicontinuum approach to the F center vibronic problem. Although this seems to explain the low-temperature ionizability, the role of 1s-like polaron states remains unclear. In addition, the 2p-based model is somewhat redundant, providing for an optical F(1s)-P(2p) absorption band that has not occurred experimentally. Thus, examining the 1s-like polaron option is very plausible, work on the matter being in progress for some time now. An extension of the bound-polaron concept applies directly to densely colored crystals in which electron transfer through tunneling between F center pairs has experimentally been established to predominate. In fact, this provides a background for developing a dynamic theory of the tunneled F-F' conversion, incorporating the phonon coupling, which has not been done so far. Detailed experiments will also be needed not only to check the theory but rather to disclose the phonon modes coupled to the electron tunneling transition. To see whether the residual ionizability does not result from factors alien to the tunneling transfer, careful measurements of the temperature and concentration dependences of the F-F' transfer rate should be made.

(v) The behavior of the electron-trapping coefficients of F and alpha centers at high color center densities and/or high excitation levels should be investigated more systematically. More theoretical efforts will be needed to derive expressions that will not discourage the experimentalists. By this time, the problem of the trapping coefficients has not been solved satisfactorily, perhaps, except for low densities. It will also be very interesting to study the trapping coefficients at low temperatures where the transition into a bound state becomes rate determining. Experiments near room temperature are also desirable in view of estimating the role of the crossover deexcitation and because of the mobility of excited and ionized F centers.

(vi) Studying nonradiative ionization, deexcitation, or transfer rates at high excitation levels is appealing by itself. It may be argued that intensity-dependent nonradiative rates leading to nonlinearities in F' related quantities can occur due to the transfigurement of vibronic potential energy surfaces in high local fields produced by charged defects.

(vii) Generally, the electronic part of the pair-interaction problem has been dealt with successfully, at least in a few cases. A large amount of experimental facts have been collected and explained in terms of simple physical models leading to a consistent quantum-mechanical outlook to the tunneling processes involved. In our opinion, this has created a good basis for taking the next step in developing a dynamic theory of the pair interactions accounting for the electron-phonon coupling. Scattered examples based on polaron concepts have already appeared in the literature.

(viii) As a result of the presumed large spread of the F' electron wavefunction, intrinsic or related quantities, such as F' thermal energy and lifetime, optical threshold energy, and electron-trapping coefficient of F centers, can all be expected to depend strongly on the environment. An example is provided by the F_A' center, another is the F' thermal energy in plastically deformed crystals. It is amazing how little has been done to study the environment effect systematically, and then utilize the F' center reciprocally as a probe to examine the environment.

(ix) The F'-dislocation interaction is of particular interest. By studying the kinetics of the pinning effect of F' centers on mobile dislocations, F' lifetimes have been deduced. These studies are, perhaps, an example of a systematic effort to tackle an environmental problem. However, the nature of the pinning mechanism is not yet clear. Although electrostatic interactions have been specifically pointed out, the overall effect may contain a significant elastic part. This appeals strongly for a detailed theoretical investigation.

(x) The bound-to-an-F-center polaron concept should be set on a more rigorous theoretical basis to check the premises of the phenomenological approach of Section 19.5. Experimentally, it will be very interesting to see just how the ir bands, attributed to bound polarons, depend on the F center density, and to measure their delay times relative to the excitation pulse by means of time-resolved spectroscopy.

(xi) The triplet F' center suspected to have occurred in magnetic-field experiments in "odd crystals" will undoubtedly stimulate further verification (Fig.69). At the same time, the theoretical premises regarding the uniqueness of singlet F' centers will have to be reexamined.

(xii) Experiments on the F-F' conversion in OH^- containing crystals have disclosed a remarkable quenching effect of the anionic impurity following some preliminary illumination at an elevated temperature to bring about a photoinduced precipitation of the F centers at impurity-neighboring sites. It has been suggested that quenching may occur due to a strong nonradiative deexcitation at the F_{OH^-} center stimulated by the negative charge on the (dipolar) impurity. Inasmuch as that impurity

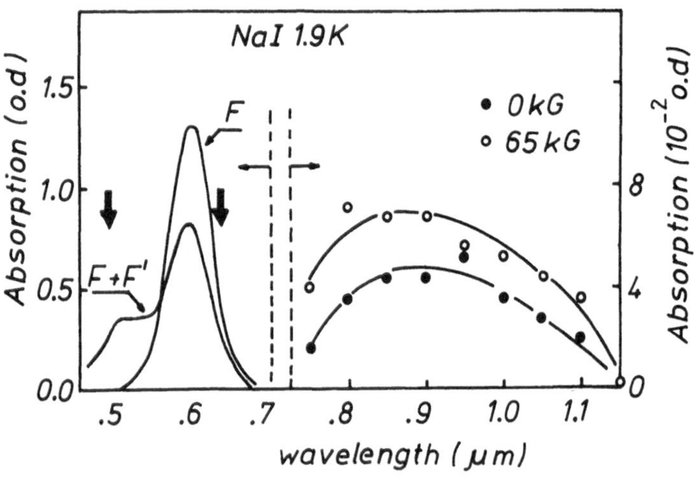

Figure 69: Left: Visible absorption spectra obtained in colored NaI by subsequent laser irradiations at wavelengths shown by arrows. Right: Infrared absorption spectrum induced by an F-F' conversion at two different magnetic fields. Courtesy of Professor U.M. Grassano. (After Baldacchini, Gallerano, Grassano, and Lüty (1983c)).

reorientates very rapidly even at the lowest temperature, the negative charge spends a considerable amount of time pointing towards the anion vacancy during the lifetime of the F* electron. It may be argued that the stimulating effect is due to the net lowering of the crossover barrier between F* and F in the average dipolar field leading to the enhancement of the nonradiative deexcitation rate in either the Franck-Condon state or the relaxed excited state. It will be very interesting to see whether any similar effect occurs with other molecular-dipole impurities, such as CN^-, as well as to work out the pertinent theory.

(xiii) It will also be important to check whether any change of the F* deexcitation rate occurs at all in crystals doped with divalent anion impurities, such as S^{2-}, or in ones doped with divalent cation impurities, such as Ca^{2+}, both forming dipolar species with corresponding anion (cation) vacancies, whose reorientational motion freezes in at low temperatures.

(xiv) Generally, the F' production by x-rays seems to be low in steady-state irradiation experiments. It would be informative to know whether that low production rate falls within the frameworks of the expectations of coloration theories. In any event, the F' role in the overall coloration process has not been appreciated comprehensively.

Table I

F' Band Optical Energies

Crystal	F' peak (eV)	F' edge (eV)	FWHM (eV)
LiF	2.0[l]		
LiCl	3.74[o]? 3.65[s]		0.40[o]?
NaF	2.64[f]	1.85±0.1[g]	
NaCl	2.4[a], 2.69[g]	1.72±0.05[g]	1.06[g]
NaBr	2.45[j]	1.55[j]	
NaI	2.44[j]	1.9[j]	
KF	1.77[i]	1.24[i]	
KCl	1.75[a], 1.65[f] 1.55[h], 1.51[g] 1.42[k], 1.51[b] 1.45[c], 1.50[m] 1.69[n]	0.90±0.03[g], 0.92[h]	1.25[c], 1.35[b] 1.39[g]
KBr	1.37[a], 1.20[g] 1.13[h], 1.21[d]	0.65±0.03[g], 0.64[h] 0.65[q]	1.26[d], 1.53[g]
KI	1.8[e], 1.68[g] 1.55[q], 1.38[t]	0.70±0.03[g], 0.65[q] 0.78[t]	0.92[g]
RbCl	1.08[g]	0.64±0.03[g]	1.2[g]
RbBr	0.85[r]	0.4[r]	0.8[r]
RbI	1.2[g]	0.55[g]	1.2[g]
CsF	1.34[p]	0.78[p]	
CsBr	1.20[g]	0.54[g]	
CsI	1.48[g]	0.55[g]	

The data listed above have been taken at the following temperatures: 173 K (KCl), 123 K (KBr), 143 K (NaCl),[a] 93 K,[b] 7.5 K,[d] 200 K,[f,m] 10 K (NaBr), 15 K (NaI),[j] LNT,[c,e,g,h,i,k] , 300 K,[n] 78 K (from spectral photoconductivity).[t] The F' bands in NaBr and NaI are very strongly temperature-dependent, unlike those in "orthodox alkali halides", such as KCl, KBr, and NaCl, which are relatively temperature-insensitive.

References:
[a]Pick (1938).
[b]Lüty (1961b).
[c]Kingsley (1961).
[d]Crandall and Mikkor (1965).
[e]Park and Faust (1966).
[f]Konrad and Neubert (1967).
[g]Lynch and Robinson (1968).
[h]Nosenzo, Reguzzoni, and Sammogia (1970).
[i]Collins and Schneider (1976).
[j]Baldacchini, Pan, and Luty (1981).
[k]Chiarotti and Grassano (1966b).
[l]Delbecq and Pringsheim (1953).
[m]Ikezawa, Hirai, and Ueta (1962).
[n]Georgiev and Todorov (1976b).
[o]Takiyama (1978).
[p]Cases, Alcala, and Orera (1982).
[q]Provoost, Debergh, and Hoebeeck (1982).
[r]Hoebeeck, Provoost, and Jacobs (1983).
[s]Takiyama, Fujita, Nojima, and Nishi (1983)
[t]Konitzer and Hersh (1966).

Table II

F' Thermal Lifetime $\tau_{F'} = \tau_0 \exp(E_{F'}/k_B T)$

Crystal	Ionization Energy $E_{F'}$ (eV)	Frequency Factor $\tau_0^{-1}(s^{-1})$
LiF	1.06 [q]	$10^{11.78}$ [q]
NaF	1.50±0.15 [p], 0.72 [q], 1.00 [s] 0.80 [x]	$10^{16.6\pm1.4}$ [p], $10^{11.27}$ [q] $10^{13.17}$ [s], $10^{10.65}$ [x]
NaCl	0.57 [a+], 0.91±0.06 [h], 0.62 [q] 0.94±0.10 [m], 1.10±0.04 [p] 0.50±0.03 [r], 1.0±0.05 [t] 0.52 [ab]	$10^{14.9\pm0.4}$ [p], $10^{12.3\pm0.5}$ [h] $10^{12.87}$ [q], $10^{13.83}$ [t], 10^{15} [r]
KF	0.77 [k]	10^{13} [k]
KCl	0.51 [a+], 0.42 [b], 0.42 [c] 0.5 [e++], 0.5±0.1 [f], 0.50 [g] 0.537±0.005 [h], 0.53±0.05 [i] 0.51±0.08 [j], 0.42±0.21 [l] 0.44±0.02 [m], 0.50±0.02 [n++] 0.60±0.04 [o], 0.35±0.02 [z] 0.40 [ab]	10^{11} [b], $10^{9.5}$ [c], $10^{10.22}$ [e] 10^{10-13} [f], $10^{9.5}$ [g], $10^{12.15}$ [i] $10^{11.4\pm0.2}$ [h], $10^{13.91}$ [o] $10^{7.48}$ [z]
KBr	0.24 [a+], 0.29 [c], 0.44 [d], 0.35 [g] 0.35±0.01 [h], 0.34±0.02 [m] 0.43±0.03 [p], 0.63 [u], 0.32 [ab]	$10^{11.1\pm0.3}$ [h], $10^{14.0\pm0.6}$ [p] $10^{11.6}$ [u], 10^{9} [d]
KI	0.47±0.02 [p]	$10^{13.7\pm0.4}$ [p]
RbCl	0.28±0.02 [m], 0.90 [v]	$10^{11.99}$ [v]
RbBr	0.14±0.02 [m], 0.14 [y]	
RbI	0.24±0.04 [m]	
CsF	0.31 [w]	

[+] Data from thermal halflife measurements. [++] Obtained using a constant γ_F/γ_α.

References:

[a] Pick (1938).
[b] Seitz (1954).
[c] Dutton and Maurer (1953).
[d] Onuki and Ohkura (1960).
[e] Ikezawa, Hirai, and Ueta (1962).
[f] Haupt (1963).
[g] Hirai and Scott (1966)
[h] Scaramelli (1966).
[i] Goldberger and Owens (1971).

[j] Terzijski and Popov (1974).

[k] Collins and Schneider (1976).

[l] Georgiev and Todorov (1976b).

[m] Dokter and Hoffmann (1976).

[n] Goode and Simpson (1977).

[o] Chandratillake et al. (1977).

[p] Chandratillake et al. (1978).

[q] Levin, Berggren, and Honnold (1967).

[r] Serpi and Serpi Macciotta (1972).

[s] Bhan (1982).

[t] Inabe and Takeuchi (1978).

[u] Ohkura (1961).

[v] Sastry and Sapru (1981).

[w] Cases, Alcala, and Orera (1982).

[x] Bhan and Rao (1982).

[y] Hoebeeck, Provoost, and Jacobs (1983).

[z] Rascon and Alvarez Rivas (1983).

[ab] Murty and Murthy (1974).

References

Aegerter, M.A. and Lüty, F. (1971a). Phys. Status Solidi B 43, 227.
Aegerter, M.A. and Lüty, F. (1971b). Phys. Status Solidi B 43, 245.
Agullo-Lopez, F. and Aguilar, M. (1979). Phys. Status Solidi B 92, K43.
Ahrenkiel, R.K. and Brown, F.C. (1964). Phys. Rev. A136, 223.
Alvarez Rivas, J.L. (1970). J. Phys. Soc. Japan 28, 997.
Amenu-Kpodo, K. and Neubert, T.J. (1965). J. Phys. Chem. Solids 26, 1615.
Apker, L. and Taft, E. (1951). Phys. Rev. 81, 698.
Arora, H.L. and Wang, S. (1969). J. Phys. Chem. Solids 30, 1649.
Baldacchini, G. and Lüty, F. (1977). Proc. Internat. Conf. on Defects in Insulating Crystals, Gatlinburg, Tennessee, p. 25.
Baldacchini, G., Gallerano, G., and Lüty, F. (1980). J. Physique (France) 41, C6-51.
Baldacchini, G., Pan, D.S., and Lüty, F. (1981). Phys. Rev. B 24, 2174.
Baldacchini, G., Gallerano, G.P., and Grassano, U.M. (1985). Nucl. Instr. & Meth. Phys. Res. A 239, 421.
Baldacchini, G., Gallerano, G.P., Grassano, U.M., and Lüty, F. (1981). Abstracts of Contributed Papers, Internat. Conf. on Defects in Insulating Crystals, Riga, p. 142.
Baldacchini, G., Gallerano, G.P., Grassano, U.M., and Lüty, F. (1983a). Paper at the Europhysics CMD Conference, Lausanne (unpublished).
Baldacchini, G., Gallerano, G.P., Grassano, U.M., and Lüty, F. (1983b). Lett. Nuovo Cim. 36, 495.
Baldacchini, G., Gallerano, G.P., Grassano, U.M., and Lüty, F. (1983c). Radiation Eff. 72, 153.
Baldacchini, G., Gallerano, G.P., Grassano, U.M., and Lüty, F. (1983d). Phys. Rev. B 27, 5039.
Bartram, R.H. and Stoneham, A.M. (1975). Solid State Commun. 17, 1593.
Bhan, S. (1982). Phys. Status Solidi B 110, 399.
Bhan, S. and Rao, S.M.D. (1982). J. Phys. C: Solid State Phys. 15, 6641.
Becker, K.H. and Pick, H. (1956). Nachr. Acad. Wiss. Gottingen Kl. IIa 7, 167.
Benci, S. and Manfredi, M. (1971). Solid State Commun. 9, 1255.
Benci, S. and Manfredi, M. (1973). Phys. Rev. B 7, 1549.
Benci, S., Fermi, F., and Manfredi, M. (1976). Solid State Commun. 18, 261.
Benci, S., Capelletti, R., Fermi, F., and Manfredi, M. (1975). J. Luminescence 11, 149.
Bennett, H.S. (1970). Phys. Rev. B 1, 1702.
Berezin, A.A. (1967). Fiz. Tverd. Tela (USSR) 9, 2756.
Berezin, A.A. (1969a). Fiz. Tverd. Tela (USSR) 11, 1587.
Berezin, A.A. (1969b). Internat. J. Quantum Chem. 3, 485.
Berezin, A.A. (1971). Optika i Spektroskopiya 31, 608.
Berezin, A.A. (1972). Phys. Status Solidi B 50, 71.
Berezin, A.A. (1976a). Fiz. Tverd. Tela (USSR) 18, 856.
Berezin, A.A. (1976b). Fiz. Tverd. Tela (USSR) 18, 858.
Berezin, A.A. (1977a). Fiz. Tverd. Tela (USSR) 19, 1480.
Berezin, A.A. (1977b). Fiz. Tverd. Tela (USSR) 19, 3372.
Berezin, A.A. (1978). Optika i spektroskopiya 44, 261.
Berezin, A.A. (1979a). Phys. Letters 72A, 48.
Berezin, A.A. (1979b). J. Phys. C: Solid State Phys. 12, L363.
Berezin, A.A. (1980). J. Phys. C: Solid State Phys. 13, L103.
Berezin, A.A. (1982). Z. Naturforschung 37a, 613.
Berezin, A.A. (1983). Phys. Letters 95A, 266.
Berezin, A.A. (1984a). Solid State Commun. 49, 87.
Berezin, A.A. (1984b). J. Chem. Phys. 81, 851.
Berezin, A.A. and Evarestov, R.A. (1971). Phys. Status Solidi B 48, 133.
Berezin, A.A. and Kirii, V.B. (1969). Fiz. Tverd. Tela (USSR) 11, 2118.
Bethe, H. and Salpeter, E.E. (1957). Quantum Mechanics of one- and two-electron atoms. Springer-Berlin.

Bichevin, V. and Käämbre, H. (1971). Phys. Status Solidi A 4, K235.
Bichevin, V. and Käämbre, H. (1987). Proc. Acad. Sci. Estonian SSR, ser. phys. math. 36, 186.
Borms, F. and Jacobs, G. (1971). Phys. Status Solidi B 43, 283.
Bosacchi, B., Fieschi, R., and Scaramelli, P. (1965). Phys. Rev. A138, 1760.
Bosi, L., Bussolati, C., and Spinolo, G. (1970). Phys. Rev. B 1, 890.
Bosi, L., Dupasquier, A., and Zappa, L. (1973). J. Physique (France) 34, C9-295.
Bosi, L., Dupasquier, A., and Zappa, L. (1975). Phys. Rev. B 11, 2485.
Bosi, L., Gagliardelli, P., and Nimis, M. (1981). Phys. Rev. B 24, 3600.
Bosi, L., Gagliardelli, P., and Nimis, M. (1982). Lett. Nuovo Cim. 35, 292.
Bosi, L., Gagliardelli, P., and Nimis, M. (1983). Phys. Status Solidi B 120, 41.
Bosi, L. and Nimis, M. (1979a). Phys. Status Solidi B 93, 183.
Bosi, L. and Nimis, M. (1979b). Phys. Status Solidi B 95, 615.
Bosi, L., Podini, P., and Spinolo, G. (1968). Phys. Rev. 175, 1133.
Brandt, R.C. (1972). Polarons in Ionic Crystals and Polar Semiconductors: J.T. Devreese, Ed., North Holland-Amsterdam, p. 227.
Brothers, A.D. and Lynch, D.W. (1968). Phys. Rev. 174, 958.
Brown, F.C. (1966). Notes of the NATO Summer School - Ghent, cited by Bosi, Podini, and Spinolo (1968).
Busse, J. and Weiss, H. (1971). Nachrichtentechnik (GDR) 21, 254.
Carlier, F. and Jacobs, G. (1978a). Phys. Status Solidi B 89, 193.
Carlier, F. and Jacobs, G. (1978b). Phys. Status Solidi B 89, K95.
Cases, R., Alcala, R., and Orera, V.M. (1982). Phys. Status Solidi B 113, K61.
Chandratillake, M.R., Newton, G.W.A., Robinson, V.J., and Rodgers, M.A.J. (1977). J. Chem. Soc. Faraday II 73, 1739.
Chandratillake, M.R., Newton, G.W.A., Patil, S.F., Robinson, V.J., and Rodgers, M.A.J. (1978). J. Chem. Soc. Faraday II 74, 480.
Chandratillake, M.R., Hamblett, I., Newton, G.W.A., Patil, S.F., Robinson, V.J., and Rodgers, M.A.J. (1978). J. Chem. Soc. Faraday II 74, 1342.
Chandratillake, M.R., Hamblett, I., Newton, G.W.A., and Robinson, V.J. (1981). J. Chem. Soc. Faraday II 77, 2319.
Cheban, A.G. (1960). Uchennie zapiski Kishinevsk. Universiteta (USSR) 55, 183.
Cheban, A.G. (1961). Optika i spektroskopiya (USSR) 10, 493.
Cheban, A.G. (1963). Optika i spektroskopiya (USSR) 14, 505.
Chiarotti, G. and Grassano, U.M. (1966a). Nuovo Cim. B 46, 78.
Chiarotti, G. and Grassano, U.M. (1966b). Phys. Rev. Lett. 16, 124.
Clark, C.D. and Newman, D.H. (1969). J. Phys. C: Solid State Phys. 2, 1726.
Collins, W.C. and Schneider, I. (1976). J. Phys. Chem. Solids 37, 917.
Comins, J.D. (1971). Phys. Status Solidi B 43, 101.
Compton, W.D. (1967). J. Physique (France) 28, C4-10.
Cordovilla, C.G. and Alvarez Rivas, J.L. (1974). J. Phys. C: Solid State Phys. 7, 3645.
Costikas, A. and Grossweiner, L.I. (1962). Phys. Rev. 126, 1410.
Crandall, R.S. (1965). Phys. Rev. 138, A1242.
Crandall, R.S. and Mikkor, M. (1965). Phys. Rev. 138, A1247.
Crawford, J.H. Jr. (1964). Phys. Rev. Lett. 12, 28.
Crippa, P.R., Paracchini, C., and Felszerfalvi, J. (1968). J. Phys. Soc. Japan 24, 92.
Christov, S.G. (1982). Phys. Rev. B 26, 6918.
Dannefaer, S., Trumpy, G., and Cotterill, R.M.J. (1974). J. Phys. C: Solid State Phys. 7, 1261.
Dechev, G., Georgiev, G., Koralov, N., Todorov, T., Gochev, A., and Georgiev, M. (1971). Proc. 5th IMEKO Symp. on Photon Detectors, Varna-Budapest, p. 247.

Delbecq, C.J., Pringsheim, P., and Yuster, P. (1951). J. Chem. Phys. 19, 574.
Delbecq, C.J. and Pringsheim, P. (1953). J. Chem. Phys. 21, 794.
Delbecq, C.J. (1963). Z. Physik 171, 560.
Dexter, D.L. (1951). Phys. Rev. 83, 1044.
Dexter, D.L., Klick, C.C., and Russell, G.A. (1955). Phys. Rev. 100, 603.
Dokter, G. and Hoffmann, H.J. (1976). Phys. Status Solidi B 76, 241.
Domanic, F. (1943). Annalen d. Physik 43, 187.
Doorn, C.Z. Van (1958). Philips Res. Reports 13, 296.
Dorendorf, H. (1951). Z. Physik 129, 317.
Dudek, F.J. and Grossweiner, L.I. (1969). J. Phys. Chem. Solids 30, 2023.
Duerig, W.H. and Markham, J.J. (1952). Phys. Rev. 88, 1043.
Dupasquier, A. (1979). Positrons in Solids. P. Hautojarvi, Ed., Springer-Berlin, p. 197.
Dutton, D. and Maurer, R. (1953). Phys. Rev. 90, 126.
Ecabert, M. and Jaccard, C. (1978). Helvetica Phys. Acta 51, 337.
Ecabert, M., Schnegg, P.A., Ruedin, Y., Aegerter, M.A., and Jaccard, C. (1972). Luminescence of Crystals, Molecules and Solutions. Ferd Williams, Ed., Plenum-New York, p. 582.
Ermakov, G.A. and Nadgornyi, E.M. (1968). Doklady Akad. Nauk USSR 181, 76.
Ermakov, G.A., Korovkin, E.V., and Soifer, Ya.M. (1974a). Fiz. Tverd. Tela (USSR) 16, 697.
Ermakov, G.A., Korovkin, E.V., and Soifer, Ya.M. (1974b). Fiz. Tverd. Tela (USSR) 16, 1756.
Faraday, B.J. and Compton, W.D. (1965). Phys. Rev. 138, A893.
Farge, Y. and Fontana, M. (1974). Perturbations electroniques et vibrationnelles localisees dans les solides ioniques. Masson-Paris.
Farge, Y., Lambert, M., and Smoluchowski, R. (1967). Phys. Rev. 159, 700.
Fedders, H., Hunger, M., and Lüty, F. (1961). J. Phys. Chem. Solids 22, 299.
Fieschi, R. and Scaramelli, P. (1966). Phys. Rev. 145, 622.
Fieschi, R. and Paracchini, C. (1969). Phys. Rev. 182, 935.
Fowler, W.B. (1964). Phys. Rev. 135, A1725.
Fowler, W.B. (1968). Physics of Color Centers. W.B. Fowler, Ed., Academic-New York, p. 54.
Fraser, S. (1974). Phys. Condensed Matter 17, 71.
Fröhlich, D. and Mahr, H. (1965). Phys. Rev. Lett. 14, 494.
Fröhlich, D. and Mahr, H. (1966a). Phys. Rev. 141, 692.
Fröhlich, D. and Mahr, H. (1966b). Phys. Rev. 148, 868.
Fuller, R.G. (1972). Point Defects in Solids. Vol. 1: General and Ionic Crystals. J.H. Crawford, Jr. and L.M. Slifkin, Eds., Plenum-New York, p. 103.
Geiger, F.E. (1955). Phys. Rev. 99, 1075.
Georgiev, M. (1984a). Phys. Rev. B 30, 7261.
Georgiev, M. (1984b). J. Luminescence 31/32, 160.
Georgiev, M. (1984c). Proc. Internat. Conf. on Defects in Insulating Crystals, Salt Lake City, Utah, p. 173.
Georgiev, M. (1985a). J. Inf. Rec. Mater. 13, 75, 177, 245.
Georgiev, M. (1985b). Rev. Mex. Fiz. 31, 221.
Georgiev, M., Gochev, A., Christov, S.G., and Kyuldjiev, A. (1982). Phys. Rev. B 26, 6936.
Georgiev, M. and Mladenov, G.M. (1986). J. Phys. Chem. Solids 47, 815.
Georgiev, M. and Staikova, M. (1987). Proc. Internat. Conf. on Luminescence, Beijing, p. 390.
Georgiev, M. and Todorov, T. (1975). Phys. Status Solidi A 27, K85.
Georgiev, M. and Todorov, T. (1976a). J. Inf. Rec. Mater. 4, 187.
Georgiev, M. and Todorov, T. (1976b). J. Physique (France) 37, C7-89.
Giuliani, G. (1969). J. Phys. Chem. Solids 30, 217.
Goldberger, G. and Owens, F.J. (1971). Phys. Rev. B 4, 3927.
Gomes, L. and Lüty, F. (1983). Bull. Amer. Phys. Soc. 28, 452.
Gomes, L. and Lüty, F. (1984). Phys. Rev. B 30, 7194.
Goode, D.H. and Simpson, J.H. (1977). Proc. Internat. Conf. on Defects

in Insulating Crystals, Gatlinburg, Tennessee, p. 151.
Gorbenko, P.K. (1964). Optika i Spektroskopiya (USSR) 16, 260.
Gorbenko, P.K. (1966). Optika i Spektroskopiya (USSR) 20, 453.
Gordan, P. and Scharmann, A. (1968). Z. Physik 217, 309.
Gourary, B.S. and Adrian, F.J. (1960). Solid State Physics. F. Seitz and D. Turnbull, Eds., Academic-New York, Vol. 10, p. 127.
Grabovskis, V.J. and Vitols, I.K. (1979). J. Luminescence 20, 337.
Gudat, A.E., Scott, A.B., and Wagner, M. (1974). J. Chem. Phys. 60, 4396.
Hagihara, T., Hayashiuchi, Y., and Okada, T. (1985). Phys. Letters 108A, 263.
Hagihara, T., Hayashiuchi, Y., and Okada, T. (1986a). Phys. Letters A 115, 385.
Hagihara, T., Hayashiuchi, Y., and Okada, T. (1986b). Abstracts, 5th Europhysics Topical LDIC Conf., Madrid, p. 263.
Hagihara, T., Hayashiuchi, Y., and Okada, T. (1986c). Abstracts, 5th Europhysics Topical LDIC Conf., Madrid, p. 393.
Hagihara, T., Inoue, Y., and Okada, T. (1982). Phys. Letters 88A, 191.
Hagihara, T., Inoue, Y., and Okada, T. (1983). Radiation Effects 72, 293.
Halperin, A., Braner, A.A., Schlesinger, M., and Kristianpoller, N. (1960). Proc. Internat. Conf. on Semiconductor Phys., Prague, p. 724.
Hardtke, F.C., Scott, A.B., and Woodley, R.E. (1960). Phys. Rev. 119, 544.
Harris, E.F., Haven, Y., and Richards, T.J. (1968). Proc. Internat. Symp. on Color Centers, Rome, p. 115.
Härtel, H. and Lüty, F. (1964a). Z. Phys. 177, 369.
Härtel, H. and Lüty, F. (1964b). Z. Phys. 182, 111.
Haupt, U. (1963). Z. Physik 176, 560.
Henry, C.H. and Slichter, C.P. (1968). Physics of Color Centers. W.B. Fowler, Ed., Academic-New York, p. 352.
Herreros, M. and Jaque, F. (1974). J. Luminescence 9, 380.
Hirai, M. (1983). Semiconductors & Insulators 5, 231.
Hirai, M. and Ueta, M. (1962). J. Phys. Soc. Japan 17, 724.
Hirai, M., Ikezawa, M., and Ueta, M. (1961). J. Phys. Soc. Japan 16, 1477.
Hirai, M., Ikezawa, M., and Ueta, M. (1962). J. Phys. Soc. Japan 17, 1483.
Hirai, M. and Scott, A.B. (1966). J. Chem. Phys. 44, 1753.
Hirai, M. and Scott, A.B. (1967). J. Chem. Phys. 46, 2896.
Hodby, J.W., Borders, J.A., and Brown, F.C. (1970). J. Phys. C: Solid State Phys. 3, 335.
Hoebeeck, G., Provoost, J., and Jacobs, G. (1983). Phys. Status Solidi B 115, K33.
Hoffmann, H.J. (1973). Phys. Status Solidi B 57, 123.
Hoffmann, H.J. (1981). Phys. Status Solidi B 107, 215.
Hoffmann, H.J., Stöckmann, F., and Tödheide-Haupt, U. (1973). Phys. Status Solidi B 56, 549.
Holzapfel, G. (1970). Z. Angewandte Physik 29, 107.
Honda, S. and Tomura, M. (1972). J. Phys. Soc. Japan 33, 1003.
Huang, K. and Rhys, A. (1950). Proc. Royal Soc. A204, 406.
Hunger, M. and Lüty, F. (1965). Phys. Letters 15, 114.
Ikezawa, M. (1964). J. Phys. Soc. Japan 19, 529.
Ikezawa, M., Hirai, M., and Ueta, M. (1962). J. Phys. Soc. Japan 17, 1474.
Inabe, K. and Takeuchi, N. (1978). Japan. J. Appl. Phys. 17, 831.
Ishii, T. and Endo, T. (1968). J. Phys. Soc. Japan 24, 524.
Jaccard, C. and Aegerter, M. (1973). Phys. Letters 44A, 391.
Jaccard, C. and Ecabert, M. (1978). Phys. Status Solidi B 87, 497.
Jaccard, C., Ruedin, Y., Aegerter, M., and Schnegg, P.A. (1972). Phys. Status Solidi B 50, 187.
Jaccard, C., Schnegg, P.A., and Aegerter, M. (1974). Proc. XVIII Ampere Congress, Nottingham, p. 219.
Jacobs, G. (1985). Phys. Status Solidi B 129, 755.
Jacobs, W., Orth, H., Putlitz, G. zu, Schafer, W., Vetter, J., Winnacker,

A., and Herlach, D. (1982). Z. Physik B 47, 95.
Jaque, F. and Agulló López, F. (1970). Phys. Rev. B 2, 4225.
Jaque, F. and Agulló López, F. (1974). Crystal Lattice Defects 5, 65.
Jenkin, G.T., Stacey, D.W., Crowder, J.G., and Hodby, J.W. (1978). J. Phys. C: Solid State Phys. 11, 1841.
Jimenez de Castro, M. and Alvarez Rivas, J.L. (1982). J. Phys. C: Solid State Phys. 15, 3019.
Kachlishvili, Z.S. (1962). Fiz. Tverd. Tela (USSR) 4, 736.
Kamada, M. and Tsutsumi, K. (1981). J. Phys. Soc. Japan 50, 3370.
Kamada, M. and Tsutsumi, K. (1985). Japan. J. Appl. Phys. 24, 15.
Kamada, M., Asai, F., and Tsutsumi, K. (1985). Japan. J. Appl. Phys. 24, 92.
Kamada, M., Furikawa, K., and Tsutsumi, K. (1981). Japan. J. Appl. Phys. 20, 71.
Kamada, M., Yoshiara, K., and Tsutsumi, K. (1984). Japan. J. Appl. Phys. 23, 286.
Kingsley, J.D. (1961). Phys. Rev. 122, 772.
Kiselev, A.A. (1963). Fiz. Tverd. Tela (USSR) 5, 1745.
Kiss, Z.J. (1970). Phys. Today 23, 42.
Kitada, T., Kakui, Y., and Tomura, M. (1968). J. Phys. Soc. Japan 25, 915. Proc. Internat. Conf. on Color Centers, Rome, p. 163.
Klick, C.C. (1972). Point Defects in Solids. Vol. 1: General and Ionic Crystals, J.H. Crawford, Jr. and L.M. Slifkin, Eds. Plenum-London, p. 291.
Kojima, K. and Nishimaki, N. (1961). J. Phys. Soc. Japan 16, 121.
Kojima, K., Nishimaki, N., and Kojima, T. (1961). J. Phys. Soc. Japan 16, 2033.
Kojima, K., Ebata, S., Tamura, A., and Kojima, T. (1976). J. Phys. Soc. Japan 40, 1397.
Kondo, Y. and Kanzaki, H. (1975). Phys. Rev. Lett. 34, 664.
Konitzer, J.D. and Hersh, H.N. (1966). J. Phys. Chem. Solids 27, 771.
Konrad, K. and Neubert, T.J. (1967). J. Chem. Phys. 47, 4946.
Korovkin, E.V. (1979). Fiz. Tverd. Tela 21, 1785.
Kouvalis, A. (1976). Phys. Rev. B 13, 2629.
Krumhansl, J.A. and Schwartz, N. (1953). Phys. Rev. 89, 1154.
Kubo, R. (1952). Phys. Rev. 86, 929.
Kubo, R. and Toyozawa, Y. (1955). Progr. Theor. Phys. 13, 160.
La, S.Y. and Bartram, R.H. (1966). Phys. Rev. 144, 670.
Lam, C.S. and Varshni, Y.P. (1978). Phys. Status Solidi B 89, 103.
Levin, H., Berggren, C.C., and Honnold, V.R. (1967). J. Phys. Chem. 71, 4228.
Link, E. and Lüty, F. (1965). Cited by Lüty (1968).
López, F.J., Aguilar, M., and Agulló-López, F. (1981). Phys. Rev. B 23, 3041.
Lüty, F. (1958). Z. Physik 153, 247.
Lüty, F. (1960). Z. Physik 160, 1.
Lüty, F. (1961a). Z. Physik 165, 17.
Lüty, F. (1961b). Halbleiterprobleme. Band VI. W. Schottky, Ed., Springer-Berlin, p. 238.
Lüty, F. (1968). Physics of Color Centers. W.B. Fowler, Ed., Academic New York, p. 182.
Lynch, D.W. and Robinson, D.A. (1968). Phys. Rev. 174, 1050.
Mahr, H. (1968). Physics of Color Centers. W.B. Fowler, Ed., Academic-New York, p. 243.
Markham, J.J. (1952). Phys. Rev. 88, 500.
Markham, J.J. (1966). F Centers in Alkali Halides. Solid State Physics. Suppl. 8. F. Seitz and D. Turnbull, Eds., Academic New York.
Markham, J.J., Platt, R.T., and Mador, I.L. (1953). Phys. Rev. 92, 597.
Martinez Negrette, M. and Ruiz Mejia, C. (1974). J. Phys. Soc. Japan 36, 1563.
Martini, F. De, Giuliani, G., and Mataloni, P. (1975). Phys. Rev. Lett. 35, 1464.
May, R. and Walker, D.C. (1973). Canad. J. Chem. 51, 2306.

Mezger, A. and Jaccard, C. (1980). Phys. Letters 79A, 118.
Mezger, A.C. and Jaccard, C. (1981). Phys. Status Solidi B 107, 689.
Mezger, A.C. and Jaccard, C. (1982). Solid State Commun. 41, 301.
Miehlich, A. (1963). Z. Physik 176, 168.
Mott, N.F. and Gurney, R.W. (1948). Electronic Processes in Ionic Crystals. Oxford.
Murayama, K., Morigaki, K. and Kanzaki, H. (1973). Solid State Commun. 13, 1197.
Murty, Y.V.G.S. and Murthy, K.R.N. (1974). J. Phys. C: Solid State Phys. 7, 1918.
Murty, Y.V.G.S. and Sucheta, N. (1982). Phys. Stat. Solidi B 109, 325.
Nadeau, J.S. (1964). J. Appl. Phys. 35, 669.
Nagarayan, T., Ramasamy, P., and Ramasamy, S. (1977). Phys. Status Solidi B 82, 75.
Nedashkovskii, A.P., Avdonin, V.P., Dugarova, L.D., Plachenov, B.T., and Savel'ev, V.P. (1974). Fiz. Tverd. Tela (USSR) 16, 3151.
Nierzewski, K.D., Todorov, T., and Georgiev, M. (1978). Phys. Status Solidi B 86, 697.
Nikolova, L., Todorov, T., Popov, D., and Terzijski, K. (1979). Phys. Stat. Sol. A 55, 333.
Nishimaki, N. and Shimanuki, S. (1982). Phys. Status Solidi B 114, K59.
Nosenzo, L., Reguzzoni, E., and Sammogia, G. (1970). Phys. Status Solidi 38, 369.
Oberly, J.J. (1951). Phys. Rev. 84, 1257.
Ohkura, H. (1961). J. Phys. Soc. Japan 16, 881.
Ohkura, H., Awane, K., and Miyamoto, S. (1960). J. Phys. Soc. Japan 15, 934.
Onaka, R. and Fujita, I. (1962). J. Quant. Spectr.& Radiat. Transfer 2, 599.
Onuki, M. and Ohkura, H. (1960). J. Phys. Soc. Japan 15, 1862.
Onuki, M. and Ohkura, H. (1961). J. Phys. Chem. Solids 22, 317.
Ostroukhov, A.A. and Tomasevich, O.F. (1958). Ukrain. Fiz. Zhur. (USSR) 3, 449.
Pan, D.S. and Luty, F. (1977). Proc. Internat. Conf. on Defects in In sulating Crystals, Gatlinburg, Tennessee, p. 329.
Park, K. (1965). Phys. Rev. 140, A1735.
Park, K. and Faust, W.L. (1966). Phys. Rev. Lett. 17, 137.
Park, K. and Hopfield, J.J. (1965). Bull. Amer. Phys. Soc. 10, 308.
Pekar, S.I. (1951). Issledovaniya po elektronnoi teorii kristallov. GITTL Moscow.
Pekar, S.I. and Tomasevich, O.F. (1951). Zh. Eksp. Teor. Fiz. (USSR) 21, 1218.
Pellaux, J.P., Sidler, T., Nouailhat, A., Aegerter, M.A. (1973). Solid State Commun. 13, 979.
Perlin, Yu.E., Cheban, A.G., and Tsukerblatt, B.S. (1961). Uch. Zap. Kishinevsk. Univer. ser. fiz. 49, 11.
Petrashen, M.I., Abarenkov, I.V., Berezin, A.A., and Evarestov, R.A. (1970a). Phys. Status Solidi 40, 9.
Petrashen, M.I., Abarenkov, I.V., Berezin, A.A., and Evarestov, R.A. (1970b). Phys. Status Solidi 40, 433.
Petroff, St. (1946). Fond Nauchni Izdaniya. Sv. Kiril Slavyanob'lgarski Universitet Varna, #52, pp. 1 60.
Petroff, St. (1950). Z. Physik 127, 443.
Philipp, H.R. and Taft, E.A. (1957). Phys. Rev. 106, 671.
Pick, H. (1938). Annalen d. Physik (Leipzig) 31, 365.
Pick, H. (1940). Annalen d. Physik (Leipzig) 37, 421.
Pick, H. (1958). Nuovo Cim. Suppl. Ser. X 7, 498.
Pick, H. (1965). Springer Tracts in Modern Physics. Vol. 38. Springer Berlin, p. 1.
Pick, H. (1972). Optical Properties of Solids. F. Abeles, Ed., North Holland Amsterdam, Chapter 9, p. 655.
Pincherle, L. (1951). Proc. Phys. Soc. Sect. A 64, 648.

Platt, R.T. and Markham, J.J. (1953). Phys. Rev. 92, 40.
Podini, P. (1966). Phys. Rev. 141, 572.
Pohl, R.W. (1937). Proc. Phys. Soc. 49, 3.
Popov, D. and Terzijski, K. (1973). Godishnik na VTUZ, ser. fiz. (Bulgaria) 10, 143.
Popov, D. and Terzijski, K. (1974). Godishnik na VUZ, tekhn. fiz. (Bulgaria) 11, 11.
Popov, D.N. and Terzijski, K.I. (1979). Phys. Status Solidi B 95, K43.
Popov, D.N. (1979). Thesis. VIMMESS Rousse (Bulgaria)(unpublished).
Porret, F. and Luty, F. (1971). Phys. Rev. Lett. 26, 843.
Provoost, J., Debergh, P., and Hoebeeck, G. (1982). Phys. Status Solidi B 113, 657.
Rabin, H. (1963). Phys. Rev. 129, 129.
Radhakrishna, S. and Chowdari, B.V.R. (1972). Phys. Status Solidi A 14, 11.
Rampacher, H. (1965). Z. Naturforschung 20a, 350.
Rascon, A. and Alvarez Rivas, J.L. (1983). J. Phys. C: Solid State Phys. 16, 241.
Raveche, H.J. (1965). J. Phys. Chem. Solids 26, 2088.
Redfield, D. (1963). Phys. Rev. 130, 914.
Reguzzoni, E. and Samoggia, G. (1968). Solid State Commun. 6, 675.
Reinberg, A.R. and Grossweiner, L.I. (1961). Phys. Rev. 122, 1734.
Rose, A. (1963). Concepts in Photoconductivity and Allied Problems. Interscience-New York.
Roth, M. and Halperin, A. (1982). J. Phys. Chem. Solids 43, 609.
Ruedin, Y. and Porret, F. (1968). Helvetica Phys. Acta 41, 1294.
Ruedin, Y., Schnegg, P.A., Jaccard, C., and Aegerter, M.A. (1972). Phys. Status Solidi B 54, 565.
Ruedin, Y., Schnegg, P.A., Jaccard, C., and Aegerter, M.A. (1973). Phys. Status Solidi B 55, 215.
Ruiz Mejia, C. (1970). Rev. Mexican. Fiz. 19, 311.
Russell, G.A. and Klick, C.C. (1956). Phys. Rev. 101, 1473.
Sastry, S.B.S. and Sapru, S. (1981). Physica B C 103B, 324.
Scaramelli, P. (1966). Nuovo Cim. 45B, 119.
Schiff, L.I. (1955). Quantum Mechanics. Mc Graw Hill-New York.
Schmid, D. and Wolf, H.C. (1962). Z. Physik 170, 455.
Schmid, D. and Zimmerman, V. (1968). Phys. Letters 27A, 459.
Schnegg, P.A., Jaccard, C., and Aegerter, M. (1973). Phys. Letters 42A, 369.
Schnegg, P.A., Jaccard, C., and Aegerter, M. (1974). Phys. Status Solidi B 63, 587.
Schnegg, P.A., Ecabert, M., Jaccard, C., and Aegerter, M.A. (1973). J. Physique Suppl. 34, C9-93.
Schnegg, P.A., Ruedin, Y., Aegerter, M.A., and Jaccard, C. (1973). Proc. XVII Congress Ampere, V. Hovi, Ed., North Holland-Amsterdam, p. 521.
Schneider, I. (1969). Phys. Rev. 177, 1324.
Schneider, I. (1970). Phys. Rev. Lett. 24, 1296.
Schneider, I. (1971). Solid State Commun. 9, 2191.
Schneider, I. and Bailey, C.E. (1969). Solid State Commun. 7, 657.
Schneider, I. and Caspari, M. (1963). Solid State Commun. 1, 9.
Schneider, I. and Caspari, M.E. (1964). Phys. Rev. 133, A1193.
Schneider, I. and Patterson, D. (1967). Phys. Rev. 164, 1136.
Schubert, M. and Vogler, K. (1980). Phys. Status Solidi B 101, 267.
Schulman, J.H. and Compton ,W.D. (1963). Color Centers in Solids. Pergamon-New York.
Schwoerer, M. and Wolf, H.C. (1963). Z. Physik 175, 457.
Seidel, H. and Wolf, H.C. (1968). Physics of Color Centers. W.B. Fowler, Ed., Academic-New York, p. 538.
Seitz, F. (1946). Revs. Modern Phys. 18, 384.
Seitz, F. (1954). Revs. Modern Phys. 26, 7.
Seitz, F. (1955). Nuovo Cim. Suppl. Ser. X 1, 95.
Serpi, A. and Serpi Macciotta, P. (1972). J. Luminescence 5, 361.

Shibaev, V.A., Vasil'ev, I.A., Plachenov, B.T., and Mochenov, M.I. (1969). Izv. Akad. Nauk (USSR), ser. fiz. 33, 1017.
Shvarts, K.K., Gotlib, V.I., and Kristapson, Ya.Zh. (1976). Opticheskie registriruyushchie sredy. Zinatne-Riga.
Smakula, A. (1930). Z. Physik 59, 603.
Smedskjaer, L., Dannefaer, S., Cotterill, R.M.J., and Trumpy, G. (1973). J. Physique (France) 34, C9-97.
Sonder, E. and Sibley, W.A. (1972). Point Defects in Solids. Vol.1: General and Ionic Crystals. J.H. Crawford, Jr. and L.M. Slifkin, Eds., Plenum-New York, p. 201.
Sonder, E., Sibley, W.A., and Mallard, W.C. (1967). Phys. Rev. 159, 755.
Stasiw, O. (1959). Elektronen und Ionenprocesse in Ionenkristallen, Springer-Berlin.
Stepien Damm, J. (1981). Abstracts of Contributed Papers. Internat. Conf. on Defects in Insulating Crystals, Riga, p. 144.
Stepien Damm, J. and Mugenski, E. (1982). Phys. Status Solidi A 73, K263.
Stiles, L.F., Jr., Fontana, M.P., and Fitchen, D.B. (1969). Solid State Commun. 7, 681.
Stiles, L.F., Fontana, M.P., and Fitchen, D.B. (1970). Phys. Rev. B 2, 2077.
Stocker, D. (1968). Z. Naturforschung 23a, 1158.
Stoneham, A.M. (1975). Theory of Defects in Solids. Clarendon-Oxford.
Strozier, J.A. and Dick, B.G. (1969). Phys. Status Solidi 31, 203.
Stumpf, H. (1961). Quantentheorie der Ionenrealkristalle. Springer-Berlin.
Swank, R.K. and Brown, F.C. (1963). Phys. Rev. 130, 34.
Taft, E. and Apker, L. (1953). Phys. Rev. 83, 479.
Takiyama, K. (1978). J. Phys. Soc. Japan 44, 1627.
Takiyama, K., Fujita, T., Nojima, H., and Nishi, M. (1983). Phys. Status Solidi B 115, K59.
Teegarden, K. and Maurer, R. (1954). Z. Physik 138, 284.
Terzijski, K. and Popov, D. (1974). Godishnik na VUZ (Bulgaria), techn. fiz. 11, 37.
Terzijski, K.I. and Popov, D.N. (1975). Fiz. Tverd. Tela (USSR) 17, 2514.
Todorov, T., Baltova, M., and Georgiev, M. (1975). Compt. Rend. Acad. Bulg. Sci. 28, 747.
Todorov, T., Koralov, N., and Georgiev, M. (1975). Optics Commun. 13, 439.
Todorov, T., Tomova, N., and Georgiev, M. (1974). Solid State Commun. 15, 1501.
Todorov, T., Dechev, G., and Georgiev, M. (1972). Solid State Commun. 11, 1731.
Todorov, T., Dechev, G., Tomova, N., and Georgiev, M. (1973). Phys. Status Solidi A 15, K123.
Tomasevich, O.F. (1951). Zh. Eksp. Teor. Fiz. (USSR) 21, 1223.
Tomasevich, O.F. (1957). Izv. Akad. Nauk (USSR), ser. fiz. 21, 74.
Tomiki, T. (1960). J. Phys. Soc. Japan 15, 488.
Tomura, M. and Kitada, T. (1967). J. Phys. Soc. Japan 22, 941.
Tomura, M., Kitada, T., and Takebayashi, M. (1965). J. Phys. Soc. Japan 20, 1531.
Tomura, M., Murase, K., Takebayashi, M., and Kitada, T. (1964). J. Phys. Soc. Japan 19, 1991.
Tsukerblat, B.S. and Cheban, A.G. (1964). Optika i Spektroskopiya (USSR) 16, 69.
Tubbs, M.R. (1973). Optics and Laser Technology 5, 155.
Tubbs, M.R. and Scrivener, G.E. (1974). J. Phot. Sci. 22, 8.
Tubbs, M.R. and Wright, D.K. (1971). Phys. Status Solidi A 7, 155.
Ueta, M. (1967). J. Phys. Soc. Japan 23, 1265.
Vassilev, Y.T., Georgiev, M., Todorov, G.S., and Todorov, T.A. (1978). Compt. Rend. Acad. Bulg. Sci. 31, 1393.
Vassilev, Y., Karamikhailova, M., Mladenova, M., and Georgiev, M. (1980). Phys. Status Solidi B 100, 463.

Walker, D.C. and May, R. (1974). Internat. J. Radiat. Phys. Chem. 6, 345.
Werner, B., Stradowski, C., and Sugier, H. (1980). Phys. Status Solidi B 99, 493.
Wild, R.L. and Brown, F.C. (1961). Phys. Rev. 121, 1296.
Wille, H. and Wahl, F. (1966). Z. Naturforschung 21a, 304.
Williams, F. (1968). Phys. Status Solidi 25, 493.
Wollbrandt, J., Bruckner, U., and Linke, E. (1983). Phys. Status Solidi A 78, 163.

AUTHOR INDEX

R

S

SUBJECT INDEX

H

I

J

K

L

M

N

O

P

Q

R

S

Lecture Notes in Mathematics

Vol. 1174: Categories in Continuum Physics, Buffalo 1982. Seminar. Edited by F.W. Lawvere and S.H. Schanuel. V, 126 pages. 1986.

Vol. 1184: W. Arendt, A. Grabosch, G. Greiner, U. Groh, H.P. Lotz, U. Moustakas, R. Nagel, F. Neubrander, U. Schlotterbeck, One-parameter Semigroups of Positive Operators. Edited by R. Nagel. X, 460 pages. 1986.

Vol. 1186: Lyapunov Exponents. Proceedings, 1984. Edited by L. Arnold and V. Wihstutz. VI, 374 pages. 1986.

Vol. 1187: Y. Diers, Categories of Boolean Sheaves of Simple Algebras. VI, 168 pages. 1986.

Vol. 1190: Optimization and Related Fields. Proceedings, 1984. Edited by R. Conti, E. De Giorgi and F. Giannessi. VIII, 419 pages. 1986.

Vol. 1191: A.R. Its, V.Yu. Novokshenov, The Isomonodromic Deformation Method in the Theory of Painlevé Equations. IV, 313 pages. 1986.

Vol. 1194: Complex Analysis and Algebraic Geometry. Proceedings, 1985. Edited by H. Grauert. VI, 235 pages. 1986.

Vol. 1203: Stochastic Processes and Their Applications. Proceedings, 1985. Edited by K. Itô and T. Hida. VI, 222 pages. 1986.

Vol. 1209: Differential Geometry, Peñíscola 1985. Proceedings. Edited by A.M. Naveira, A. Ferrández and F. Mascaró. VIII, 306 pages. 1986.

Vol. 1214: Global Analysis – Studies and Applications II. Edited by Yu.G. Borisovich and Yu.E. Gliklikh. V, 275 pages. 1986.

Vol. 1218: Schrödinger Operators, Aarhus 1985. Seminar. Edited by E. Balslev. V, 222 pages. 1986.

Vol. 1227: H. Helson, The Spectral Theorem. VI, 104 pages. 1986.

Vol. 1229: O. Bratteli, Derivations, Dissipations and Group Actions on C*-algebras. IV, 277 pages. 1986.

Vol. 1236: Stochastic Partial Differential Equations and Applications. Proceedings, 1985. Edited by G. Da Prato and L. Tubaro. V, 257 pages. 1987.

Vol. 1237: Rational Approximation and its Applications in Mathematics and Physics. Proceedings, 1985. Edited by J. Gilewicz, M. Pindor and W. Siemaszko. XII, 350 pages. 1987.

Vol. 1250: Stochastic Processes – Mathematics and Physics II. Proceedings 1985. Edited by S. Albeverio, Ph. Blanchard and L. Streit. VI, 359 pages. 1987.

Vol. 1251: Differential Geometric Methods in Mathematical Physics. Proceedings, 1985. Edited by P.L. García and A. Pérez-Rendón. VII, 300 pages. 1987.

Vol. 1255: Differential Geometry and Differential Equations. Proceedings, 1985. Edited by C. Gu, M. Berger and R.L. Bryant. XII, 243 pages. 1987.

Vol. 1256: Pseudo-Differential Operators. Proceedings, 1986. Edited by H.O. Cordes, B. Gramsch and H. Widom. X, 479 pages. 1987.

Vol. 1258: J. Weidmann, Spectral Theory of Ordinary Differential Operators. VI, 303 pages. 1987.

Vol. 1260: N.H. Pavel, Nonlinear Evolution Operators and Semigroups. VI, 285 pages. 1987.

Vol. 1263: V.L. Hansen (Ed.), Differential Geometry. Proceedings, 1985. XI, 288 pages. 1987.

Vol. 1267: J. Lindenstrauss, V.D. Milman (Eds), Geometrical Aspects of Functional Analysis. Seminar. VII, 212 pages. 1987.

Vol. 1269: M. Shiota, Nash Manifolds. VI, 223 pages. 1987.

Vol. 1270: C. Carasso, P.-A. Raviart, D. Serre (Eds), Nonlinear Hyperbolic Problems. Proceedings, 1986. XV, 341 pages. 1987.

Vol. 1272: M.S. Livšic, L.L. Waksman, Commuting Nonselfadjoint Operators in Hilbert Space. III, 115 pages. 1987.

Vol. 1273: G.-M. Greuel, G. Trautmann (Eds), Singularities, Representation of Algebras, and Vector Bundles. Proceedings, 1985. XIV, 383 pages. 1987.

Lecture Notes in Physics

Vol. 277: Molecular Dynamics and Relaxation Phenomena in Glasses. Proceedings, 1985. Edited by Th. Dorfmüller and G. Williams. VII, 218 pages. 1987.

Vol. 278: The Physics of Phase Space. Proceedings, 1986. Edited by Y.S. Kim and W.W. Zachary. IX, 449 pages. 1987.

Vol. 279: Symmetries and Semiclassical Features of Nuclear Dynamics. Proceedings, 1986. Edited by A.A. Raduta. VI, 465 pages. 1987.

Vol. 280: Field Theory, Quantum Gravity and Strings II. Proceedings, 1985/86. Edited by H.J. de Vega and N. Sánchez. V, 245 pages. 1987.

Vol. 281: Ph. Blanchard, Ph. Combe, W. Zheng, Mathematical and Physical Aspects of Stochastic Mechanics. VIII, 171 pages. 1987.

Vol. 282: F. Ehlotzky (Ed.), Fundamentals of Quantum Optics II. Proceedings, 1987. X, 289 pages. 1987.

Vol. 283: M. Yussouff (Ed.), Electronic Band Structure and Its Applications. Proceedings, 1986. VIII, 441 pages. 1987.

Vol. 284: D. Baeriswyl, M. Droz, A. Malaspinas, P. Martinoli (Eds.), Physics in Living Matter. Proceedings, 1986. V, 180 pages. 1987.

Vol. 285: T. Paszkiewicz (Ed.), Physics of Phonons. Proceedings, 1987. X, 486 pages. 1987.

Vol. 286: R. Alicki, K. Lendi, Quantum Dynamical Semigroups and Applications. VIII, 196 pages. 1987.

Vol. 287: W. Hillebrandt, R. Kuhfuß, E. Müller, J.W. Truran (Eds.), Nuclear Astrophysics. Proceedings. IX, 347 pages. 1987.

Vol. 288: J. Arbocz, M. Potier-Ferry, J. Singer, V.Tvergaard, Buckling and Post-Buckling. VII, 246 pages. 1987.

Vol. 289: N. Straumann, Klassische Mechanik. XV, 403 Seiten. 1987.

Vol. 290: K.T. Hecht, The Vector Coherent State Method and Its Application to Problems of Higher Symmetries. V, 154 pages. 1987.

Vol. 291: J.L. Linsky, R.E. Stencel (Eds.), Cool Stars, Stellar Systems, and the Sun. Proceedings, 1987. XIII, 537 pages. 1987.

Vol. 292: E.-H. Schröter, M. Schüssler (Eds.), Solar and Stellar Physics. Proceedings, 1987. V, 231 pages. 1987.

Vol. 293: Th. Dorfmüller, R. Pecora (Eds.), Rotational Dynamics of Small and Macromolecules. Proceedings, 1986. V, 249 pages. 1987.

Vol. 294: D. Berényi, G. Hock (Eds.), High-Energy Ion-Atom Collisions. Proceedings, 1987. VIII, 540 pages. 1988.

Vol. 295: P. Schmüser, Feynman-Graphen und Eichtheorien für Experimentalphysiker. VI, 217 Seiten. 1988.

Vol. 296: M. Month, S. Turner (Eds.), Frontiers of Particle Beams. XII, 700 pages. 1988.

Vol. 297: A. Lawrence (Ed.), Comets to Cosmology. X, 415 pages. 1988.

Vol. 298: M. Georgiev, F' Centers in Alkali Halides. XI, 287 pages. 1988.